AF702868

Support Vector Machines Applications

Yunqian Ma • Guodong Guo
Editors

Support Vector Machines Applications

Springer

Editors
Yunqian Ma
Honeywell
Golden Valley, MN, USA

Guodong Guo
West Virginia University
Morgantown, WV, USA

ISBN 978-3-319-02299-4 ISBN 978-3-319-02300-7 (eBook)
DOI 10.1007/978-3-319-02300-7
Springer Cham Heidelberg New York Dordrecht London

Library of Congress Control Number: 2014930758

© Springer International Publishing Switzerland 2014
This work is subject to copyright. All rights are reserved by the Publisher, whether the whole or part of the material is concerned, specifically the rights of translation, reprinting, reuse of illustrations, recitation, broadcasting, reproduction on microfilms or in any other physical way, and transmission or information storage and retrieval, electronic adaptation, computer software, or by similar or dissimilar methodology now known or hereafter developed. Exempted from this legal reservation are brief excerpts in connection with reviews or scholarly analysis or material supplied specifically for the purpose of being entered and executed on a computer system, for exclusive use by the purchaser of the work. Duplication of this publication or parts thereof is permitted only under the provisions of the Copyright Law of the Publisher's location, in its current version, and permission for use must always be obtained from Springer. Permissions for use may be obtained through RightsLink at the Copyright Clearance Center. Violations are liable to prosecution under the respective Copyright Law.
The use of general descriptive names, registered names, trademarks, service marks, etc. in this publication does not imply, even in the absence of a specific statement, that such names are exempt from the relevant protective laws and regulations and therefore free for general use.
While the advice and information in this book are believed to be true and accurate at the date of publication, neither the authors nor the editors nor the publisher can accept any legal responsibility for any errors or omissions that may be made. The publisher makes no warranty, express or implied, with respect to the material contained herein.

Printed on acid-free paper

Springer is part of Springer Science+Business Media (www.springer.com)

Preface

The support vector machine (SVM) has been developed in the statistical learning theory and later adopted in machine learning, statistics, and signal processing. SVM has been applied to solve a variety of practical problems. Recent advances have seen diverse developments and applications of the SVM. This book would like to summarize these progresses in robotics, computer vision, pattern recognition, computer security, neural and medical image analysis, soft biometrics, etc. The selected studies will present how to formulate their application, use basic SVM algorithm, and extend SVM algorithm for their application. It will motivate the readers to think about their own applications and come up with solutions using related SVM techniques.

Chapter 1 presents an SVM application in robotics. They modified the basic SVM method to an augmented-SVM approach, which combines gradient observations with the standard observations of function values, then applies in robotics for better performance. Chapter 2 proposes a simplified multi-class SVM approach and applies for multi-class pattern recognition problems. Chapter 3 proposes two novel approaches for transfer learning, inductive transfer learning and transductive transfer learning, where a domain adaptation kernelized SVM was developed with two extensions. They evaluate the performance of the proposed method in a biochemical process with real-world datasets. Chapter 4 discusses the SVM for security applications in adversarial environments, e.g., malware detection, intrusion detection, and spam filtering. Chapter 5 presents the application of the SVMs for image categorization, i.e., classifying visual images into different object categories.

Chapter 6 describes an SVM application in neuroimage analysis. Neuroimage analysis is to characterize the group difference captured by the SVMs with anatomically interpretable patterns, providing insights into the unknown mechanism of the brain. The authors introduce the SVM-based methods to the neuroimage analysis. Discriminative patterns are decoded from the SVM through distinctive feature selection, SVM decision boundary interpretation, and discriminative learning of generative models. Chapter 7 studies the SVMs for the imbalanced data problem, where an imbalanced set of samples are used for training the SVMs. They use in two biomedical applications, the abnormal ECG beat annotation and detection of abnormal regions in colonoscopic images, where the imbalanced data problem exists. Chapter 8 presents the applications of the SVMs for extracting and recognizing soft biometrics, such as human age, gender, and ethnicity from facial images. Soft biometrics is different from the traditional biometrics, such as face recognition or iris, fingerprint recognition. It can be used in business intelligence. The author focuses on the SVMs to learn an estimator or recognizer to extract the soft biometrics. A combination of SVM regression and classifiers has also been developed for age estimation.

We would like to sincerely thank the contributors of each chapter. The book would not be possible to be edited without their great contributions. Hopefully these excellent research works and related reviews give readers a broad and deep representation of recent development of the SVM algorithms and applications.

Golden Valley, MN, USA — Yunqian Ma
Morgantown, WV, USA — Guodong Guo

Contents

Chapter 1
Augmented-SVM for Gradient Observations with Application to Learning Multiple-Attractor Dynamics

Ashwini Shukla and Aude Billard

Abstract In this chapter we present a new formulation that exploits the principle of support vector machine (SVM). This formulation—*Augmented-SVM* (A-SVM)—aims at combining gradient observations with the standard observations of function values (integer labels in classification problems and real values in regression) within a single SVM-like optimization framework. The presented formulation adds onto the existing SVM by enforcing constraints on the gradient of the classifier/regression function. The new constraints modify the original SVM dual, whose optimal solution then results in a new class of support vectors (SV). We present our approach in the light of a particular application in robotics, namely, learning a nonlinear dynamical system (DS) with multiple attractors. Nonlinear DS have been used extensively for encoding robot motions with a single attractor placed at a predefined target where the motion is required to terminate. In this chapter, instead of insisting on a single attractor, we focus on combining several such DS with distinct attractors, resulting in a multi-stable DS. While exploiting multiple attractors provides more flexibility in recovering from unseen perturbations, it also increases the complexity of the underlying learning problem. We address this problem by augmenting the standard SVM formulation with gradient-based constraints derived from the individual DS. The new SV corresponding to the gradient constraints ensure that the resulting multi-stable DS incurs minimum deviation from the original dynamics and is stable at each of the attractors within a finite region of attraction. We show, via implementations on a simulated ten degrees of freedom mobile robotic platform, that the model is capable of real-time motion generation and is able to adapt on-the-fly to perturbations.

A. Shukla (✉) • A. Billard
École Polytechnique Fédérale de Lausanne (EPFL), Lausanne, 1015, Switzerland
e-mail: ashwini.shukla@epfl.ch; aude.billard@epfl.ch

Y. Ma and G. Guo (eds.), *Support Vector Machines Applications*,
DOI 10.1007/978-3-319-02300-7_1,
© Springer International Publishing Switzerland 2014

1.1 Introduction

Dynamical systems (DS) have proved to be a promising framework for encoding and generating complex motions. A major advantage of representing motion using DS-based models [3, 10, 15, 16] is the ability to counter perturbations by virtue of the fact that re-planning of trajectories is instantaneous. These are generative schemes that define the flow of trajectories in state space $\mathbf{x} \in \mathbb{R}^N$ by means of a nonlinear dynamical function $\dot{\mathbf{x}} = f(\mathbf{x})$. DS with single stable attractors have been used in pick and place tasks to control for both the motion of the end-effector [2, 7, 12] and the placement of the fingers on an object [17]. Assuming a single attractor, and hence a single grasping location on the object, constrains considerably the applicability of these methods to realistic grasping problems. A DS composed of multiple stable attractors provides an opportunity to encode different ways to reach and grasp an object. Recent neuro-physiological results [5] have shown that a DS-based modeling best explains the trajectories followed by humans while switching between several reaching targets. From a robotics viewpoint, a robot controlled using a DS with multiple attractors would be able to switch online across grasping strategies. This may be useful, e.g., when one grasping point becomes no longer accessible due to a sudden change in the orientation of the object or the appearance of an obstacle along the current trajectory. Here we present a method using which one can combine—in a single dynamical system—multiple dynamics directed toward different attractors.

The dynamical function $f(\mathbf{x})$ is usually estimated using nonlinear regression functions such as Gaussian Process Regression (GPR) [11], Gaussian Mixture Regression (GMR) [7], and Locally Weighted Projection Regression (LWPR) [13]. However, all of these works modeled DS with a single attractor. While [7, 10] ensure global stability at the attractor, other approaches result in unstable DS with spurious attractors.

Stability at multiple targets has been addressed to date largely through neural networks approaches. The Hopfield network and variants offered a powerful means to encode several stable attractors in the same system to provide a form of content-addressable memory [4, 9]. The dynamics to reach these attractors was, however, not controlled for, nor was the partitioning of the state space that would send the trajectories to each attractor. Echo-state networks provide alternative ways to encode various complex dynamics [6]. Although they have proved to be universal estimators, their ability to generalize in untrained regions of state space remains unverified. Also, the key issue of global stability of the learned dynamics is achieved using heuristic rules. To our knowledge, this is the first attempt at learning simultaneously a partitioning of the state space and an embedding of multiple dynamical systems with separate regions of attractions (ROAs) and distinct attractors.

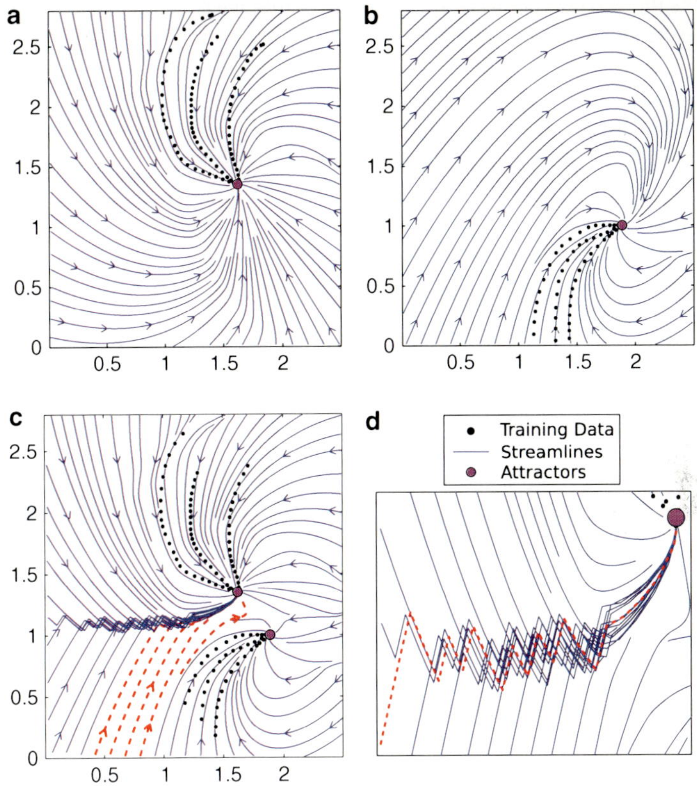

Fig. 1.1 Combining motions using naive SVM classification-based switching. (**a**, **b**)—Two different dynamics with distinct attractors which are to be combined. (**c**)—Employing a simple switching scheme leads to crossing over of some trajectories shown in *red*. (**d**)—Zoomed in around the boundary, showing the fast switching near the boundary

1.2 Identifying Dynamic Constraints

A naive approach to building a multi-attractor DS would be to first partition the space and then learn a DS in each partition separately. This would unfortunately rarely result in the desired compound system. Consider, for instance, two DS with distinct attractors, as shown in Fig. 1.1a, b. First, we build an Support Vector Machine (SVM) classifier to separate data points of the first DS, labeled $+1$, from data points of the other DS, labeled -1. We then estimate each DS separately using any of the techniques reviewed in the previous section. Let $h : \mathbb{R}^N \mapsto \mathbb{R}$ denote the classifier function that separates the state space $\mathbf{x} \in \mathbb{R}^N$ into two regions with labels $y_i \in \{+1, -1\}$. Also, let the two DS be $\dot{\mathbf{x}} = f_{y_i}(\mathbf{x})$ with stable attractors at $\mathbf{x}^*_{y_i}$. The combined DS is then given by $\dot{\mathbf{x}} = f_{\text{sgn}(h(\mathbf{x}))}(\mathbf{x})$. Figure 1.1c shows the trajectories resulting from this approach. Due to the nonlinearity of the dynamics, trajectories initialized in one region cross the boundary and converge to the attractor located in the opposite region. In other words, each region partitioned by the SVM hyperplane is not a region of attraction for its attractor. In a real-world scenario where the attractors represent grasping points on an object and the trajectories are to be followed by robots, crossing over may take the trajectories towards kinematically unreachable regions. Also, as shown in Fig. 1.1d, trajectories that encounter the boundary may switch rapidly between different dynamics leading to jittery motion.

To ensure that the trajectories do not cross the boundary and remain within the region of attraction of their respective attractors, one could adopt a more informed approach in which each of the original DS is modulated such that the generated trajectories always move away from the classifier boundary. Recall that by construction, the absolute value of the classifier function $h(\mathbf{x})$ increases as one moves away from the classification hyperplane. The gradient $\nabla h(\mathbf{x})$ is hence positive, respectively negative, as one moves inside the region of the positive, respectively negative, class. We can exploit this observation to deflect selective components of the velocity signal from the original DS along, respectively opposite to, the direction $\nabla h(\mathbf{x})$. Concretely, if $\dot{\mathbf{x}}_O = f_{\mathrm{sgn}(h(\mathbf{x}))}(\mathbf{x})$ denotes the velocity obtained from the original DS and

$$\lambda(\mathbf{x}) = \begin{cases} \max\left(\varepsilon, \nabla h(\mathbf{x})^T \dot{\mathbf{x}}_O\right) & \text{if } h(\mathbf{x}) > 0 \\ \min\left(-\varepsilon, \nabla h(\mathbf{x})^T \dot{\mathbf{x}}_O\right) & \text{if } h(\mathbf{x}) < 0 \end{cases}, \tag{1.1}$$

the modulated dynamical system is given by

$$\dot{\mathbf{x}} = \tilde{f}(\mathbf{x}) = \lambda(\mathbf{x})\nabla h(\mathbf{x}) + \dot{\mathbf{x}}_\perp . \tag{1.2}$$

Here, ε is a small positive scalar and $\dot{\mathbf{x}}_\perp = \dot{\mathbf{x}}_O - \left(\frac{\nabla h(\mathbf{x})^T \dot{\mathbf{x}}_O}{\|\nabla h(\mathbf{x}\|^2}\right)\nabla h(\mathbf{x})$ is the component of the original velocity perpendicular to ∇h. This results in a vector field that flows along increasing values of the classifier function in the regions of space where $h(\mathbf{x}) > 0$ and along decreasing values for $h(\mathbf{x}) < 0$. As a result, the trajectories move away from the classification hyperplane and converge to a point located in the region where they were initialized. Such modulated systems have been used extensively for estimating stability regions of interconnected power networks [8] and are known as *quasi gradient systems* [1]. If $h(\mathbf{x})$ is upper bounded,[1] all trajectories converge to one of the stationary points $\{\mathbf{x} : \nabla h(\mathbf{x}) = 0\}$ and $h(\mathbf{x})$ is a Lyapunov function of the overall system [1, Proposition 1]. Figure 1.2 shows the result of applying the above modulation to our pair of DS. As expected, it forces the trajectories to flow along the gradient of the function $h(\mathbf{x})$. Although this solves the problem of "crossing-over" the boundary, the trajectories obtained are deficient in two major ways. They depart heavily from the original dynamics and do not terminate at the desired attractors. This is due to the fact that the function $h(\mathbf{x})$ used to modulate the DS was designed

[1] SVM classifier function is bounded if the Radial Basis Function (RBF) is used as kernel.

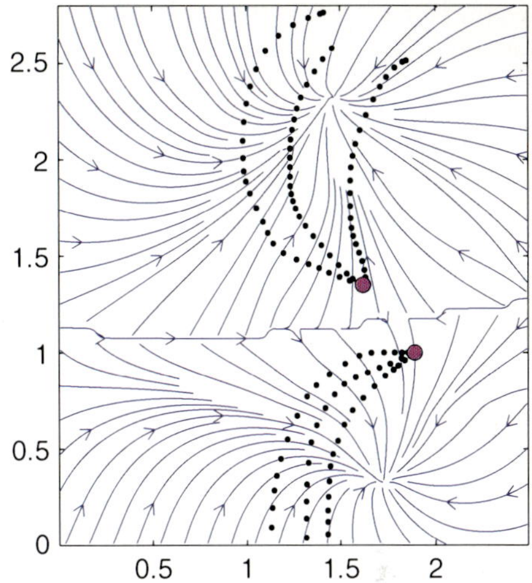

Fig. 1.2 Trajectories obtained by modulating the two original DS with an SVM classifier function. The resulting trajectories flow along directions which register an increase in the value of the classifier function which in turn leads to the bifurcation at the SVM decision boundary

solely for classification and contained no information about the dynamics of the two original DS. In other words, the vector field given by $\nabla h(\mathbf{x})$ was not aligned with the flow of the training trajectories and the stationary points of the modulation function did not coincide with the desired attractors.

In subsequent sections, we show how we can learn a new modulation function which takes into account the three issues we highlighted in this preliminary discussion. We will seek a system that (a) ensures strict classification across ROA for each DS, (b) follows closely the dynamics of each DS in each ROA, and (c) ensures that all trajectories in each ROA reach the desired attractor. Satisfying requirements (a) and (b) above is equivalent to performing classification and regression simultaneously. We take advantage of the fact that the optimization in support vector classification and support vector regression has the same form to phrase our problem in a single constrained optimization framework. In the next sections, we show that in addition to the usual SVM support vectors (SVs), the resulting modulation function is composed of an additional class of SVs. We analyze geometrically the effect of these new support vectors on the resulting dynamics. While this preliminary discussion considered solely binary classification, we will now extend the problem to multi-class classification.

1.3 Problem Formulation

The N-dimensional state space of the system represented by $\mathbf{x} \in \mathbb{R}^N$ is partitioned into M different classes, one for each of the M motions to be combined. We collect trajectories in the state space, yielding a set of P data points $\{\mathbf{x}_i; \dot{\mathbf{x}}_i; l_i\}_{i=1\ldots P}$

where $l_i \in \{1,2,\cdots,M\}$ refers to the class label of each point.[2] To learn the set of modulation functions $\{h_m(\mathbf{x})\}_{m=1\ldots M}$, we proceed recursively. We learn each modulation function in a one-vs-all classifier scheme and then compute the final modulation function $\tilde{h}(\mathbf{x}) = \max_{m=1\cdots M} h_m(\mathbf{x})$. In the multi-class setting, the behavior of avoiding boundaries is obtained if the trajectories move along *increasing* values of the function $\tilde{h}(\mathbf{x})$. To this effect, the deflection term $\lambda(\mathbf{x})$ presented in the binary case Eq. (1.1) becomes $\lambda(\mathbf{x}) = \max\left(\varepsilon, \nabla\tilde{h}(\mathbf{x})^T\dot{\mathbf{x}}_O\right); \forall\mathbf{x} \in \mathbb{R}^N$. Next, we describe the procedure for learning a single $h_m(\mathbf{x})$ function.

We follow the classical SVM formulation and lift the data into a higher dimensional feature space through the mapping $\phi : \mathbb{R}^N \mapsto \mathbb{R}^F$ where F denotes the dimension of the feature space. We also assume that each function $h_m(\mathbf{x})$ is linear in feature space, i.e., $h_m(\mathbf{x}) = \mathbf{w}^T\phi(\mathbf{x}) + b$ where $\mathbf{w} \in \mathbb{R}^F, b \in \mathbb{R}$. We label the current ($m$-th) motion class as positive and all others negative such that the set of labels for the current subproblem is given by

$$y_i = \begin{cases} +1 & \text{if } l_i = m \\ -1 & \text{if } l_i \neq m \end{cases}; \qquad i = 1\cdots P.$$

Also, the set indexing the positive class is then defined as $\mathcal{I}_+ = \{i : i \in [1,P]; l_i = m\}$. With this, we formalize the three constraints explained in Sect. 1.2 as:

Classification — Each point must be classified correctly yields P constraints

$$y_i\left(\mathbf{w}^T\phi(\mathbf{x}_i) + b\right) \geq 1 \quad \forall i = 1\ldots P. \tag{1.3}$$

Lyapunov Constraint — The gradient of the modulation function must have a positive component along the velocities at the data points. This ensures that the modulated flow is aligned with the training trajectories. We have the constraint

$$\nabla h_m(\mathbf{x}_i)^T\hat{\dot{\mathbf{x}}}_i = \mathbf{w}^T\mathrm{J}(\mathbf{x}_i)\hat{\dot{\mathbf{x}}}_i \geq 0 \quad \forall i \in \mathcal{I}_+ \tag{1.4}$$

where $\mathrm{J} \in \mathbb{R}^{F\times N}$ is the Jacobian matrix given by $\mathrm{J} = \left[\nabla\phi_1(\mathbf{x})\nabla\phi_2(\mathbf{x})\cdots\nabla\phi_F(\mathbf{x})\right]^T$ and $\hat{\dot{\mathbf{x}}}_i = \dot{\mathbf{x}}_i/\|\dot{\mathbf{x}}_i\|$ is the normalized velocity at the i-th data point.

Stability — The gradient of the modulation function must vanish at the attractor $\mathbf{x}^*$ of the positive class. This constraint can be expressed as

$$\nabla h_m(\mathbf{x}^*)^T\mathbf{e}_i = \mathbf{w}^T\mathrm{J}(\mathbf{x}^*)\mathbf{e}_i = 0 \quad \forall i = 1\ldots N \tag{1.5}$$

where the set of vectors $\{\mathbf{e}_i\}_{i=1\cdots N}$ is the canonical basis of $\mathbb{R}^N$.

[2] Boldfaced fonts represent vectors. $\mathbf{x}_i$ denotes the i-th vector and x_i denotes the i-th element of vector $\mathbf{x}$.

Incorporating Gradient Observations with Augmented-SVM

Existing variants of the SVM methodology aim at learning **(a)** Classifiers—which satisfy certain *inequality* constraints and **(b)** Regressors—which satisfy *equality* constraints at the data points. Also, in both cases, the constraints are solely defined on the scalar function value. In the A-SVM framework presented in the next sections, we will combine both inequality and equality constraints within the same optimization framework. Moreover, the constraints will be enforced not only on the function value but also on its *gradient*.

1.3.1 *Primal and Dual Forms*

As in the standard SVM [14], we optimize for maximal margin between the positive and negative classes, subject to constraints (1.3)–(1.5) above. This can be formulated as:

$$\begin{aligned} &\underset{\mathbf{w},\xi_i}{\textbf{minimize}} \quad \frac{1}{2}\|\mathbf{w}\|^2 + C\sum_{i\in\mathcal{I}_+}\xi_i \\ &\textbf{subject to} \quad \left.\begin{array}{rr} y_i\left(\mathbf{w}^T\phi(\mathbf{x}_i)+b\right)\geq 1 & \forall i=1\cdots P \\ \mathbf{w}^T\mathrm{J}(\mathbf{x}_i)\hat{\mathbf{x}}_i+\xi_i>0 & \forall i\in\mathcal{I}_+ \\ \xi_i>0 & \forall i\in\mathcal{I}_+ \\ \mathbf{w}^T\mathrm{J}(\mathbf{x}^*)\mathbf{e}_i=0 & \forall i=1\cdots N \end{array}\right\}. \end{aligned} \tag{1.6}$$

Here $\xi_i \in \mathbb{R}$ are slack variables that relax the Lyapunov constraint in Eq. (1.4). We retain these in our formulation to accommodate noise in the data representing the dynamics. $C \in \mathbb{R}_+$ is a penalty parameter for the slack variables. The Lagrangian for the above problem can be written as

$$\begin{aligned} \mathcal{L}(\mathbf{w},b,\alpha,\beta,\gamma) = \frac{1}{2}\|\mathbf{w}\|^2 + C\sum_{i\in\mathcal{I}_+}\xi_i - \sum_{i\in\mathcal{I}_+}\mu_i\xi_i - \sum_{i=1}^{P}\alpha_i\left(y_i(\mathbf{w}^T\phi(\mathbf{x}_i)+b)-1\right) \\ - \sum_{i\in\mathcal{I}_+}\beta_i\left(\mathbf{w}^T\mathrm{J}(\mathbf{x}_i)\hat{\mathbf{x}}_i+\xi_i\right) + \sum_{i=1}^{N}\gamma_i\mathbf{w}^T\mathrm{J}(\mathbf{x}^*)\mathbf{e}_i \end{aligned} \tag{1.7}$$

where $\alpha_i, \beta_i, \mu_i, \gamma_i$ are the Lagrange multipliers with $\alpha_i, \beta_i, \mu_i \in \mathbb{R}_+$, and $\gamma_i \in \mathbb{R}$. Employing a similar analysis as in the standard SVM, we derive the dual by setting the derivatives of the Lagrangian w.r.t all the variables and multipliers to zero, we get

$$\frac{\partial \mathcal{L}}{\partial \mathbf{w}} = \mathbf{w} - \sum_{i=1}^{P} \alpha_i y_i \phi(\mathbf{x}_i) - \sum_{i \in \mathcal{I}_+} \beta_i \mathrm{J}(\mathbf{x}_i)\hat{\mathbf{x}}_i + \sum_{i=1}^{N} \gamma_i \mathrm{J}(\mathbf{x}^*)\mathbf{e}_i = 0.$$
$$\Rightarrow \mathbf{w} = \sum_{i=1}^{P} \alpha_i y_i \phi(\mathbf{x}_i) + \sum_{i \in \mathcal{I}_+} \beta_i \mathrm{J}(\mathbf{x}_i)\hat{\mathbf{x}}_i - \sum_{i=1}^{N} \gamma_i \mathrm{J}(\mathbf{x}^*)\mathbf{e}_i; \tag{1.8}$$

$$\frac{\partial \mathcal{L}}{\partial b} = \sum_{i=1}^{P} \alpha_i y_i = 0; \tag{1.9}$$

$$\frac{\partial \mathcal{L}}{\partial \xi_i} = C - \beta_i - \mu_i = 0 \qquad \forall i \in \mathcal{I}_+. \tag{1.10}$$

Combining (1.10) with the constraints that all the Lagrange multipliers β_i and μ_i be positive, we obtain

$$0 \le \beta_i \le C \qquad \forall i \in \mathcal{I}_+. \tag{1.11}$$

Using (1.8), (1.9), and (1.10) in (1.7) we get the dual objective function to be maximized as

$$\hat{\mathcal{L}}(\alpha,\beta,\gamma) = \sum_{i=1}^{P} \alpha_i - \frac{1}{2}\mathbf{w}^T\mathbf{w}. \tag{1.12}$$

Note that although the dual has the same general form as the dual in the standard SVM formulation, it differs in the expression of the term $\mathbf{w}$. Expanding using (1.8) we have

$$\begin{aligned}
\mathbf{w}^T\mathbf{w} = & \left(\sum_{i=1}^{P} \alpha_i y_i \phi(\mathbf{x}_i)^T + \sum_{i \in \mathcal{I}_+} \beta_i \hat{\mathbf{x}}_i^T \mathrm{J}(\mathbf{x}_i)^T - \sum_{i=1}^{N} \gamma_i \mathbf{e}_i^T \mathrm{J}(\mathbf{x}^*)^T\right) \\
& \left(\sum_{j=1}^{P} \alpha_j y_j \phi(\mathbf{x}_j) + \sum_{j \in \mathcal{I}_+} \beta_j \mathrm{J}(\mathbf{x}_j)\hat{\mathbf{x}}_j - \sum_{j=1}^{N} \gamma_j \mathrm{J}(\mathbf{x}^*)\mathbf{e}_j\right) \\
= & \sum_{i=1}^{P}\left(\sum_{j=1}^{P} \alpha_i y_i \alpha_j y_j \phi(\mathbf{x}_i)^T \phi(\mathbf{x}_j) + \sum_{j \in \mathcal{I}_+} \alpha_i y_i \beta_j \phi(\mathbf{x}_i)^T \mathrm{J}(\mathbf{x}_j)\hat{\mathbf{x}}_j - \sum_{j=1}^{N} \alpha_i y_i \gamma_j \phi(\mathbf{x}_i)^T \mathrm{J}(\mathbf{x}^*)\mathbf{e}_j\right) \\
& + \sum_{i \in \mathcal{I}_+}\left(\sum_{j=1}^{P} \beta_i \alpha_j y_j \hat{\mathbf{x}}_i^T \mathrm{J}(\mathbf{x}_i)^T \phi(\mathbf{x}_j) + \sum_{j \in \mathcal{I}_+} \beta_i \beta_j \hat{\mathbf{x}}_i^T \mathrm{J}(\mathbf{x}_i)^T \mathrm{J}(\mathbf{x}_j)\hat{\mathbf{x}}_j - \sum_{j=1}^{N} \beta_i \gamma_j \hat{\mathbf{x}}_i^T \mathrm{J}(\mathbf{x}_i)^T \mathrm{J}(\mathbf{x}^*)\mathbf{e}_j\right) \\
& - \sum_{i=1}^{N}\left(\sum_{j=1}^{P} \gamma_i \alpha_j y_j \mathbf{e}_i^T \mathrm{J}(\mathbf{x}^*)^T \phi(\mathbf{x}_j) + \sum_{j \in \mathcal{I}_+} \gamma_i \beta_j \mathbf{e}_i^T \mathrm{J}(\mathbf{x}^*)^T \mathrm{J}(\mathbf{x}_j)\hat{\mathbf{x}}_j - \sum_{j=1}^{N} \gamma_i \gamma_j \mathbf{e}_i^T \mathrm{J}(\mathbf{x}^*)^T \mathrm{J}(\mathbf{x}^*)\mathbf{e}_j\right).
\end{aligned} \tag{1.13}$$

Rewriting in matrix form,

$$\mathbf{w}^T\mathbf{w} = \begin{bmatrix}\alpha^T & \beta^T & \gamma^T\end{bmatrix}\begin{bmatrix} \mathrm{K} & \mathrm{G} & -\mathrm{G}_* \\ \mathrm{G}^T & \mathrm{H} & -\mathrm{H}_* \\ -\mathrm{G}_*^T & -\mathrm{H}_*^T & \mathrm{H}_{**}\end{bmatrix}\begin{bmatrix}\alpha \\ \beta \\ \gamma\end{bmatrix} \tag{1.14}$$

where $\mathrm{K} \in \mathbb{R}^{P\times P}, \mathrm{G} \in \mathbb{R}^{P\times|\mathcal{I}_+|}, \mathrm{G}_* \in \mathbb{R}^{P\times N}, \mathrm{H} \in \mathbb{R}^{|\mathcal{I}_+|\times|\mathcal{I}_+|}, \mathrm{H}_* \in \mathbb{R}^{|\mathcal{I}_+|\times N}, \mathrm{H}_{**} \in \mathbb{R}^{N\times N}$ are given by

$$\left.\begin{array}{llll} [\mathrm{K}]_{ij} = y_i y_j \phi(\mathbf{x}_i)^T\phi(\mathbf{x}_j) ; & [\mathrm{H}]_{ij} = \hat{\mathbf{x}}_i^T \mathrm{J}(\mathbf{x}_i)^T \mathrm{J}(\mathbf{x}_j)\hat{\mathbf{x}}_j \\ [\mathrm{G}]_{ij} = y_i \phi(\mathbf{x}_i)^T \mathrm{J}(\mathbf{x}_j)\hat{\mathbf{x}}_j ; & [\mathrm{H}_*]_{ij} = \hat{\mathbf{x}}_i^T \mathrm{J}(\mathbf{x}_i)^T \mathrm{J}(\mathbf{x}^*)\mathbf{e}_j \\ [\mathrm{G}_*]_{ij} = y_i \phi(\mathbf{x}_i)^T \mathrm{J}(\mathbf{x}^*)\mathbf{e}_j ; & [\mathrm{H}_{**}]_{ij} = \mathbf{e}_i^T \mathrm{J}(\mathbf{x}^*)^T \mathrm{J}(\mathbf{x}^*)\mathbf{e}_j \end{array}\right\}. \tag{1.15}$$

where $[.]_{ij}$ denotes the i,j-th entry of the corresponding matrix. Further using the relations (1.19) and (1.20) from Appendix 1, we can rewrite the above block matrices in terms of the kernel function and data:

$$\left.\begin{array}{llll} [\mathrm{K}]_{ij} = y_i y_j k(\mathbf{x}_i,\mathbf{x}_j) & ; & [\mathrm{H}]_{ij} = \hat{\mathbf{x}}_i^T \frac{\partial^2 k(\mathbf{x}_i,\mathbf{x}_j)}{\partial \mathbf{x}_i \partial \mathbf{x}_j}\hat{\mathbf{x}}_j \\ [\mathrm{G}]_{ij} = y_i \left(\frac{\partial k(\mathbf{x}_i,\mathbf{x}_j)}{\partial \mathbf{x}_j}\right)^T \hat{\mathbf{x}}_j & ; & [\mathrm{H}_*]_{ij} = \hat{\mathbf{x}}_i^T \frac{\partial^2 k(\mathbf{x}_i,\mathbf{x}^*)}{\partial \mathbf{x}_i \partial \mathbf{x}^*}\mathbf{e}_j \\ [\mathrm{G}_*]_{ij} = y_i \left(\frac{\partial k(\mathbf{x}_i,\mathbf{x}^*)}{\partial \mathbf{x}^*}\right)^T \mathbf{e}_j & ; & [\mathrm{H}_{**}]_{ij} = \mathbf{e}_i^T \frac{\partial^2 k(\mathbf{x}^*,\mathbf{x}^*)}{\partial \mathbf{x}^* \partial \mathbf{x}^*}\mathbf{e}_j \end{array}\right\}. \tag{1.16}$$

These can be further expanded given a choice of the kernel. Expansions for the RBF and the nonhomogeneous polynomial kernel are given in Appendix 2.

Using Eqs. (1.9), (1.11), (1.12), and (1.14) the dual optimization problem can be stated as

$$\underset{\alpha,\beta,\gamma}{\textbf{minimize}}\ \frac{1}{2}\begin{bmatrix}\alpha^T \beta^T \gamma^T\end{bmatrix}\begin{bmatrix} \mathrm{K} & \mathrm{G} & -\mathrm{G}_* \\ \mathrm{G}^T & \mathrm{H} & -\mathrm{H}_* \\ -\mathrm{G}_*^T & -\mathrm{H}_*^T & \mathrm{H}_{**}\end{bmatrix}\begin{bmatrix}\alpha \\ \beta \\ \gamma\end{bmatrix} - \alpha^T\bar{\mathbf{1}}$$

subject to

$$\left.\begin{array}{ll} 0 \le \alpha_i & \forall i = 1\cdots P \\ 0 \le \beta_i \le C & \forall i \in \mathcal{I}_+ \\ \sum_{i=1}^P \alpha_i y_i = 0 & \end{array}\right\}. \tag{1.17}$$

Note that the Lagrange multipliers γ_i are completely unconstrained as they correspond to the *equality* constraints in the primal. Also, since the matrices K, H, and H_{**} are symmetric, the overall Hessian matrix for the resulting quadratic program is also symmetric. In our implementation, we use the MATLAB® *quadprog* solver to solve this quadratic program. We initialize the iterations by setting α_i as the solution to the standard SVM classification problem. All β_i and γ_i are set to zeros. Once the optimal solution for the above problem is obtained, the modulation function can be written as

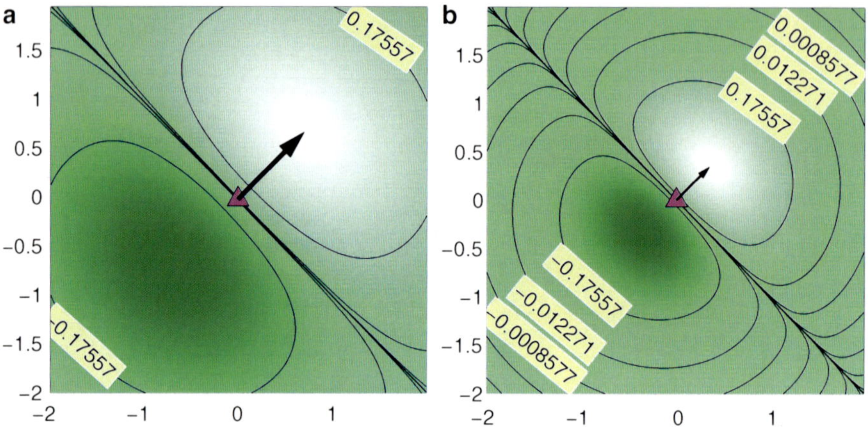

Fig. 1.3 Isocurves of $f(\mathbf{x}) = \hat{\dot{\mathbf{x}}}_i^T \frac{\partial k(\mathbf{x},\mathbf{x}_i)}{\partial \mathbf{x}_i}$ at $\mathbf{x}_i = [0\,0]^T, \hat{\dot{\mathbf{x}}}_i = [\frac{1}{\sqrt{2}}\,\frac{1}{\sqrt{2}}]^T$ for the RBF kernel with width σ. (**a**) $\sigma = 1$. (**b**) $\sigma = 0.5$

$$
\begin{aligned}
h(\mathbf{x}) &= \mathbf{w}^T \phi(\mathbf{x}) + b \\
&= \sum_{i=1}^{P} \alpha_i y_i \phi(\mathbf{x}_i)^T \phi(\mathbf{x}) + \sum_{i\in\mathcal{I}_+} \beta_i \hat{\dot{\mathbf{x}}}_i^T \mathrm{J}(\mathbf{x}_i)^T \phi(\mathbf{x}) - \sum_{i=1}^{N} \gamma_i \mathbf{e}_i^T \mathrm{J}(\mathbf{x}^*)^T \phi(\mathbf{x}) + b \\
&= \sum_{i=1}^{P} \alpha_i y_i k(\mathbf{x}, \mathbf{x}_i) + \sum_{i\in\mathcal{I}_+} \beta_i \dot{\mathbf{x}}_i^T \frac{\partial k(\mathbf{x}, \mathbf{x}_i)}{\partial \mathbf{x}_i} - \sum_{i=1}^{N} \gamma_i \mathbf{e}_i^T \frac{\partial k(\mathbf{x}, \mathbf{x}^*)}{\partial \mathbf{x}^*} + b
\end{aligned}
\tag{1.18}
$$

This modulation function has noticeable similarities with the standard SVM classifier function. The first summation term on the right-hand side is composed of the α support vectors (α-SV) which act as support to the classification hyperplane. The second term entails a new class of support vectors that perform a linear combination of the normalized velocity $\hat{\dot{\mathbf{x}}}_i$ at the training data points $\mathbf{x}_i$. These β support vectors (β-SVs) collectively contribute to the fulfilment of the Lyapunov constraint in Eq. (1.4) by introducing a positive slope in the modulation function value along the directions $\hat{\dot{\mathbf{x}}}_i$. Figure 1.3 shows the influence of a single β-SV for the RBF kernel $k(\mathbf{x}_i, \mathbf{x}_j) = e^{-1/2\sigma^2 \|\mathbf{x}_i - \mathbf{x}_j\|^2}$ with $\mathbf{x}_i$ at the origin and $\hat{\dot{\mathbf{x}}}_i = [\frac{1}{\sqrt{2}}\,\frac{1}{\sqrt{2}}]^T$. Observe that the smaller the kernel width σ, the steeper the slope. The third summation term is a nonlinear bias, which does not depend on the chosen support vectors, and performs a local modification around the desired attractor $\mathbf{x}^*$ to ensure that the modulation function has a local maximum at that point. b is the constant bias which normalizes the classification margins as -1 and $+1$. We calculate its value by making use of the fact that for all the data points $\mathbf{x}_i$ chosen as α-SV, we must have $y_i h_m(\mathbf{x}_i) = 1$. We use the average of the values obtained from different support vectors.

Figure 1.4 illustrates the effects of the support vectors in a 2D example by progressively adding them and overlaying the resulting DS flow in each case.

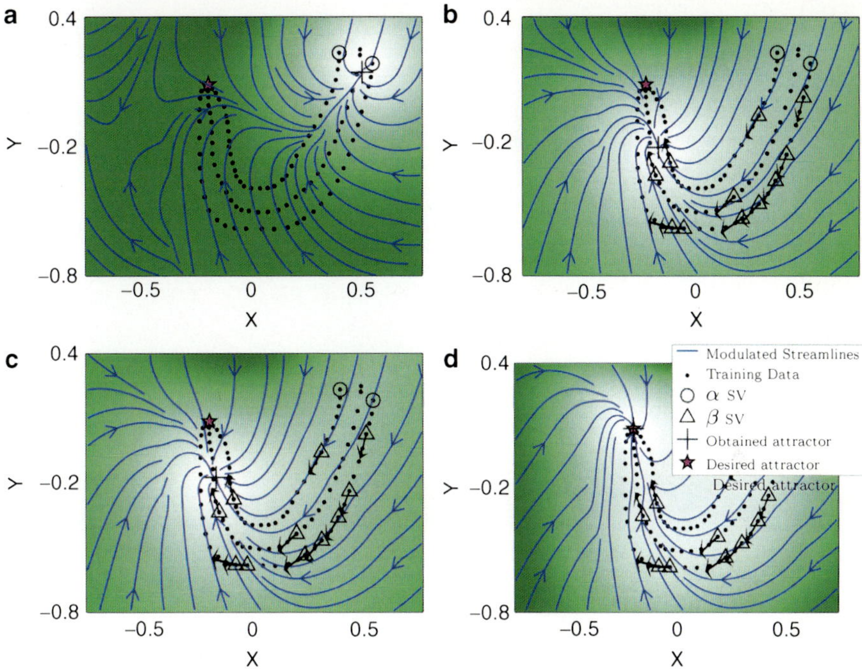

Fig. 1.4 Progressively adding support vectors to highlight their effect on shaping the dynamics of the motion. (**a**) α—SVs largely affect classification. (**b**)—β-SVs guide the flow of trajectories along their respective associated directions $\hat{\mathbf{x}}_i$ shown by *arrows*. (**c**, **d**) The two γ terms force the local maximum of the modulation function to coincide with the desired attractor along the X and Y axes respectively

The value of the modulation function $h_m(\mathbf{x})$ is shown by the color plot (white indicates high values). As the β-SVs are added—as shown in Fig. 1.4b—they *force* the flow of trajectories along their associated directions. In Fig. 1.4c, d, adding the two γ terms shifts the location of the maximum of the modulation function to coincide with the desired attractor. Once all the SVs have been taken into account, the streamlines of the resulting DS achieve the desired criteria, i.e., they follow the training trajectories and terminate at the desired attractor.

1.4 Application Examples

In this section, we validate the presented A-SVM model on 2D (synthetic) data and on a robotic simulated experiment using a seven degrees of freedom (DOF) KUKA-LWR arm mounted on a three-DOF Omnirob base to catch falling objects. A video of the robotic experiment—simulated and real—is provided at the project url http://asvm.epfl.ch. Next, we present a cross-validation analysis of the error

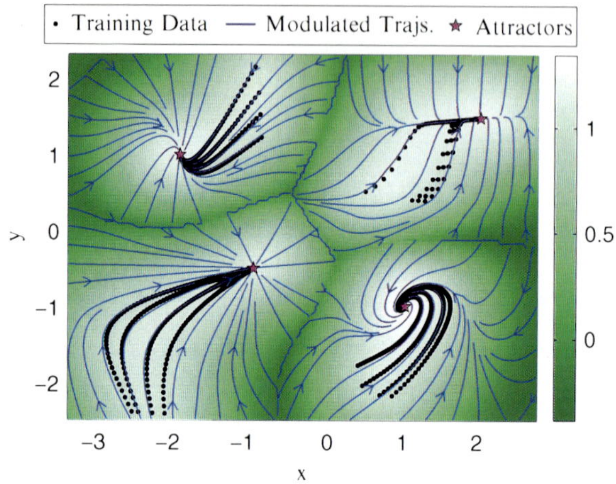

Fig. 1.5 Resulting flow of the synthetic four-attractor example. Color plot depicts the value of the one-vs-all classifier function which has local maximums at all the four attractors

introduced by the modulation in the original dynamics. A sensitivity analysis of the region of attraction of the resulting dynamical system with respect to the model parameters is also presented. We used the RBF kernel for all the results presented in this section. As discussed in Sect. 1.2, the RBF kernel is advantageous as it ensures that the function $h_m(\mathbf{x})$ is bounded. To generate an initial estimate of each individual dynamical system, we used the technique proposed in Khansari-Zadeh and Billard [7].

1.4.1 2D Example

Figure 1.5 shows a synthetic example with four motion classes, each generated from a different closed form dynamics and containing 160 data points. The color plot indicates the value of the combined modulation function $\tilde{h}(\mathbf{x}) = \max_{m=1\cdots M} h_m(\mathbf{x})$ where each of the functions $h_m(\mathbf{x})$ is learned using the presented A-SVM technique. A total of nine support vectors were obtained which is $<10\,\%$ of the number of training data points. The trajectories obtained after modulating the original dynamical systems flow along increasing values of the modulation function, thereby bifurcating towards different attractors at the region boundaries. Unlike the dynamical system in Fig. 1.2, the flow here is aligned with the training trajectories and terminates at the desired attractors. To recall, this is made possible thanks to the additional constraints Eqs. (1.4) and (1.5) in our formulation.

In a second example, we tested the ability of our model to accommodate a higher density of attractors. We created eight synthetic dynamics by capturing motion data using a screen mouse. Figure 1.6 shows the resulting eight-attractor system.

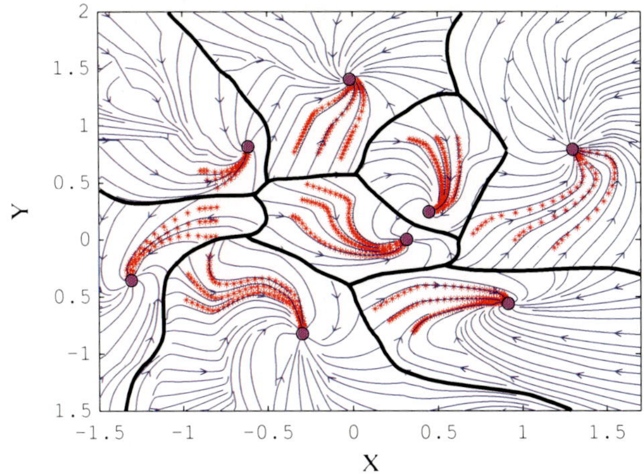

Fig. 1.6 A DS with eight attractors learned using only a few representative trajectories (*black dots*) from each DS. The bifurcation boundaries, as well as the dynamics of each DS need to be estimated from these trajectories

1.4.2 Error Analysis

As formulated in Eq. (1.6), the Lyapunov constraints admit some slack, which allows the modulation to introduce slight deviations from the original dynamics. Here we statistically analyze this error via fivefold cross validation. In the four-attractor problem presented above, we generate a total of ten trajectories per motion class and use 2:3 training to testing ratio for cross validation.

We calculate the average percentage error between the original velocity (read off from the data) and the modulated velocity [calculated using Eq. (1.2)] for the m-th class as $e_m = \left\langle \frac{\|\dot{\mathbf{x}}_i - \tilde{f}(\mathbf{x}_i)\|}{\|\dot{\mathbf{x}}_i\|} \times 100 \right\rangle_{i:l_i=m}$ where $< . >$ denotes average over the indicated range. Figure 1.7a shows the cross-validation error (mean and standard deviation over the fivefolds) for a range of values of kernel width. The general trend revealed here is that for each class of motion, there exists a band of optimum values of the kernel width for which the testing error is the smallest. The region covered by this band of optimal values may vary depending on the relative location of the attractors and other data points. In Fig. 1.5, motion classes 2 (upper left) and 4 (upper right) are better fitted and show less sensitivity to the choice of kernel width than classes 1 (lower left) and 3 (lower right). We will show later in this section that this is correlated with the distance between the attractors. A comparison of testing and training errors for the least error case is shown in Fig. 1.7b. We see that the testing errors for all the classes in the best case scenario are less than 1 %.

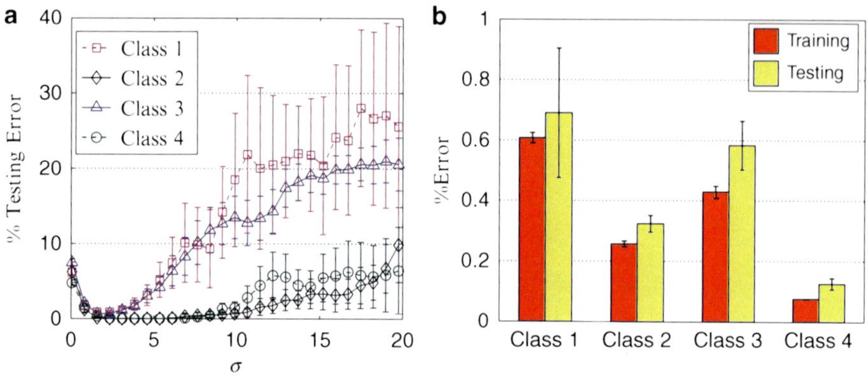

Fig. 1.7 Error analysis for the synthetic four-attractor example. (**a**) Cross-validation error. (**b**) Best case errors

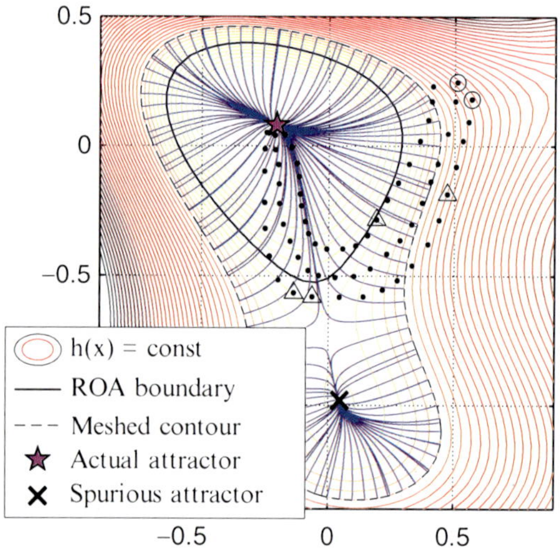

Fig. 1.8 Test trajectories starting from several points on an isocurve (*dotted line*) to determine spurious attractors

1.4.3 Sensitivity Analysis

The partitioning of space created by our method results in M ROA for each of our M attractors. To assess the size of these regions and the existence of spurious attractors, we adopt an empirical approach. For each class, we compute the isosurfaces of the corresponding modulation function $h_m(\mathbf{x})$ in the range $[0, h_m(\mathbf{x}^*)]$. These hypersurfaces incrementally span the volume of the m-th region around its attractor. We mesh each of these test surfaces and compute trajectories starting from the obtained mesh-points, looking for spurious attractors. h_{ROA} is the isosurface of maximal value that encloses no spurious attractor and marks the ROA of the corresponding motion dynamics. We use the example in Fig. 1.4 to illustrate this process. Figure 1.8 shows a case where one spurious attractor is detected using a

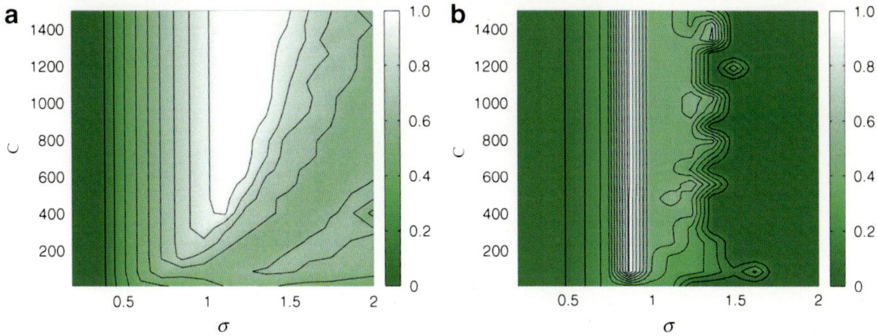

Fig. 1.9 Variation of r_{ROA} with varying model parameters. (**a**) $d_{att} = 1.0$. (**b**) $d_{att} = 0.2$

larger test surface (*dotted line*) whereas the actual ROA (*solid line*) is smaller. Once h_{ROA} is calculated, we define the size of ROA as $r_{ROA} = (h(\mathbf{x}^*) - h_{ROA})/h(\mathbf{x}^*)$. $r_{ROA} = 0$ when no trajectory except those originating at the attractor itself lead to the attractor. $r_{ROA} = 1$ when the ROA is bounded by the isosurface $h(\mathbf{x}) = 0$. The size of the r_{ROA} is affected by both the choice of kernel width and the distance across nearby attractors. This is illustrated in Fig. 1.9 using data points from class 1 of Fig. 1.5 and translating the attractors so that they are either very far apart (left, distance $d_{att} = 1.0$) or very close to one another (right, $d_{att} = 0.2$). As expected, r_{ROA} increases as we reach the optimal range of parameters. Furthermore, when the attractors are farther apart, high values of r_{ROA} are obtained for a larger range of values of the kernel width, i.e., the model is less sensitive to the chosen kernel width. With smaller distance between the attractors (Fig. 1.9b), only a small deviation from the optimum kernel width results in a considerable loss in r_{ROA}, exhibiting high sensitivity to the model parameter.

1.4.4 3D Example

We validated our method on a real-world 3D problem. The attractors here represent manually labeled grasping points on a pitcher. The 3D model of the object was taken from the ROS IKEA object library. We use the seven-DOF KUKA-LWR arm mounted on the three-DOF KUKA-Omnirob base for executing the modulated Cartesian trajectories in simulation. We control all ten DOF of the robot using the damped least square inverse kinematics. Training data for this implementation was obtained by recording the end-effector positions $\mathbf{x}_i \in \mathbb{R}^3$ from kinesthetic demonstrations of reach-to-grasp motions directed towards these grasping points, yielding a three-class problem (see Fig. 1.10a). Each class was represented by 75 data points. Figure 1.10b shows the isosurfaces $h_m(\mathbf{x}) = 0; m \in \{1, 2, 3\}$ learned using the presented method. Figure 1.11a, b show the robot executing two trajectories when started from two different locations and converging to a different attractor (grasping point). Figure 1.10c shows the flow of motion around the object. Note

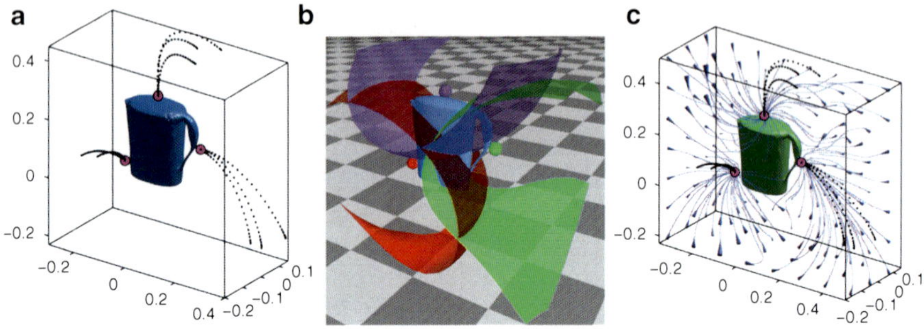

Fig. 1.10 3D Experiment. (**a**) shows training trajectories for three manually chosen grasping points. (**b**) shows the isosurfaces $h_m(\mathbf{x}) = 0; m = 1, 2, 3$ along with the locations of the corresponding attractors. (**c**) shows the complete flow of motion

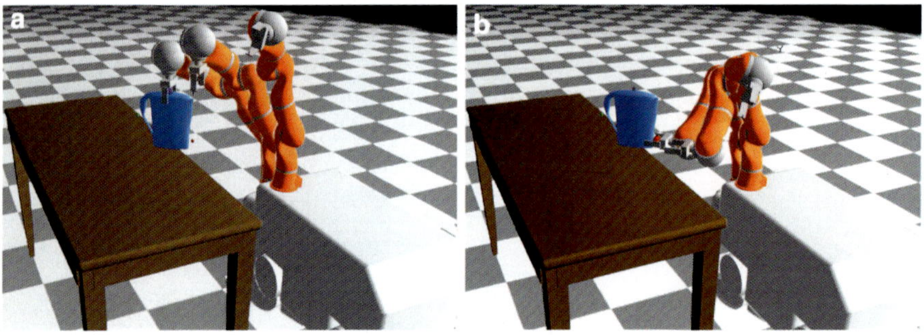

Fig. 1.11 Ten-DOF mobile robot executes the generated trajectories starting from different positions and hence converging to different grasping points (attractors). (**a**) Trajectory 1. (**b**) Trajectory 2

that the time required to generate each trajectory point is $O(S)$ where S denotes the total number of support vectors in the model. In this particular example with a total of 18 SVs, the trajectory points were generated at 1,000 Hz which is well suited for real-time control. Such a fast generative model allows the robot to switch on-the-fly between the attractors and adapt to real-time perturbations in the object or the end-effector pose, without any re-planning or re-learning. Results for another object—a champagne glass with two attractors—are shown in Fig. 1.12. We performed high speed experiments in which the glass is falling and the robot needs to catch it at one of the two attractors. This requires real-time adaptation to the constantly changing position and orientation of the object. The robot might need to switch between the attractors and move the end-effector toward the chosen attractor. Figure 1.13 shows the experiments in simulation and with the real KUKA robot. Full videos explaining the A-SVM methodology and these experiments are available at http://asvm.epfl.ch/download.php.

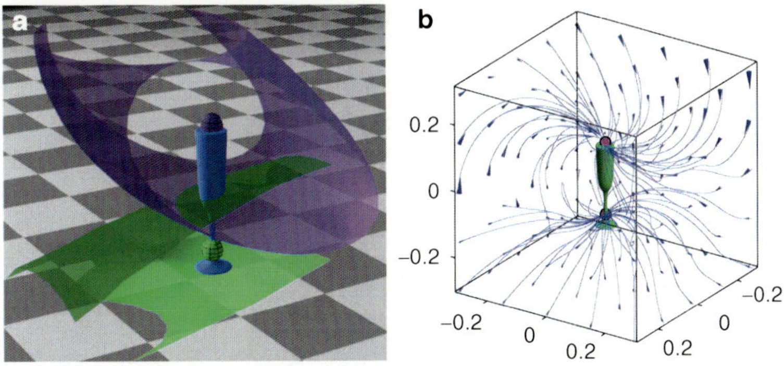

Fig. 1.12 (**a**) Two attractors placed on a champagne glass and their corresponding classification surfaces. (**b**) Complete flow of motion around the object

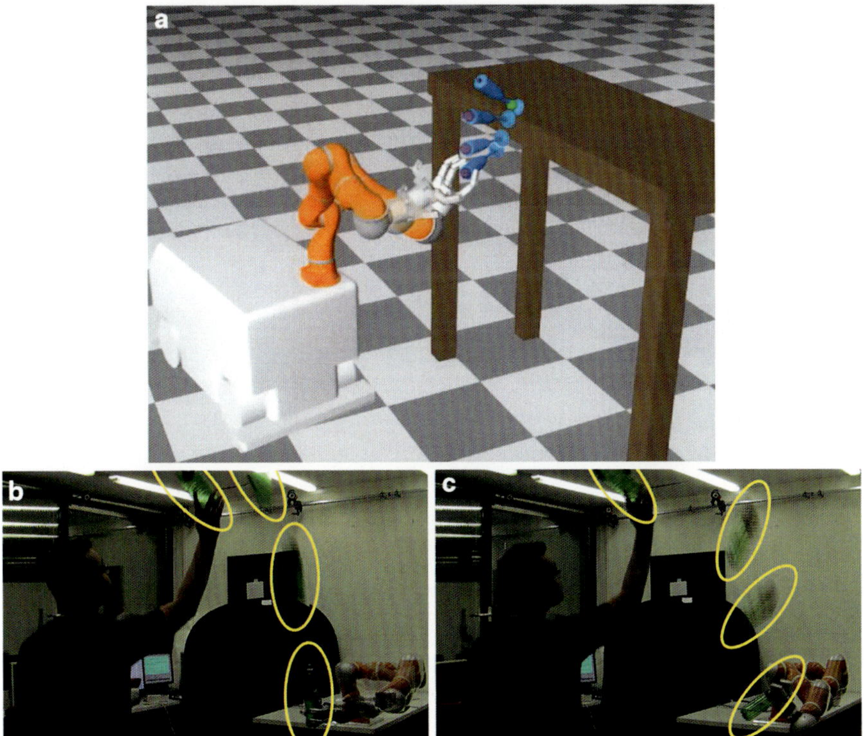

Fig. 1.13 (**a**)—Simulation experiment of catching a falling object with the ten-DOF KUKA-Omnirob. The robot switches between attractors (*green* to *magenta*) as the object falls down. (**b**, **c**)—Real seven-DOF KUKA arm catching the falling object at different grasping points (attractors) in different throwing situations

1.5 Conclusions

We presented the A-SVM model for combining nonlinear dynamical systems through a partitioning of the space. We reformulated the optimization framework of SVM to encapsulate gradient-based constraints that ensure accurate reproduction of the dynamics of motion. The new set of constraints result in a new class of support vectors that exploit partial derivatives of the kernel function to align the flow of trajectories with the training data. The resulting model behaves as a multi-stable DS with attractors at the desired locations. Each of the classified regions is forward invariant w.r.t the learned DS. This ensures that the trajectories do not cross over region boundaries. We validated the presented method on synthetic motions in 2D and 3D grasping motions on real objects. Results show that even though spurious attractors may occur, in practice they can be avoided by a careful choice of model parameters through grid search. The applicability of the method for real-time control of a ten-DOF robot was also demonstrated.

Acknowledgments This work was supported by EU Project *First-MM* (FP7/2007–2013) under grant agreement number *248258*. The authors would also like to thank Prof. François Margot for his insightful comments on the technical material.

Appendix 1: Kernel Derivatives

For scalar variables $x_i, x_j \in \mathbb{R}$ and any feature transformation $\phi : \mathbb{R} \mapsto \mathbb{R}^F$ we define a valid Mercer kernel as $k(x_i,x_j) \equiv \phi(x_i)^T \phi(x_j)$. If $'$ denotes the derivative w.r.t the state variable, then the identities $\phi'(x_i)^T \phi(x_j) = \frac{\partial k(x_i,x_j)}{\partial x_i}$ and $\phi'(x_i)^T \phi'(x_j) = \frac{\partial^2 k(x_i,x_j)}{\partial x_i \partial x_j}$ follow directly from the definition of the kernel. We can rewrite these identities for vector variables $\mathbf{x}_i, \mathbf{x}_j \in \mathbb{R}^N$ by taking the derivative w.r.t one of the components (say n-th) as $\left(\frac{\partial \phi(\mathbf{x}_i)}{\partial \mathbf{x}(n)}\right)^T \phi(\mathbf{x}_j) = \frac{\partial k(\mathbf{x}_i,\mathbf{x}_j)}{\partial \mathbf{x}_i(n)}$. Expanding the first vector term we get

$$\Rightarrow \left[\frac{\partial \phi_1(\mathbf{x}_i)}{\partial \mathbf{x}(n)}, \frac{\partial \phi_2(\mathbf{x}_i)}{\partial \mathbf{x}(n)}, \cdots, \frac{\partial \phi_F(\mathbf{x}_i)}{\partial \mathbf{x}(n)}\right] \phi(\mathbf{x}_j) = \frac{\partial k(\mathbf{x}_i, \mathbf{x}_j)}{\partial \mathbf{x}_i(n)}.$$

Stacking the above equation in rows for $n = 1 \ldots N$, we get

$$\begin{bmatrix} \frac{\partial \phi_1(\mathbf{x}_i)}{\partial \mathbf{x}(1)} & \frac{\partial \phi_2(\mathbf{x}_i)}{\partial \mathbf{x}(1)} & \cdots & \frac{\partial \phi_F(\mathbf{x}_i)}{\partial \mathbf{x}(1)} \\ \frac{\partial \phi_1(\mathbf{x}_i)}{\partial \mathbf{x}(2)} & \frac{\partial \phi_2(\mathbf{x}_i)}{\partial \mathbf{x}(2)} & \cdots & \frac{\partial \phi_F(\mathbf{x}_i)}{\partial \mathbf{x}(2)} \\ \vdots & \vdots & \vdots & \vdots \\ \frac{\partial \phi_1(\mathbf{x}_i)}{\partial \mathbf{x}(N)} & \frac{\partial \phi_2(\mathbf{x}_i)}{\partial \mathbf{x}(N)} & \cdots & \frac{\partial \phi_F(\mathbf{x}_i)}{\partial \mathbf{x}(N)} \end{bmatrix} \phi(\mathbf{x}_j) = \begin{bmatrix} \frac{\partial k(\mathbf{x}_i,\mathbf{x}_j)}{\partial \mathbf{x}_i(1)} \\ \frac{\partial k(\mathbf{x}_i,\mathbf{x}_j)}{\partial \mathbf{x}_i(2)} \\ \vdots \\ \frac{\partial k(\mathbf{x}_i,\mathbf{x}_j)}{\partial \mathbf{x}_i(N)} \end{bmatrix}$$

$$\Rightarrow \mathrm{J}(\mathbf{x}_i)^T \phi(\mathbf{x}_j) = \frac{\partial k(\mathbf{x}_i, \mathbf{x}_j)}{\partial \mathbf{x}_i} \tag{1.19}$$

where J denotes the standard Jacobian matrix for a vector valued function. Similarly, by writing the derivatives w.r.t (n,m)-th dimension and putting them as the corresponding element of a Hessian matrix we get

$$\mathrm{J}(\mathbf{x}_i)^T \mathrm{J}(\mathbf{x}_j) = \frac{\partial^2 k(\mathbf{x}_i, \mathbf{x}_j)}{\partial \mathbf{x}_i \partial \mathbf{x}_j}. \tag{1.20}$$

Appendix 2: Specific Kernel Expansions

The above formulation is generic and can be applied to any kernel. Here we give the RBF kernel-specific expressions for the block matrices in (1.15).

RBF Kernel

$$[\mathrm{K}]_{ij} = y_i y_j k(\mathbf{x}_i, \mathbf{x}_j) = y_i y_j e^{-d\|\mathbf{x}_i - \mathbf{x}_j\|^2}$$

$$[\mathrm{G}]_{ij} = y_i \left(\frac{\partial k(\mathbf{x}_i, \mathbf{x}_j)}{\partial \mathbf{x}_j} \right)^T \hat{\mathbf{x}}_j = -2d y_i e^{-d\|\mathbf{x}_i - \mathbf{x}_j\|^2} (\mathbf{x}_j - \mathbf{x}_i)^T \hat{\mathbf{x}}_j$$

Replacing $\mathbf{x}_j$ by $\mathbf{x}^*$ in the above equation we get

$$[\mathrm{G}_*]_{ij} = y_i \left(\frac{\partial k(\mathbf{x}_i, \mathbf{x}^*)}{\partial \mathbf{x}^*} \right)^T \mathbf{e}_j = -2d y_i e^{-d\|\mathbf{x}_i - \mathbf{x}^*\|^2} (\mathbf{x}^* - \mathbf{x}_i)^T \mathbf{e}_j$$

$$\begin{aligned}[\mathrm{H}]_{ij} &= \hat{\mathbf{x}}_i^T \frac{\partial^2 k(\mathbf{x}_i, \mathbf{x}_j)}{\partial \mathbf{x}_i \partial \mathbf{x}_j} \hat{\mathbf{x}}_j = \hat{\mathbf{x}}_i^T \left[\frac{\partial}{\partial \mathbf{x}_i} \left\{ -2d e^{-d\|\mathbf{x}_i - \mathbf{x}_j\|^2} (\mathbf{x}_j - \mathbf{x}_i) \right\} \right] \hat{\mathbf{x}}_j \\ &= 2d e^{-d\|\mathbf{x}_i - \mathbf{x}_j\|^2} \left[\hat{\mathbf{x}}_i^T \hat{\mathbf{x}}_j - 2d \left\{ \hat{\mathbf{x}}_i^T (\mathbf{x}_i - \mathbf{x}_j) \right\} \left\{ (\mathbf{x}_i - \mathbf{x}_j)^T \hat{\mathbf{x}}_j \right\} \right].\end{aligned}$$

Again, replacing $\mathbf{x}_j$ by $\mathbf{x}^*$,

$$[\mathrm{H}_*]_{ij} = \hat{\mathbf{x}}_i^T \frac{\partial^2 k(\mathbf{x}_i, \mathbf{x}^*)}{\partial \mathbf{x}_i \partial \mathbf{x}^*} \mathbf{e}_j = 2d e^{-d\|\mathbf{x}_i - \mathbf{x}^*\|^2} \left[\hat{\mathbf{x}}_i^T \mathbf{e}_j - 2d \left\{ \hat{\mathbf{x}}_i^T (\mathbf{x}_i - \mathbf{x}^*) \right\} \left\{ (\mathbf{x}_i - \mathbf{x}^*)^T \mathbf{e}_j \right\} \right].$$

Replacing $\mathbf{x}_i$ also by $\mathbf{x}^*$,

$$[\mathrm{H}_{**}]_{ij} = \mathbf{e}_i^T \frac{\partial^2 k(\mathbf{x}^*, \mathbf{x}^*)}{\partial \mathbf{x}^* \partial \mathbf{x}^*} \mathbf{e}_j = 2d \left(\mathbf{e}_i^T \mathbf{e}_j \right).$$

Polynomial Kernel

$$[\mathrm{K}]_{ij} = y_i y_j k(\mathbf{x}_i, \mathbf{x}_j) = y_i y_j (\mathbf{x}_i^T \mathbf{x}_j + 1)^d$$

$$[\mathrm{G}]_{ij} = y_i \left(\frac{\partial k(\mathbf{x}_i, \mathbf{x}_j)}{\partial \mathbf{x}_j} \right)^T \hat{\mathbf{x}}_j = y_i d (\mathbf{x}_i^T \mathbf{x}_j + 1)^{d-1} \mathbf{x}_i^T \hat{\mathbf{x}}_j.$$

Replacing $\mathbf{x}_j$ by $\mathbf{x}_*$ in the above equation we get

$$[\mathrm{G}_*]_{ij} = y_i \left(\frac{\partial k(\mathbf{x}_i, \mathbf{x}_*)}{\partial \mathbf{x}_*} \right)^T \mathbf{e}_j = y_i d \left(\mathbf{x}_i^T \mathbf{x}_* + 1 \right)^{d-1} \mathbf{x}_i^T \mathbf{e}_j.$$

$$\begin{aligned}
[\mathrm{H}]_{ij} =& \hat{\mathbf{x}}_i^T \frac{\partial^2 k(\mathbf{x}_i, \mathbf{x}_j)}{\partial \mathbf{x}_i \partial \mathbf{x}_j} \hat{\mathbf{x}}_j \\
&= \hat{\mathbf{x}}_i^T \left[\frac{\partial}{\partial \mathbf{x}_i} \left\{ d(\mathbf{x}_i^T \mathbf{x}_j + 1)^{d-1} \mathbf{x}_i \right\} \right] \hat{\mathbf{x}}_j \\
&= \hat{\mathbf{x}}_i^T \left[d(\mathbf{x}_i^T \mathbf{x}_j + 1)^{d-1} \frac{\partial \mathbf{x}_i}{\partial \mathbf{x}_i} + \mathbf{x}_i \frac{\partial}{\partial \mathbf{x}_i} \left\{ d(\mathbf{x}_i^T \mathbf{x}_j + 1)^{d-1} \right\} \right] \hat{\mathbf{x}}_j \\
&= \hat{\mathbf{x}}_i^T \left[d(\mathbf{x}_i^T \mathbf{x}_j + 1)^{d-1} \mathrm{I}_N + d(d-1)(\mathbf{x}_i^T \mathbf{x}_j + 1)^{d-2} \mathbf{x}_i \mathbf{x}_j^T \right] \hat{\mathbf{x}}_j \\
&= d(\mathbf{x}_i^T \mathbf{x}_j + 1)^{d-2} \left[(\mathbf{x}_i^T \mathbf{x}_j + 1) \hat{\mathbf{x}}_i^T \hat{\mathbf{x}}_j + (d-1) \left(\hat{\mathbf{x}}_i^T \mathbf{x}_i \right) \left(\mathbf{x}_j^T \hat{\mathbf{x}}_j \right) \right].
\end{aligned}$$

Again, replacing $\mathbf{x}_j$ by $\mathbf{x}_*$,

$$\begin{aligned}
[\mathrm{H}_*]_{ij} &= \hat{\mathbf{x}}_i^T \frac{\partial^2 k(\mathbf{x}_i, \mathbf{x}_*)}{\partial \mathbf{x}_i \partial \mathbf{x}_*} \mathbf{e}_j \\
&= d(\mathbf{x}_i^T \mathbf{x}_* + 1)^{d-2} \left[(\mathbf{x}_i^T \mathbf{x}_* + 1) \hat{\mathbf{x}}_i^T \mathbf{e}_j + (d-1) \left(\hat{\mathbf{x}}_i^T \mathbf{x}_i \right) \left(\mathbf{x}_*^T \mathbf{e}_j \right) \right].
\end{aligned}$$

Replacing $\mathbf{x}_i$ also by $\mathbf{x}_*$,

$$\begin{aligned}
[\mathrm{H}_{**}]_{ij} &= \mathbf{e}_i^T \frac{\partial^2 k(\mathbf{x}_*, \mathbf{x}_*)}{\partial \mathbf{x}_*^2} \mathbf{e}_j \\
&= d(\mathbf{x}_*^T \mathbf{x}_* + 1)^{d-2} \left[(\mathbf{x}_*^T \mathbf{x}_* + 1) \mathbf{e}_i^T \mathbf{e}_j + (d-1) \left(\mathbf{e}_i^T \mathbf{x}_* \right) \left(\mathbf{x}_*^T \mathbf{e}_j \right) \right].
\end{aligned}$$

References

1. Chiang, H., Chu, C.: A systematic search method for obtaining multiple local optimal solutions of nonlinear programming problems. IEEE Trans. Circ. Syst. I Fundam. Theory Appl. **43**(2), 99–109 (1996)
2. Dixon, K., Khosla, P.: Trajectory representation using sequenced linear dynamical systems. In: Proceedings of 2004 IEEE International Conference on Robotics and Automation, 2004 (ICRA'04), vol. 4, pp. 3925–3930. IEEE (2004)
3. Ellekilde, L., Christensen, H.: Control of mobile manipulator using the dynamical systems approach. In: Proceedings of 2009 IEEE International Conference on Robotics and Automation, 2009 (ICRA'09), pp. 1370–1376. IEEE (2009)
4. Fuchs, A., Haken, H.: Pattern recognition and associative memory as dynamical processes in a synergetic system. I. Translational invariance, selective attention, and decomposition ofscenes. Biol. Cybern. **60**, 17–22 (1988). http://dl.acm.org/citation.cfm?id=56852.56854
5. Hoffmann, H.: Target switching in curved human arm movements is predicted by changing a single control parameter. Exp. Brain Res. **208**(1), 73–87 (2011)
6. Jaeger, H., Lukosevicius, M., Popovici, D., Siewert, U.: Optimization and applications of echo state networks with leaky-integrator neurons. Neural Netw. **20**(3), 335–352 (2007)
7. Khansari-Zadeh, S.M., Billard, A.: Learning stable non-linear dynamical systems with Gaussian mixture models. IEEE Trans. Robot. **27**(5), 943–957 (2011). http://lasa.epfl.ch/khansari
8. Lee, J.: Dynamic gradient approaches to compute the closest unstable equilibrium point for stability region estimate and their computational limitations. IEEE Trans. Automat. Contr. **48**(2), 321–324 (2003)
9. Michel, A., Farrell, J.: Associative memories via artificial neural networks. IEEE Contr. Syst. Mag. **10**(3), 6–17 (1990). doi:10.1109/37.55118
10. Pastor, P., Hoffmann, H., Asfour, T., Schaal, S.: Learning and generalization of motor skills by learning from demonstration. In: Proceedings of 2009 IEEE International Conference on Robotics and Automation, 2009 (ICRA '09), pp. 763–768 (2009). doi:10.1109/ROBOT.2009.5152385
11. Rasmussen, C.: Gaussian processes in machine learning. In: Advanced Lectures on Machine Learning, pp. 63–71. Springer, Berlin (2004)
12. Reimann, H., Iossifidis, I., Schöner, G.: Autonomous movement generation for manipulators with multiple simultaneous constraints using the attractor dynamics approach. In: Proceedings of 2011 IEEE International Conference on Robotics and Automation, 2011 (ICRA), pp. 5470–5477. IEEE (2011)
13. Schaal, S., Atkeson, C., Vijayakumar, S.: Scalable techniques from nonparametric statistics for real time robot learning. Appl. Intell. **17**(1), 49–60 (2002)
14. Schölkopf, B., Smola, A.: Learning with Kernels: Support Vector Machines, Regularization, Optimization, and Beyond. MIT Press, Cambridge (2001)
15. Schöner, G., Dose, M.: A dynamical systems approach to task-level system integration used to plan and control autonomous vehicle motion. Robot. Auton. Syst. **10**(4), 253–267 (1992)
16. Schöner, G., Dose, M., Engels, C.: Dynamics of behavior: theory and applications for autonomous robot architectures. Robot. Auton. Syst. **16**(2), 213–245 (1995)
17. Shukla, A., Billard, A.: Coupled dynamical system based armhand grasping model for learning fast adaptation strategies. Robot. Auton. Syst. **60**(3), 424–440 (2012). doi:10.1016/j.robot.2011.07.023. http://www.sciencedirect.com/science/article/pii/S0921889011001576

Chapter 2
Multi-Class Support Vector Machine

Zhe Wang and Xiangyang Xue

Abstract Support vector machine (SVM) was initially designed for binary classification. To extend SVM to the multi-class scenario, a number of classification models were proposed such as the one by Crammer and Singer (J Mach Learn Res 2:265–292, 2001). However, the number of variables in Crammer and Singer's dual problem is the product of the number of samples (l) by the number of classes (k), which produces a large computational complexity. This chapter sorts the existing classical techniques for multi-class SVM into the indirect and direct ones and further gives the comparison for them in terms of theory and experiments. Especially, this chapter exhibits a new Simplified Multi-class SVM (SimMSVM) that reduces the size of the resulting dual problem from $l \times k$ to l by introducing a relaxed classification error bound. The experimental discussion demonstrates that the SimMSVM approach can greatly speed up the training process, while maintaining a competitive classification accuracy.

2.1 Introduction

Support vector machine (SVM) originally separates the binary classes ($k = 2$) with a maximized margin criterion [6]. However, real-world problems often require the discrimination for more than two categories. Thus, the multi-class pattern recognition has a wide range of applications including optical character recognition [27], intrusion detection [18], speech recognition [11], and bioinformatics [2]. In practice,

Z. Wang (✉)
Department of Computer Science and Engineering, East China University of Science and Technology, Shanghai 200237, China
e-mail: wangzhe@ecust.edu.cn

X. Xue
School of Computer Science, Fudan University, Shanghai 200433, China
e-mail: xyxue@fudan.edu.cn

Y. Ma and G. Guo (eds.), *Support Vector Machines Applications*,
DOI 10.1007/978-3-319-02300-7_2,
© Springer International Publishing Switzerland 2014

the multi-class classification problems ($k > 2$) are commonly decomposed into a series of binary problems such that the standard SVM can be directly applied. Two representative ensemble schemes are one-versus-rest (1VR) [36] and one-versus-one (1V1) [21] approaches. Both one-versus-rest and one-versus-one are special cases of the Error Correcting Output Codes (ECOC) [8] which decomposes the multi-class problem into a predefined set of binary problems. A main issue of this approach is to construct a good ECOC matrix. Another proposals [4, 7, 12, 38] are to directly address the multi-class problem in one single optimization processing. This kind of models combines multiple binary-class optimization problems into one single objective function and simultaneously achieves classification of multiple classes. However, a larger computational complexity is required for the size of the resulting Quadratic Programming (QP) problem.

Moreover, Szedmak et al. [33] proposed a multi-class model for L1-norm SVM. In their formulation, the one-versus-rest framework is used. A potential problem with one-versus-rest is that when the number of classes is large, each binary classification becomes highly unbalanced. The unbalanced classification problem occurs when there are many more samples of some classes than others. In this case, standard classifiers tend to be overwhelmed by the large-scale classes and ignore the small ones. The SVM algorithm is constructing a separating hyperplane with the maximal margin. Since only support vectors are used for classification and many majority samples far from the decision boundary can be removed, SVM can be more accurate on moderately unbalanced data. However, SVM is sensitive to high unbalanced classification since it is prone to generating a classifier that has a strong estimation bias towards the majority class and would give a bad accuracy in the classification performance for the minority class, which is discussed in the work of Chawla et al. [5] and Tang et al. [34]. Wang and Shen [37] proposed a method that circumvents the difficulties of one-versus-rest by treating multiple classes jointly. Suykens and Vandewalle [32] extended the LS-SVM methodology [31] to the multi-class case. A drawback of LS-SVM is that its solution is constructed from most of the training examples, which is referred to as the non-sparseness problem. Xia and Li [39] presented a new multi-class LS-SVM algorithm whose solution is sparse in the weight coefficient of support vectors. Fung and Mangasarian [10] followed the PSVM (proximal SVM) [9] idea and applied it to the multi-class problem. This approach is closely aligned with the one-versus-rest method. For each decomposed subproblem, the solution is similar to its binary case (PSVM) which classifies new samples by assigning them to the closer of the two parallel planes that are pushed apart as far as possible.

This chapter revisits the main multi-class methods including indirect and direct ones for SVM. Especially, in this chapter we introduce a new Simplified Multi-class SVM (SimMSVM) [13]. The SimMSVM gives a direct solution for training multi-class predictors. Following Crammer and Singer's work, SimMSVM introduces a relaxed classification error bound. By doing so, the resulting dual problem only involves l variables, where l is the size of training samples. That is, solving a single l-variable QP is enough for a multi-class classification task. We then investigate the theoretical properties of the loss function of SimMSVM, including the Fisher consistency issue. We also give the experimental comparisons and find that the new simplified model achieves significant speed-up compared with its original model [7].

2.2 Indirect Multi-Class SVM

2.2.1 SVM Classification

SVM seeks to find the optimal separating hyperplane between binary classes by following the maximized margin criterion [6]. Given training vectors $\mathbf{x}_i \in \mathcal{R}^d$, $i = 1,\dots,l$, in two classes, and the label vector $\mathbf{y} \in \{1,-1\}^l$, the support vector technique requires the solution of the following optimization problem:

$$\begin{aligned} \min_{\mathbf{w}\in\mathcal{H},b\in\mathcal{R},\xi_i\in\mathcal{R}} \quad & \frac{1}{2}\mathbf{w}^T\mathbf{w} + C\sum_{i=1}^{l}\xi_i \\ \text{subject to} \quad & y_i(\mathbf{w}^T\varphi(\mathbf{x}_i) + b) \geq 1 - \xi_i, \\ & \xi_i \geq 0, \quad i = 1,\dots,l, \end{aligned} \tag{2.1}$$

where $\mathbf{w} \in \mathcal{R}^d$ is the weight vector, $C \in \mathcal{R}_+$ is the regularization constant, and the mapping function φ projects the training data into a suitable feature space $\mathcal{H}$ so as to allow for nonlinear decision surfaces. A common method to solve (2.1) is through its dual:

$$\begin{aligned} \min_{\alpha\in\mathcal{R}^l} \quad & \frac{1}{2}\alpha^T(\mathbf{K}\odot(\mathbf{y}\mathbf{y}^T))\alpha - \mathbf{e}^T\alpha \\ \text{subject to} \quad & \mathbf{0} \leq \alpha \leq C\mathbf{e}, \quad \mathbf{y}^T\alpha = 0, \end{aligned} \tag{2.2}$$

where $\mathbf{e} \in \mathcal{R}^l$ is a vector of all 1's, $\mathbf{K} \in \mathcal{R}^{l\times l}$ is the kernel matrix, and $\odot$ denotes the Hadamard–Schur (elementwise) product. Symbols in bold represent matrices or vectors, as particular case, the symbol $\mathbf{0}$ represents the vector with all components set to 0.

Instead of (2.1), Mangasarian and Musicant [25] proposed a new SVM formulation:

$$\begin{aligned} \min_{\mathbf{w}\in\mathcal{H},b\in\mathcal{R},\xi_i\in\mathcal{R}} \quad & \frac{1}{2}(\mathbf{w}^T\mathbf{w} + b^2) + C\sum_{i=1}^{l}\xi_i \\ \text{subject to} \quad & y_i(\mathbf{w}^T\varphi(\mathbf{x}_i) + b) \geq 1 - \xi_i, \\ & \xi_i \geq 0, \quad i = 1,\dots,l, \end{aligned} \tag{2.3}$$

whose dual becomes a bound-constrained problem:

$$\begin{aligned} \min_{\alpha \in \mathcal{R}^l} \quad & \frac{1}{2}\alpha^T \left((\mathbf{K} + \mathbf{e}\mathbf{e}^T) \odot (\mathbf{y}\mathbf{y}^T) \right) \alpha - \mathbf{e}^T \alpha \\ \text{subject to} \quad & \mathbf{0} \leq \alpha \leq C\mathbf{e}. \end{aligned} \tag{2.4}$$

The motivation behind (2.3) is that, it is supposed easier to handle its dual problem (2.4) without linear constraint. Hsu and Lin [15] demonstrate that (2.3) is an acceptable formulation in terms of generalization errors though an additional term $b^2/2$ is added to the objective. This method is implemented in the software BSVM [16] which is proved to have fast convergence in the literature [15].

2.2.2 *One-Versus-Rest Approach*

The one-versus-rest (1VR) approach [36] constructs k separate binary classifiers for k-class classification. The m-th binary classifier is trained using the data from the m-th class as positive examples and the remaining $k-1$ classes as negative examples. During test, the class label is determined by the binary classifier that gives maximum output value. A major problem of the one-versus-rest approach is the imbalanced training set. Suppose that all classes have an equal size of training examples, the ratio of positive to negative examples in each individual classifier is $\frac{1}{k-1}$. In this case, the symmetry of the original problem is lost.

2.2.3 *One-Versus-One Approach*

Another classical approach for multi-class classification is the one-versus-one (1V1) or pairwise decomposition [20]. It evaluates all possible pairwise classifiers and thus induces $k(k-1)/2$ individual binary classifiers. Applying each classifier to a test example would give one vote to the winning class. A test example is labeled to the class with the most votes. The size of classifiers created by the one-versus-one approach is much larger than that of the one-versus-rest approach. However, the size of QP in each classifier is smaller, which makes it possible to train fast. Moreover, compared with the one-versus-rest approach, the one-versus-one method is more symmetric. Platt et al. [28] improved the one-versus-one approach and proposed a method called Directed Acyclic Graph SVM (DAGSVM) that forms a tree-like structure to facilitate the testing phase. As a result, it takes only $k-1$ individual evaluations to decide the label of a test example.

2.3 Direct Multi-Class SVM

2.3.1 *Weston and Watkins' Multi-Class SVM*

Instead of creating several binary classifiers, a more natural way is to distinguish all classes in one single optimization processing, as proposed by Vapnik [36], Weston and Watkins [38], and Bredensteiner and Bennett [4]. For a k-class problem, these methods design a single objective function for training all k-binary SVMs simultaneously and maximize the margins from each class to the remaining ones. Here, we take the method by Weston and Watkins as an example. Given a labeled training set represented by $\{(\mathbf{x}_1, y_1), \ldots, (\mathbf{x}_l, y_l)\}$ of cardinality l, where $\mathbf{x}_i \in \mathcal{R}^d$ and $y_i \in \{1, \ldots, k\}$, the formulation proposed in [38] is given as follows:

$$\begin{aligned} \min_{\mathbf{w}_m \in \mathcal{H}, \mathbf{b} \in \mathcal{R}^k, \xi \in \mathcal{R}^{l \times k}} \quad & \frac{1}{2} \sum_{m=1}^{k} \mathbf{w}_m^T \mathbf{w}_m + C \sum_{i=1}^{l} \sum_{t \neq y_i} \xi_{i,t} \\ \text{subject to} \quad & \mathbf{w}_{y_i}^T \varphi(\mathbf{x}_i) + b_{y_i} \geq \mathbf{w}_t^T \varphi(\mathbf{x}_i) + b_t + 2 - \xi_{i,t}, \\ & \xi_{i,t} \geq 0, \\ & i = 1, \ldots, l, \ t \in \{1, \ldots, k\} \setminus y_i. \end{aligned} \tag{2.5}$$

The resulting decision function is

$$\operatorname{argmax}_m f_m(\mathbf{x}) = \operatorname{argmax}_m (\mathbf{w}_m^T \varphi(\mathbf{x}) + b_m). \tag{2.6}$$

The main disadvantage of this approach is that the computational time may be very high due to the enormous size of the resulting QP.

2.3.2 *Crammer and Singer's Multi-Class SVM*

Crammer and Singer [7] presented another "all-together" approach by solving the following optimization problem:

$$\begin{aligned} \min_{\mathbf{w}_m \in \mathcal{H}, \xi \in \mathcal{R}^l} \quad & \frac{1}{2} \sum_{m=1}^{k} \mathbf{w}_m^T \mathbf{w}_m + C \sum_{i=1}^{l} \xi_i \\ \text{subject to} \quad & \mathbf{w}_{y_i}^T \varphi(\mathbf{x}_i) - \mathbf{w}_t^T \varphi(\mathbf{x}_i) \geq 1 - \delta_{y_i,t} - \xi_i, \\ & i = 1, \ldots, l, \ t \in \{1, \ldots, k\}, \end{aligned} \tag{2.7}$$

where $\delta_{i,j}$ is the Kronecker delta (defined as 1 for $i = j$ and as 0 otherwise). The resulting decision function is

$$\text{argmax}_m f_m(\mathbf{x}) = \text{argmax}_m \mathbf{w}_m^T \varphi(\mathbf{x}). \tag{2.8}$$

Note that the constraints $\xi_i \geq 0, i = 1, \ldots, l$, are implicitly indicated in the margin constraints of (2.7) when t equals y_i. In addition, (2.7) focuses on classification rule (2.8) without the bias terms b_m, $m = 1, \ldots, k$. A nonzero b_m can be easily modeled by adding an additional constant feature to each $\mathbf{x}$ (see, e.g., [26]).

2.3.3 *Simplified Multi-Class SVM (SimMSVM)*

2.3.3.1 Rationale of SimMSVM

Although Crammer and Singer's multi-class SVM gives a compact set of constraints, the number of variables in its dual problem is still $l \times k$ [7]. This value may explode even for small datasets. For instance, an English letter recognition problem with 2,600 samples (100 samples per letter) would require solving a QP of size 2,600 × 26, which will result in a large computational complexity. Here, we follow Crammer and Singer's work and further introduce a simplified method named SimMSVM for relaxing its constraints [13]. By doing so, solving one single l-variable QP is enough for a multi-class classification task.

Before we describe the new direct multi-class SVM method [13], we first compare the loss function of the above two "all-together" approaches. For a training example $\mathbf{x}_i$, we let

$$\zeta_{i,m} = 1 - f_{y_i}(\mathbf{x}_i) + f_m(\mathbf{x}_i), \tag{2.9}$$

for $m \in \{1, \ldots, k\} \setminus y_i$. If $\zeta_{i,m} > 0$, it depicts the pairwise margin violation between the true class y_i and some other class m. In Weston and Watkins' work, their loss function adds up all positive margin violation ($\zeta_{i,m} > 0$),

$$\xi_i^{(1)} = \sum_{m \neq y_i} [\zeta_{i,m}]_+, \tag{2.10}$$

where $[\cdot]_+ \equiv \max(\cdot, 0)$. In the original work proposed by Weston and Watkins [38], the term ζ adopts the "2" rather than "1" as follows:

$$\zeta_{i,m} = 2 - f_{y_i}(\mathbf{x}_i) + f_m(\mathbf{x}_i),$$

Here in order to compare the work proposed by Weston and Watkins [38] with the other methods consistently, we scale the "2" into "1," i.e. adopt the Eq. (2.9). As for Crammer and Singer's approach, sample loss counts only the maximum positive margin violation

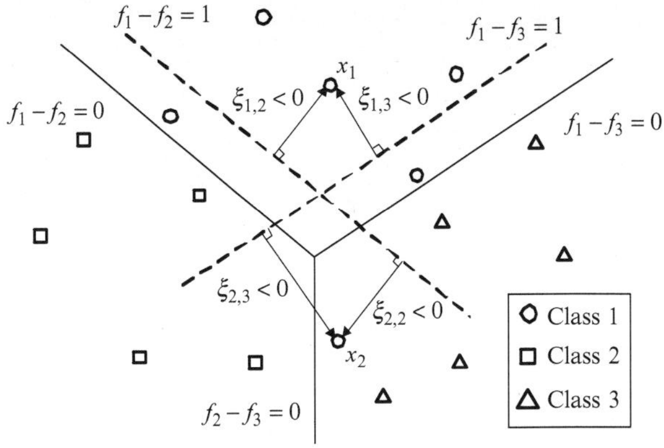

Fig. 2.1 Multi-class classification visualization. Class 1, 2, and 3 are represented by *circles*, *rectangles*, and *triangles*, respectively. The *bold lines* represent some possible class boundaries. The two *dash lines* are two positive margins for the first class. The pairwise margin violations for two examples from the first class, $\zeta_{1,2}$ and $\zeta_{1,3}$ for $\mathbf{x}_1$, and $\zeta_{2,2}$ and $\zeta_{2,3}$ for $\mathbf{x}_2$, are depicted in the figure

$$\xi_i^{(2)} = \left[\max_{m \neq y_i} \zeta_{i,m} \right]_+ . \tag{2.11}$$

Figure 2.1 gives a multi-class classification graphical illustration, where three classes are represented as circles, rectangles, and triangles, respectively. The bold lines represent some possible decision boundaries. We plot the two positive pairwise margins for the first class (shown in dash lines). For a correctly classified example $\mathbf{x}_1$, we have $\xi_1^{(1)} = 0$ and $\xi_1^{(2)} = 0$, i.e., no loss counted, since both $\zeta_{1,2}$ and $\zeta_{1,3}$ are negative. On the other hand, for an example $\mathbf{x}_2$ that violates two margin bounds ($\zeta_{2,2}, \zeta_{2,3} > 0$), both methods generate a loss.

In order to reduce the problem size, the number of constraints should be proportional to l instead of $l \times k$. To construct the multi-class predictors of SimMSVM, we introduce the following relaxed bound

$$\xi_i^{(3)} = \left[\frac{1}{k-1} \sum_{m=1, m \neq y_i}^{k} \zeta_{i,m} \right]_+ . \tag{2.12}$$

That is, we suffer a loss which is linearly proportional to the average of directed distances between an example and its pairwise margins. For the above example shown in Fig. 2.1, the loss (2.12) yields no loss for $\mathbf{x}_1$ as the other two loss functions.

To clearly analyze the SimMSVM, we compare the three loss functions (2.10)–(2.12). The loss (2.10) adds up all positive margin violation ($\zeta_{i,m} > 0$). The loss (2.11) counts only the maximum positive margin violation. The loss (2.12) represents a linear proportion to the average of directed distances between a sample and its pairwise margins. Thus, the relationship between these losses is

Table 2.1 The rate of f_{y_i} be the maximum value of $f_1, \ldots, f_k$ for the ten datasets

Dataset	Iris	Wine	Glass	Vowel	Segment
Rate	98.67%	99.44 %	89.25 %	100.00 %	99.70 %
Dataset	Waveform	DNA	Satimage	Letter	USPS
Rate	94.20 %	98.45 %	99.66 %	99.97 %	99.93 %

$$\xi_i^{(1)} \geq \xi_i^{(2)} \geq \xi_i^{(3)}. \tag{2.13}$$

Figure 2.1 gives a visual example for the three loss functions (2.10)–(2.12). According to their definition, the corresponding losses for the sample x_2 that violates two margin bounds in Fig. 2.1 are

$$\begin{aligned} \xi_2^{(1)} &= \zeta_{2,2} + \zeta_{2,3} \\ \xi_2^{(2)} &= \max\{\zeta_{2,2}, \zeta_{2,3}\} = \zeta_{2,3} \\ \xi_2^{(3)} &= (\zeta_{2,2} + \zeta_{2,3})/2 \end{aligned}$$

From the inequality (2.13), the relationship in Fig. 2.1 for the losses is

$$\xi_2^{(1)} \geq \xi_2^{(2)} \geq \xi_2^{(3)}.$$

In order to clearly explore the chosen loss function, we rewrite the relaxed loss function (2.12) as

$$\xi_i = \left[1 - f_{y_i}(\mathbf{x}_i) + \frac{1}{k-1} \sum_{m \neq y} f_m(\mathbf{x}_i) \right]_+ . \tag{2.14}$$

In other words, the SimMSVM will yield no loss for a training example $\mathbf{x}_i$ if $f_{y_i}(\mathbf{x}_i) \geq 1 + \frac{1}{k-1} \sum_{m \neq y_i} f_m(\mathbf{x}_i)$. Although we cannot guarantee if $\xi_i = 0$ that $f_{y_i}(\mathbf{x}_i) > f_m(\mathbf{x}_i)$ for all $m \neq y_i$, it is most likely that f_{y_i} will be the maximum value of $f_1, \ldots, f_k$. To further explore this statement, we empirically study the problem in all of the ten datasets. Table 2.1 gives the rates that f_{y_i} be the maximum value of $f_1, \ldots, f_k$ and thus validate the claim about the loss function of SimMSVM. In this sense, the (2.14) is reasonable as a classification criterion.

In theory, the relaxed loss function (2.14) may incur more classification error since it fails to upper-bound the $0-1$ loss. In practice, we admit some tolerable classification error in order to obtain a significant speed-up in the training process. The experimental result in Sect. 2.3.4 demonstrate that, SimMSVM gives competitive accuracies with the other multi-class SVM approaches on real world datasets. On the other hand, the multi-class hinge losses of Weston and Watkins [38], Crammer and Singer [7] both satisfy as the upper bound of $0-1$ loss. However, in practice, they all involve a QP of size $l \times k$ and thus cause a high computational complexity for a relatively large number of classes.

2.3.3.2 Architecture of SimMSVM

Induced by the above error bound, the unbiased primal problem of SimMSVM comes up as follows:

$$\begin{aligned} \min_{\mathbf{w}_m \in \mathcal{H}, \xi \in \mathcal{R}^l} \quad & \frac{1}{2}\sum_{m=1}^{k} \mathbf{w}_m^T \mathbf{w}_m + C\sum_{i=1}^{l} \xi_i \\ \text{subject to} \quad & \mathbf{w}_{y_i}^T \varphi(\mathbf{x}_i) - \frac{1}{k-1}\sum_{m \neq y_i} \mathbf{w}_m^T \varphi(\mathbf{x}_i) \geq 1 - \xi_i, \\ & \xi_i \geq 0, \quad i = 1, \ldots, l. \end{aligned} \tag{2.15}$$

Note that the number of margin constraints (2.15) has already been reduced to l. To solve the optimization problem (2.15) we use the Karush—Kuhn—Tucker (KKT) theorem. We add a dual set of variables, one for each constraint and get the Lagrangian of the optimization problem

$$\begin{aligned} L(\mathbf{W}, \xi, \alpha, \lambda) = & \frac{1}{2}\sum_{m=1}^{k} \mathbf{w}_m^T \mathbf{w}_m + C\sum_{i=1}^{l} \xi_i - \sum_{i=1}^{l} \lambda_i \xi_i \\ & - \sum_{i=1}^{l} \alpha_i \left(\mathbf{w}_{y_i}^T \varphi(\mathbf{x}_i) - \frac{1}{k-1}\sum_{m \neq y_i} \mathbf{w}_m^T \varphi(\mathbf{x}_i) - 1 + \xi_i \right). \end{aligned} \tag{2.16}$$

By differentiating the Lagrangian with respect to the primal variables, we obtain the following constraints that the variables have to fulfill in order to be an optimal solution:

$$\frac{\partial L}{\partial \mathbf{w}_m} = 0 \Longleftrightarrow \mathbf{w}_m = \sum_{i:y_i = m} \alpha_i \varphi(\mathbf{x}_i) - \frac{1}{k-1}\sum_{i:y_i \neq m} \alpha_i \varphi(\mathbf{x}_i), \tag{2.17}$$

$$\frac{\partial L}{\partial \xi} = 0 \Longleftrightarrow C\mathbf{e} = \lambda + \alpha, \tag{2.18}$$

subject to the constraints $\alpha \geq \mathbf{0}$ and $\lambda \geq \mathbf{0}$.

By elimination of $\mathbf{w}_m$, ξ, and λ, the dual problem of (2.15) is reduced to the following succinct form

$$\begin{aligned} \min_{\alpha \in \mathcal{R}^l} \quad & \frac{1}{2}\alpha^T \mathbf{G} \alpha - \mathbf{e}^T \alpha \\ \text{subject to} \quad & \mathbf{0} \leq \alpha \leq C\mathbf{e}. \end{aligned} \tag{2.19}$$

The Hession $\mathbf{G}$ is an $l \times l$ matrix with its entries

$$G_{i,j} = \begin{cases} \frac{k}{k-1} K_{i,j}, & \text{if } y_i = y_j, \\ \frac{-k}{(k-1)^2} K_{i,j}, & \text{if } y_i \neq y_j, \end{cases} \tag{2.20}$$

where we abbreviate the kernel value $\kappa(\mathbf{x}_i, \mathbf{x}_j) \equiv \varphi(\mathbf{x}_i)^T \varphi(\mathbf{x}_j)$ for $K_{i,j}$. Furthermore, let $\mathbf{V}$ be an $l \times k$ matrix defined as:

$$V_{i,j} = \begin{cases} 1, & \text{if } y_i = j, \\ \frac{-1}{k-1}, & \text{if } y_i \neq j. \end{cases} \tag{2.21}$$

We have

$$\mathbf{G} = \mathbf{K} \odot (\mathbf{V}\mathbf{V}^T). \tag{2.22}$$

Since the kernel matrix $\mathbf{K}$ and $\mathbf{V}\mathbf{V}^T$ are both positive semi-definite, so is their Hadamard product $\mathbf{G}$. In the next section, we show that, up to a change of variables, (2.19) is equivalent to the dual problem of Crammer and Singer's approach with an additional constraint.

In order to avoid the division operation in the kernel computation of (2.20), we further let $\bar{\alpha} = \frac{k}{(k-1)^2}\alpha$. Therefore, we rewrite (2.19) as

$$\begin{aligned} \min_{\bar{\alpha} \in \mathcal{R}^l} \quad & \frac{1}{2}\bar{\alpha}^T \bar{\mathbf{G}} \bar{\alpha} - \mathbf{e}^T \bar{\alpha} \\ \text{subject to} \quad & 0 \leq \bar{\alpha} \leq \bar{C}, \end{aligned} \tag{2.23}$$

where $\bar{C} = \frac{k}{(k-1)^2}C$, and

$$\bar{G}_{i,j} = \begin{cases} (k-1)K_{i,j}, & \text{if } y_i = y_j, \\ -K_{i,j}, & \text{if } y_i \neq y_j. \end{cases} \tag{2.24}$$

As for the decision function, from (2.17) we obtain

$$\begin{aligned} f_m^*(\mathbf{x}) &= \sum_{i:y_i=m} \alpha_i^* \kappa(\mathbf{x}_i, \mathbf{x}) - \frac{1}{k-1} \sum_{i:y_i \neq m} \alpha_i^* \kappa(\mathbf{x}_i, \mathbf{x}) \\ &= \frac{k}{k-1} \sum_{i:y_i=m} \alpha_i^* \kappa(\mathbf{x}_i, \mathbf{x}) - \underbrace{\frac{1}{k-1} \sum_{i=1}^{l} \alpha_i^* \kappa(\mathbf{x}_i, \mathbf{x})}_{\text{the same for all } f_m^*}. \end{aligned} \tag{2.25}$$

Finally, the resulting decision function can be simplified as

$$\arg\max_m f_m^*(\mathbf{x}) = \arg\max_m \sum_{i:y_i=m} \alpha_i^* \kappa(\mathbf{x}_i, \mathbf{x}). \tag{2.26}$$

Thus, each class model is built upon its own support vectors.

Table 2.2 Dataset description

Dataset	#Training data (l)	#Test data	#Class (k)	#Attr (d)
Iris	150	0	3	4
Wine	178	0	3	13
Glass	214	0	6	9
Vowel	528	0	11	10
Segment	2,310	0	7	19
Waveform	5,000	0	3	21
DNA	2,000	1,186	3	180
Satimage	4,435	2,000	6	36
Letter	15,000	5,000	26	16
USPS	7,291	2,007	10	256

2.3.4 *Example: A Comparative Study*

In this section, we validate the accuracy and efficiency of the new SimMSVM algorithm on several multi-class pattern recognition problems including **Iris**, **Wine**, **Glass**, **Vowel**, **Segment**, and **Waveform** from the **UCI** repository [1], **DNA**, **Satimage**, **Letter** from **Statlog** collection [19], and **USPS** [17]. For all the datasets except **DNA**, we linearly scale each attribute to be in the range $[-1, 1]$. Table 2.2 gives a detailed description about these datasets. Here we compare the new SimMSVM with some classical multi-class algorithms in terms of classification and computation time. Moreover, we also analyze the scaling behavior of the SimMSVM with the sample size and the number of classes in terms of training time. All the experiments are performed using the tool of BSVM [16] on a 1.50 GHz Intel Core^{TM}2 Duo PC running Windows XP with 1 GB main memory.

2.3.4.1 Classification Performance Comparisons

The tool of BSVM is used in the experiments. BSVM is developed to efficiently solve the bound-constrained problem (2.4). We find that the only difference between (2.19) and (2.4) is the Hessian of the objective function. To proceed the SimMSVM algorithm and take advantage of its sophisticated implementations, we only need to modify the kernel evaluation part of BSVM. For the one-versus-one (1V1) and one-versus-rest (1VR) approaches, we formulate their binary subproblems by (2.3) and solve each of them using BSVM. For the two so-called all-together methods proposed by Weston and Watkins and Crammer and Singer, their major disadvantage remains to the enormous size of the optimization problems. To tackle these difficulties, Hsu and Lin proposed decomposition methods for these two approaches, respectively. The implementations are integrated into BSVM as well.

In conclusion, we adopt BSVM for three reasons: (1) BSVM uses a simple working set selection which leads to a faster convergence for difficult cases.

The use of a special implementation of the optimization solver allows BSVM to stably identify bounded variables. (2) The implementations of Weston and Watkins' approach [38] and Crammer and Singer's approach [7] are integrated into BSVM. We can formulate the binary subproblems of the one-versus-one and one-versus-rest approaches and solve each of them using BSVM. (3) For the SimMSVM algorithm, we just modify the kernel evaluation part of BSVM. Furthermore, it should be stated that as described in Hsu and Lin [14], the BSVM software includes a decomposition implementation of Crammer and Singer's approach Crammer and Singer [7] for multi-class problems. The other three approaches (one-versus-one, one-versus-rest, and Weston and Watkins' approach) are also based on BSVM for the fair comparison. Therefore, the experiments of this manuscript are all implemented in a fair condition.

The most important criterion for evaluating the performance of the multi-class SVM is the classification accuracy. For all the five methods, we use the gaussian RBF kernel

$$\kappa(\mathbf{x}_i, \mathbf{x}_j) = \exp\left(-\gamma\|\mathbf{x}_i - \mathbf{x}_j\|^2\right), \quad \gamma \geq 0.$$

We follow the strategy used in Hsu and Lin [14] and use model selection to get the optimal parameters. The hyper-parameter space is explored on a two-dimensional grid with $\gamma = [10^{-2}, 10^{-1}, 10^{0}, 10^{1}]$ and the regularization parameter $C = [10^{-1}, 10^{0}, 10^{1}, 10^{2}, 10^{3}]$. We use two criteria to estimate the generalized accuracy.

For the used **Iris**, **Wine**, **Glass**, **Vowel**, **Segment**, and **Waveform**, we randomly subdivide each dataset into ten disjoint subsets. Then for each procedure, one of the subsets is used for testing and the others for training, where the tenfold cross-validation strategy is employed in the training set so as to select the optimal parameter C and γ. In order to evaluate the variability over different random splits for each dataset, the whole procedure above is repeated five times and the final experimental results are averaged. For the used **DNA**, **Satimage**, **Letter**, and **USPS** where both training and testing sets are available, we randomly draw the subsets of the original training set for training in order to evaluate the variability over different choices of the training set on each dataset. Here the 70 % data points are randomly chosen from the original training set, where the tenfold cross-validation is used for parameter selection on the selected training set. Then we train on the 70 % dataset with the optimal pair (C, γ) and predict for the test set. The whole procedure above is repeated five times and the results are averaged. We present the comparison results in Table 2.3, where "W&W" denotes the method proposed by Weston and Watkins and "C&S" indicates Crammer and Singers approach.

Table 2.3 shows the optimal classification accuracy and their corresponding number of support vectors (SV). Both the average standard deviation of the classification accuracy and the number of SV are also reported in Table 2.3 so as to evaluate the statistical significance of the cross-validation results and the number of SV are not integers since they are the average values of the tenfold cross-validation. From Table 2.3, we can find that the SimMSVM gives competitive accuracies with

Table 2.3 Classification accuracy (%) and the number of SV comparison between 1VR, 1V1, W&W, C&S, and SimMSVM (best rates boldfaced)

	1VR		1V1		W&W		C&S		SimMSVM	
Dataset	ACC	SV	ACC	SV	YACC	YSV	ACC	SV	ACC	SV
Iris	95.47	43.0	95.47	40.1	96.00	33.3	95.33	35.7	**96.53**	18.0
	±5.62	±17.5	±5.46	±24.5	±5.04	±18.8	±5.08	±23.6	±4.52	±15.4
Wine	96.82	64.2	**97.76**	70.8	97.53	64.9	97.06	64.0	97.18	101.6
	±4.15	±25.9	±3.74	±17.4	±3.58	±28.4	±3.41	±36.4	±3.80	±15.0
Glass	65.90	140.7	68.67	128.8	67.81	141.3	69.05	152.8	**71.62**	136.6
	±9.69	±10.7	±10.50	±8.2	±10.02	±19.2	±10.77	±17.1	±10.36	±4.2
Vowel	97.88	348.2	**99.31**	346.0	98.65	312.8	98.88	367.1	99.08	403.9
	±2.37	±5.5	±1.27	±4.8	±1.61	±60.8	±1.61	±64.0	±1.24	±4.1
Segment	96.73	483.5	97.23	444.4	97.07	338.5	97.31	626.1	**97.33**	1106.3
	±0.93	±43.9	±0.96	±53.7	±0.87	±57.8	±0.83	±384.8	±0.91	±16.5
Waveform	86.38	1843.6	**87.06**	2374.6	87.00	2187.1	86.75	2061.2	86.64	2870.0
	±1.77	±180.7	±1.16	±302.1	±1.17	±285.3	±1.52	±149.9	±1.32	±8.5
DNA	93.79	808.4	94.69	734.4	94.77	708.2	**94.95**	660.2	93.81	458.6
	±0.59	±10.3	±0.32	±84.4	±0.34	±72.6	±0.31	±37.3	±0.23	±4.7
Satimage	89.20	1312.2	**91.21**	1233.2	90.72	1122.8	90.60	1467.4	90.57	2731.0
	±0.44	±18.9	±0.50	±23.8	±0.31	±34.8	±0.24	±735.3	±0.41	±8.6
Letter	94.01	4464.4	**97.04**	5439.4	96.52	3541.8	96.39	3424.8	96.37	8126.2
	±0.24	±13.9	±0.16	±22.7	±0.21	±30.6	±0.17	±21.1	±0.14	±27.3
USPS	93.49	1348.6	94.96	1247.2	94.83	974.4	94.80	2486.2	**95.00**	3349.2
	±0.27	±14.4	±0.22	±10.8	±0.41	±20.2	±0.52	±138.2	±0.37	±17.5

the method by Crammer and Singer. In detail, we observe that: (1) not all the 1VR and 1V1 approaches outperform "all together" approaches (i.e., W&W, C&S, and SimMSVM) in the used ten datasets; (2) all the five implemented methods give the comparable results. The experimental results in the previous work such as Hsu and Lin [14] also suggest that no one is statistically better than the others for the 1VR and 1V1 approaches and all together approaches.

Since the test time of the W&W, C&S, and SimMSVM depends on the number of SV, we analyze their decision function expressions. Suppose that there are kernel function $\kappa(\mathbf{x}_i, \mathbf{x})$ and let k and n_{SV} be the number of class and SVs. The computational complexity of Eq. (2.25) (i.e. $\arg\max_m f_m^*(\mathbf{x}) = \arg\max_m \sum_{i:y_i=m} \alpha_i^* \kappa(\mathbf{x}_i, \mathbf{x})$) is $O(n_{SV})$ since each SV is operated once for testing. According to Hsu and Lin [14], Eq. (2.6) (i.e., $\arg\max_m f_m(\mathbf{x}) = \arg\max_m(\mathbf{w}_m^T \varphi(\mathbf{x}) + b_m)$) can be rewritten as

$$\arg\max_m f_m(\mathbf{x}) = \arg\max_m \sum_{i=1}^{l} (c_i^m A_i - \alpha_i^m)(\kappa(\mathbf{x}_i, \mathbf{x}) + 1)$$

where $A_i = \sum_m \alpha_i^m$, $c_i^m = 1$, if $y_i = m$ and $c_i^m = 0$, otherwise. Furthermore, Eq. (2.8) (i.e., $\arg\max_m f_m(\mathbf{x}) = \arg\max_m \mathbf{w}_m^T \varphi(\mathbf{x})$) can be rewritten as

$$\arg\max_m f_m(\mathbf{x}) = \arg\max_m \sum_{i=1}^{l} \alpha_i^m \kappa(\mathbf{x}_i, \mathbf{x})$$

Therefore, the computational complexity of Eqs. (2.6) and (2.8) is $O(kn_{SV})$. A quantitative test time comparison of the used datasets among these methods is also given in Table 2.4. From this table, the SimMSVM has a comparable test time to the multi-class SVM methods W&W and C&S.

2.3.4.2 Computational Complexity Comparisons

We have claimed that the SimMSVM achieves a significant speed-up compared with the method by Crammer and Singer for its simplified model. To demonstrate it, we report the average training time of the above model selection procedures. Due to the limit of the space here, we select the six out of ten datasets for comparison, e.g., **Vowel**, **Segment**, **Waveform**, **Satimage**, **Letter**, and **USPS**. For the six datasets, we report the average cross-validation time of the five competitive methods as shown in Figs. 2.2 and 2.3. Each row of Figs. 2.2 and 2.3 has the parameter γ with the value 10^{-2}, 10^{-1}, 10^{0}, and 10^{1}, respectively, and the C varies from 10^{-1} to 10^{3} in a log scale. For the parameter pair (C, γ) for each learning algorithm, we report the average time on ten runs of the cross-validation. Most of the standard deviation values are below 0.05 in the experiments.

Figures 2.2 and 2.3 show that the SimMSVM has the best computational cost for most of the parameter combinations. For example, the 18 out of 20

Table 2.4 Testing time (in second) and the number of SV comparison between W&W, C&S, and SimMSVM

Dataset	W&W			C&S			SimMSVM		
(#Class)	Time	SV	$\frac{\text{Time}}{\text{SV}}$	Time	SV	$\frac{\text{Time}}{\text{SV}}$	Time	SV	$\frac{\text{Time}}{\text{Sv}}$
Iris	2.44×10^{-4}	33.3	7.31×10^{-6}	3.69×10^{-4}	35.7	10.34×10^{-6}	0.63×10^{-4}	18.0	3.52×10^{-6}
(3)	$\pm 1.34 \times 10^{-4}$	± 18.8		$\pm 2.44 \times 10^{-4}$	± 23.6		$\pm 0.54 \times 10^{-4}$	± 15.4	
Wine	5.31×10^{-4}	64.9	8.19×10^{-6}	7.48×10^{-4}	64.0	11.69×10^{-6}	4.08×10^{-4}	101.6	4.01×10^{-6}
(3)	$\pm 2.30 \times 10^{-4}$	± 28.4		$\pm 4.26 \times 10^{-4}$	± 36.4		$\pm 0.64 \times 10^{-4}$	± 15.0	
Glass	35.40×10^{-4}	141.3	25.07×10^{-6}	44.30×10^{-4}	152.8	28.97×10^{-6}	6.75×10^{-4}	136.6	4.94×10^{-6}
(6)	$\pm 4.89 \times 10^{-4}$	± 19.2		$\pm 4.91 \times 10^{-4}$	± 17.1		$\pm 0.21 \times 10^{-4}$	± 4.2	
Vowel	3.87×10^{-2}	312.8	12.37×10^{-5}	4.83×10^{-2}	367.1	13.17×10^{-5}	0.49×10^{-2}	403.9	1.21×10^{-5}
(11)	$\pm 0.76 \times 10^{-2}$	± 60.8		$\pm 0.84 \times 10^{-2}$	± 64.0		$\pm 0.00 \times 10^{-2}$	± 4.1	
Segment	1.12×10^{-1}	338.5	32.98×10^{-5}	2.34×10^{-1}	626.1	37.32×10^{-5}	0.60×10^{-1}	1106.3	5.39×10^{-5}
(7)	$\pm 0.19 \times 10^{-1}$	± 57.8		$\pm 1.44 \times 10^{-1}$	± 384.8		$\pm 0.01 \times 10^{-1}$	± 16.5	
Waveform	0.52	2187.1	2.4×10^{-4}	0.72	2061.2	3.4×10^{-4}	0.34	2870.0	1.2×10^{-4}
(3)	± 0.07	± 285.3		± 0.05	± 149.9		± 0.00	± 8.5	
DNA	0.63	708.2	9×10^{-4}	0.82	660.2	12×10^{-4}	0.28	458.6	6×10^{-4}
(3)	± 0.01	± 72.6		± 0.01	± 37.3		± 0.00	± 4.7	
Satimage	2.66	1122.8	24×10^{-4}	3.07	1467.4	21×10^{-4}	1.28	2731.0	5×10^{-4}
(6)	± 0.31	± 34.8		± 0.24	± 735.3		± 0.41	± 8.6	
Letter	107.1	3541.8	30×10^{-3}	103.7	3424.8	30×10^{-3}	9.5	8126.2	1×10^{-3}
(26)	± 0.6	± 30.6		± 0.8	± 21.1		± 0.0	± 27.3	
USPS	4.19	974.4	43×10^{-4}	16.46	2486.2	66×10^{-4}	1.59	3349.2	4×10^{-4}
(10)	± 0.07	± 20.2		± 0.17	± 138.2		± 0.01	± 17.5	

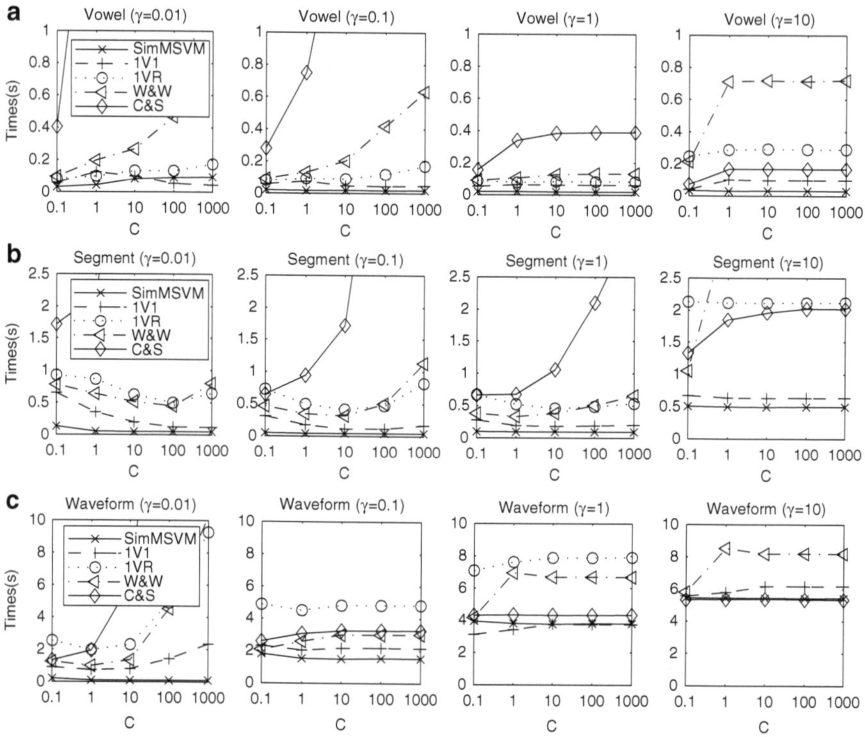

Fig. 2.2 Cross-validation time (in seconds) comparison as a function of C for $\gamma = 0.01,\ 0.1,\ 1,$ and 10 on (**a**) Vowel dataset, (**b**) Segment dataset, (**c**) Waveform dataset

parameter combinations in **Vowel** dataset by the SimMSVM outperform the other methods. All the parameter combinations in **Satimage** dataset by SimMSVM lead to better performance. Totally, the 81.7 % combinations by SimMSVM are the most efficient one for the used six datasets. In fact, it gives a significant speed-up from Crammer and Singer's approach (95.8 % combinations by SimMSVM perform better). Although the BSVM has applied decomposition methods for the two "all-together" approaches, their training speed remains slow especially for the relatively small γ. Since we find that the optimal parameters (C, γ) are in various ranges for different classification cases in the experiments, it is necessary to carry out more parameter pairs to obtain the optimal model. If the training speed is an important issue, the SimMSVM could be an option.

2.3.4.3 Speed vs. Data Size and Number of Class

In this subsection, we give a detailed analysis for the effect of the number of classes and the training set size on the SimMSVM. We compare the SimMSVM with the other multi-class SVM approaches on two datasets with the large training samples:

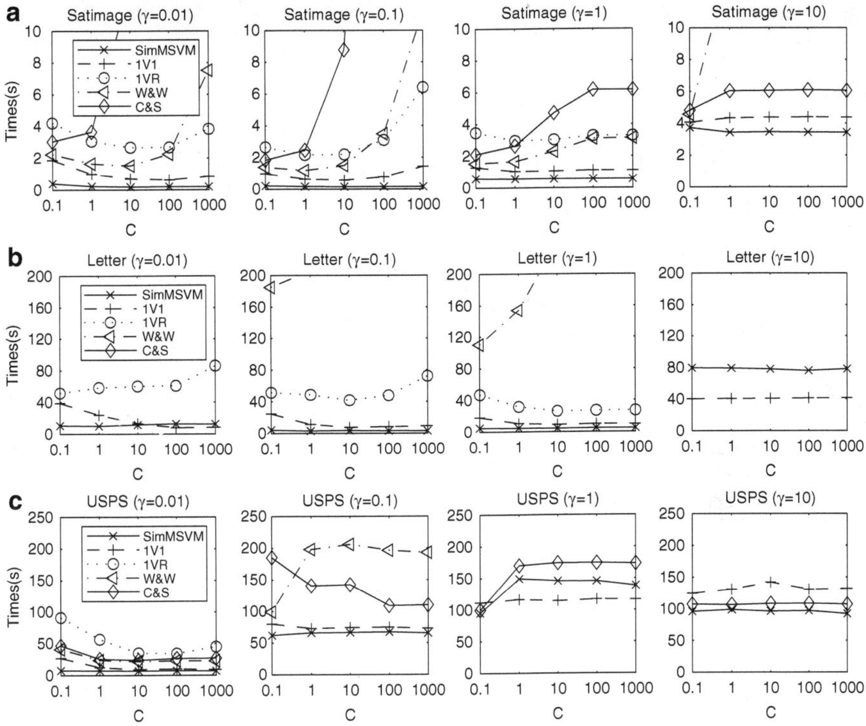

Fig. 2.3 Cross-validation time (in seconds) comparison as a function of C for $\gamma = 0.01$, 0.1, 1, and 10 on (**a**) Satimage dataset, (**b**) Letter dataset, (**c**) USPS dataset

Letter and **USPS**. First, we restrict the analysis to binary-class case for the first two classes, then proceed to a multi-class analysis by gradually adding a new class. As described in Figs. 2.2 and 2.3, different kernel parameters (C, γ) would cause different computational time. For fairness, we choose the default parameter values of the BSVM package for the comparison: $C = 1, \gamma = 1/k$. Figure 2.4 summarizes the results.

For the two "all-together" multi-class SVM approaches, we have to solve an $(l \times k)$-variable QP. For the one-versus-rest approach, instead we need to solve k l-variable QPs. As for the SimMSVM, we solve a single l-variable QP, which is not affected by k. Figure 2.4 validates that the training time of SimMSVM scales slowly as the number of classes and dataset increase. We show an increasing training set size by keeping the number of class number ($k = 26$ for Letter and $k = 10$ for USPS) constant and adding training samples to each class. Each learning algorithm is repeated with five times and we report the average training time in Fig. 2.5. We further show an increase of class number by keeping the training set size constant. In the experiments, we set the constant training set size to be 5,000

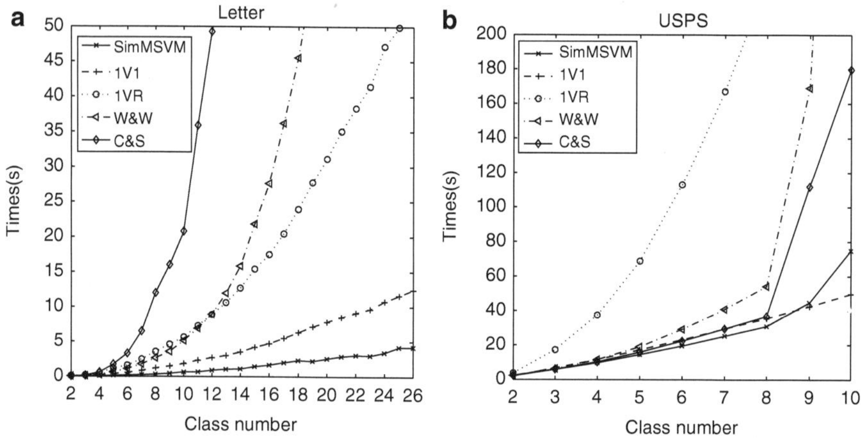

Fig. 2.4 Training time (seconds) vs. class number on (**a**) Letter dataset, (**b**) USPS dataset

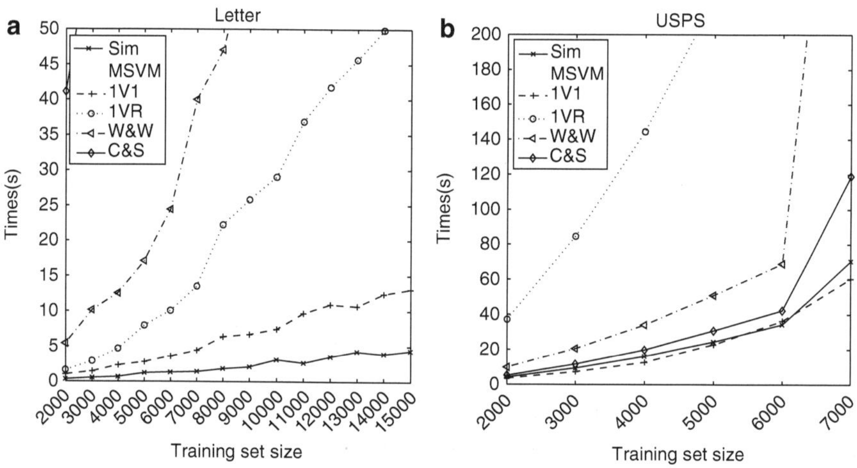

Fig. 2.5 Training time (seconds) vs. increasing training set size by keeping constant the class number on (**a**) Letter dataset, (**b**) USPS dataset

for Letter and 3,000 for USPS. The number of the training samples of each class is reduced accordingly while the class number increases. Figure 2.6 summarizes the results.

From Figs. 2.4, 2.5, and 2.6, it can be found that the SimMSVM takes the least computational cost in most cases except the **USPS**. In the **USPS**, the SimMSVM has a comparable computational cost to that of the one-versus-one method. The reason for this phenomenon is that the optimization for the dual problem of SVM with the 1V1 is much lower than that of the SimMSVM in the large-scale case. In this case, the advantage of the SimMSVM might be counteracted with that of the optimization

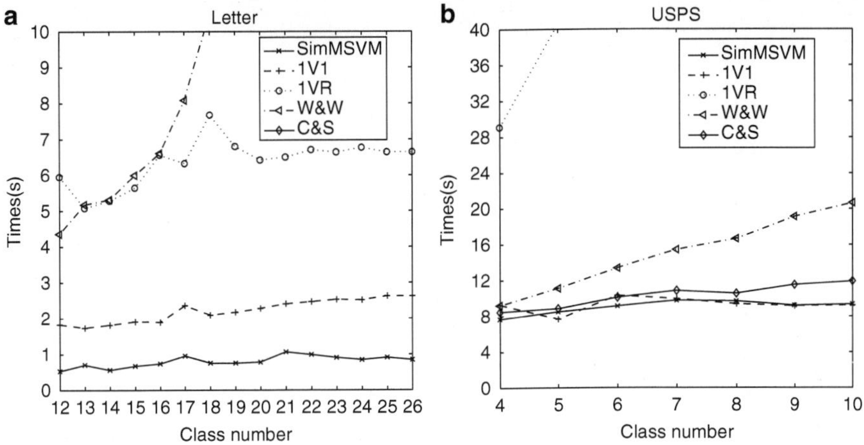

Fig. 2.6 Training time (seconds) vs. increasing class number by keeping constant the training set size on (**a**) Letter dataset, (**b**) USPS dataset

for the traditional SVM dual problem with 1V1. Compared with the other three algorithms (one-versus-rest, the C&S, and the W&W), the SimMSVM here has a significant advantage, which can demonstrate the efficiency of SimMSVM.

2.4 Discussions for SimMSVM

2.4.1 Relation to Binary SVM

In binary case where $k = 2$, from (2.17), the SimMSVM approach will produce $\mathbf{w}_1$ and $\mathbf{w}_2$ where $\mathbf{w}_1 = -\mathbf{w}_2$. Let $\mathbf{w} = 2\mathbf{w}_1$ and target $+1$ for the positive class and -1 negative, (2.15) is equivalent to

$$\begin{aligned} \min_{\mathbf{w}\in\mathcal{H},\xi\in\mathcal{R}^l} \quad & \frac{1}{2}\mathbf{w}^T\mathbf{w} + 2C\sum_{i=1}^{l}\xi_i \\ \text{subject to} \quad & y_i\mathbf{w}^T\mathbf{x}_i \geq 1 - \xi_i, \\ & \xi_i \geq 0, \quad i = 1,\ldots,l, \end{aligned} \tag{2.27}$$

which is the unbiased binary SVM formulation.

Lee et al. [22] proposed a variant of multi-class SVM by modifying the target values so that the positive class has target $+1$ and the negative class has target $-\frac{1}{k-1}$. For notational convenience, we define $\mathbf{v}_m$ for $m = 1,\ldots,k$ as a k-dimensional vector with 1 in the m-th coordinate and $-\frac{1}{k-1}$ elsewhere. If the label of a training example $\mathbf{x}_i$ is coded in vector form as $\mathbf{v}_{y_i}$, the margin constraints of (2.15) can be rewritten as

$$\mathbf{v}_{y_i}^T \mathbf{W}^T \varphi(\mathbf{x}_i) \geq 1 - \xi_i, \tag{2.28}$$

where $\mathbf{W} = [\mathbf{w}_1, \ldots, \mathbf{w}_k]$. Thus, it can be found that SimMSVM can be degenerated into the unbiased binary SVM.

Note that the transpose of the label vector $\mathbf{v}_{y_i}$ is exactly the i-th row of $\mathbf{V}$ defined in (2.21). From (2.22), the Hessian $\mathbf{G}$ contains not only a similarity measure between data samples but also their label information. In addition, we find that SimMSVM can be easily extended to the multi-label classification framework in which an instance may be simultaneously relevant to several labels. Multi-label classification is useful for text categorization, multimedia retrieval, and many other areas. In SimMSVM, we solve a multi-label classification problem by using the same dual formulation (2.19) except for a different representation of the label vector $\mathbf{v}_{y_i}$. For example, if a training example is relevant to r out of k labels simultaneously, we can set its label vector so that the relevant labels have target $+\frac{1}{r}$ and the other labels $-\frac{1}{k-r}$, where the values of the k labels are taken into two categories with $+\frac{1}{r}$ and $-\frac{1}{k-r}$. We then compute the Hessian $\mathbf{G}$ from (2.21) and solve the dual problem (2.19) to obtain the optimal α^*.

The resulting discriminant function is given as follows:

$$f_m^*(\mathbf{x}) = \sum_{i=1}^{l} v_{y_i,m} \alpha_i^* \kappa(\mathbf{x}_i, \mathbf{x}), \tag{2.29}$$

where $v_{y_i,m}$ is the m-th element of $\mathbf{v}_{y_i}$. Since the values of the k labels are taken into two categories with $+\frac{1}{r} > 0$ and $-\frac{1}{k-r} < 0$, the label set Y for the test sample $\mathbf{x}$ is determined as:

$$Y = \{m | f_m^*(\mathbf{x}) > 0, m \in \{1 \ldots k\}\} \tag{2.30}$$

2.4.2 Relation to Crammer and Singer's Multi-Class SVM

In [7], the dual problem is given as follows:

$$\begin{aligned}
\min_{\alpha \in \mathcal{R}^{l \times k}} \quad & \frac{1}{2} \sum_{i=1}^{l} \sum_{j=1}^{l} \sum_{m=1}^{k} K_{i,j} \alpha_{i,m} \alpha_{j,m} - \sum_{i=1}^{l} \alpha_{i,y_i} \\
\text{subject to} \quad & \sum_{m=1}^{k} \alpha_{i,m} = 0, \\
& \alpha_{i,m} \leq 0 \quad \text{for all } m \neq y_i, \\
& \alpha_{i,y_i} \leq C, \\
& i = 1, 2, \ldots, l, \quad m = 1, \ldots, k.
\end{aligned} \tag{2.31}$$

Let $\mathbf{M}$ be the $l \times k$ dual matrix whose (i,m)-th element is the dual variable $\alpha_{i,m}$, and let

$$\alpha = [\alpha_{1,y_1}, \alpha_{2,y_2}, \ldots, \alpha_{l,y_l}]^T. \tag{2.32}$$

We then rewrite the dual objective function of (2.31) as

$$\frac{1}{2}\mathrm{tr}(\mathbf{M}^T\mathbf{K}\mathbf{M}) - \mathbf{e}^T\alpha. \tag{2.33}$$

From the linear constraint of (2.31), every k dual variables associated with the same x_i follows the sum-to-zero constraint. If $\alpha_{i,y_i} = 0$ for some $\mathbf{x}_i$, we have $\alpha_{i,m} = 0$ for all $m \neq y_i$. We carry over the notion of support vectors to the multi-class setting, and define support vectors as examples with $[\alpha_{i,1}, \ldots, \alpha_{i,k}] \neq \mathbf{0}$. A support vector $\mathbf{x}_i$ plays a positive role in its class model since $\alpha_{i,yi} > 0$, while punishing some other classes for those $\alpha_{i,m} < 0$. In order to reduce the size of dual variables, we further add to (2.31) a seemingly aggressive constraint that all those $\alpha_{i,m}$ $(m \neq y_i)$ corresponding to the i-th sample are of the same value (share equal punishment):

$$\alpha_{i,m} = -\frac{1}{k-1}\alpha_{i,y_i}, \text{ for all } m \neq y_i. \tag{2.34}$$

Thus, the sum-to-zero constraint of (2.31) is satisfied accordingly. As for the objective function of (2.31), from (2.34) we have

$$\sum_{m=1}^{k} \alpha_{i,m}\alpha_{j,m} = \begin{cases} \frac{k}{k-1}\alpha_i\alpha_j, & \text{if } y_i = y_j, \\ \frac{-k}{(k-1)^2}\alpha_i\alpha_j, & \text{if } y_i \neq y_j, \end{cases} \tag{2.35}$$

where we simply denote α_{i,y_i} as α_i, for $i = 1, \ldots, l$.

To sum up, Crammer and Singer's dual formulation (2.31) with additional constraint (2.34) is exactly the same as (2.19), which is the dual of SimMSVM.

2.4.3 Relation to One-Class SVM

There are also One-class SVMs, which solve an unsupervised learning problem related to probability density estimation. Two approaches that extend the SVM methodology have been proposed in [30,35]. The dual problem of the support vector domain description (SVDD) method described in Tax and Duin, [35] using RBF kernel is as follows:

$$\begin{aligned} \min_{\alpha \in \mathcal{R}^l} \quad & \frac{1}{2}\alpha^T\mathbf{K}\alpha - \mathbf{e}^T\alpha \\ \text{subject to} \quad & \mathbf{e}^T\alpha = 0, \\ & 0 \leq \alpha \leq C. \end{aligned} \tag{2.36}$$

If we ignore the equality constraint of (2.36), the only difference between (2.19) and (2.36) is the Hessian of the quadratic objective function. In fact, SVDD treats all training samples involved as one class, and the support vectors (those $\mathbf{x}_i$ with $\alpha_i > 0$) usually appear at the boundaries of data distribution. The next lemma proves that the Hessian $\mathbf{G} = \mathbf{K} \odot (\mathbf{P}\mathbf{P}^T)$ of (2.19) will endow the SimMSVM with discrimination power between all classes.

Lemma 1. *By solving the dual optimization problem (2.19), the decision value for the m-th class* $f_m(\mathbf{x}) = \sum_{i:y_i=m} \alpha_i \kappa(\mathbf{x_i}, \mathbf{x})$ *is not trivially 0 when the RBF kernel is adopted, i.e., there is at least one support vector for each class.*

Proof. Prove by contradiction. From the Karush—Kuhn—Tucker optimality condition of (2.19), a vector α is a stationary point of (2.19) if and only if there are two nonnegative vectors λ and μ such that

$$\begin{aligned} &\mathbf{G}\alpha - \mathbf{e} = \lambda - \mu, \\ &\lambda^T \alpha = 0, \ \mu^T(C\mathbf{e} - \alpha) = 0, \\ &\lambda \geq \mathbf{0}, \ \mu \geq \mathbf{0}. \end{aligned}$$

We rewrite this condition as

$$\begin{aligned} &(\mathbf{G}\alpha)_i > 1 \quad \text{if } \alpha_i = 0, \\ &(\mathbf{G}\alpha)_i = 1 \quad \text{if } 0 < \alpha_i < C, \\ &(\mathbf{G}\alpha)_i < 1 \quad \text{if } \alpha_i = C. \end{aligned}$$

Assume that the m-th class has no support vectors, i.e., $\alpha_i = 0$ for all i where $y_i = m$. Since (2.20) indicates that $G_{ij} \leq 0$ for all $y_i \neq y_j$ when using a gaussian RBF kernel, we have $(\mathbf{G}\alpha)_i = \sum_{j:y_j \neq m} G_{ij}\alpha_j \leq 0$, which contradicts the above KKT condition. □

2.4.4 Fisher Consistency Issue

In binary case where the class label $y \in \{-1, +1\}$, we denote $P_+(\mathbf{x}) = P(Y=1|\mathbf{X}=\mathbf{x})$. Then a binary loss function $\xi(f(\mathbf{x}), y)$ is Fisher consistent if the minimizer of $E[\xi(f(\mathbf{X}), Y)|\mathbf{X} = \mathbf{x})]$ is the same as $\text{sign}(P_+(\mathbf{x}) - \frac{1}{2})$. In other words, Fisher consistency requires that the loss function should yield the Bayes decision boundary asymptotically [3, 23, 24]. In the multi-class setting where $y \in \{1, \ldots, k\}$, we let $P_m(\mathbf{x}) = P(Y = j|\mathbf{X} = \mathbf{x})$. Suppose $\xi(\mathbf{f}(\mathbf{x}), y)$ is a multi-class loss function, we denote the minimizer of $E(\xi(\mathbf{f}(\mathbf{X}, Y)|\mathbf{X} = \mathbf{x})$ as $\mathbf{f}^* = (f_1^*, \ldots, f_k^*)$. Fisher consistency requires $\text{argmax}_m f_m^* = \text{argmax}_m P_m$.

In the next lemma, we show that the loss (2.14) is Fisher inconsistent. For remedy, we introduce additional constraints to force the loss function of SimMSVM to be Fisher consistent.

Lemma 2. *The loss (2.14) is Fisher inconsistent when $k \geq 3$.*

Proof. From the above Fisher consistency definition, we have

$$E\left\{\left[1 - f_Y(\mathbf{X}) + \frac{1}{k-1}\sum_{M\neq Y} f_M(\mathbf{X})\right]_+\right\}$$
$$= E\left\{\sum_{y=1}^{k} P_y(\mathbf{X})\left[1 - f_y(\mathbf{X}) + \frac{1}{k-1}\sum_{m\neq y} f_m(\mathbf{X})\right]_+\right\}.$$

For any fixed $\mathbf{X} = \mathbf{x}$, the goal is to minimize

$$\sum_{y=1}^{k} P_y(\mathbf{x})\left[1 - f_y(\mathbf{x}) + \frac{1}{k-1}\sum_{m\neq y} f_m(\mathbf{x})\right]_+ . \tag{2.37}$$

We let

$$g_y(\mathbf{x}) = f_y(\mathbf{x}) - \frac{1}{k-1}\sum_{m\neq y} f_m(\mathbf{x}), \tag{2.38}$$

for $y = 1,\ldots,k$. Therefore, (2.37) is equivalent to

$$\sum_{y=1}^{k} P_y(\mathbf{x})\left[1 - g_y(\mathbf{x})\right]_+ . \tag{2.39}$$

Clearly, $\sum_{y=1}^{k} g_y(\mathbf{x}) = 0$. From the proof of **Lemma 1** in [24], we know that the minimizer $\mathbf{g}^*$ satisfies $g_m^* \leq 1$, $\forall m = 1,\ldots,k$. Thus, the problem (2.37) reduces to

$$\begin{aligned} \max_{\mathbf{g}} \quad & \sum_{m=1}^{k} P_m(\mathbf{x}) g_m(\mathbf{x}) \\ \text{subject to} \quad & \sum_{m=1}^{k} g_m(\mathbf{x}) = 0, \\ & g_m(\mathbf{x}) \leq 1, \quad m = 1,\ldots,k. \end{aligned} \tag{2.40}$$

It is easy to verify that the solution of (2.40) satisfies $g_m^*(\mathbf{x}) = -(k-1)$ if $m = \mathrm{argmin}_j P_j(\mathbf{x})$ and 1 otherwise. We then rewrite (2.38) as

$$f_y(\mathbf{x}) = \frac{k-1}{k}\left(g_y(\mathbf{x}) + A\right), \tag{2.41}$$

where $A = \frac{1}{k-1}\sum_{m=1}^{k} f_m(\mathbf{x})$. We have

$$\operatorname{argmax}_m f^*(\mathbf{x}) = \operatorname{argmax}_m g^*(\mathbf{x}). \tag{2.42}$$

Therefore, the loss function (2.14) is not Fisher consistent when $k \geq 3$. □

Although Fisher consistency is a desirable condition, a consistent loss may not always lead to better classification accuracy [15, 24, 29]. However, if Fisher consistency is required for the loss function of SimMSVM, one alternative solution is to add further constraints

$$(k-1)f_y(\mathbf{x}) - \sum_{m \neq y} f_m(\mathbf{x}) + 1 \geq 0, \tag{2.43}$$

for $y = 1, \ldots, k$. These constraints (2.43) amount to $g_m(\mathbf{x}) \geq -\frac{1}{k-1}$. We have the following Lemma:

Lemma 3. *The maximizer* $\mathbf{g}^*$ *of* $E[g_Y(\mathbf{X})|\mathbf{X} = \mathbf{x}]$, *subject to* $\sum_{m=1}^{k} g_m(\mathbf{x}) = 0$ *and* $-\frac{1}{k-1} \leq g_m(\mathbf{x}) \leq 1$, $\forall m$, *satisfies the following:* $g_m^*\mathbf{x}) = 1$ *if* $m = \operatorname{argmax}_m P_m(\mathbf{x})$ *and* $-\frac{1}{k-1}$ *otherwise.*

For the detailed proof, please see Lemma 5 in [24]. □

Lemma 3 justifies that, with additional constraints (2.43), the loss function of SimMSVM is Fisher consistent. Fisher consistency is an attractive property for a loss function, the related Fisher consistency issue can be studied detailedly in the future work.

2.5 Conclusion

In this chapter, we discuss the main approaches for the multi-class SVM and especially introduce a new SimMSVM algorithm that directly solves a multi-class classification problem. Through modifying Crammer and Singer's multi-class SVM by introducing a relaxed classification error bound, the SimMSVM reduces the size of the dual variables from $l \times k$ to l, where l and k are the size of training data and the number of classes, respectively. We here prove that the dual formulation of the proposed SimMSVM is exactly the dual of Crammer and Singer's approach with an additional constraint. The experimental evaluations on real-world datasets show that the new SimMSVM approach can greatly speed-up the training process and achieve competitive or better classification accuracies.

References

1. Asuncion, A., Newman, D.: UCI machine learning repository. http://archive.ics.uci.edu/ml/datasets.html (2007)
2. Baldi, P., Pollastri, G.: A machine-learning strategy for protein analysis. IEEE Intell. Syst. **17**(2), 28–35 (2002)
3. Bartlett, P., Jordan, M., McAuliffe, J.: Convexity, classification, and risk bounds. J. Am. Stat. Assoc. **101**, 138–156 (2006)
4. Bredensteiner, E., Bennett, K.: Multicategory classification by support vector machines. Comput. Optim. Appl. **12**, 53–79 (1999)
5. Chawla, N.V., Japkowicz, N., Kolcz, A.: Editorial: special issue on learning from imbalanced data sets. ACM SIGKDD Explor. **6**(1), 1–6 (2004)
6. Cortes, C., Vapnik, V.: Support vector networks. Mach. Learn. **20**(3), 273–297 (1995)
7. Crammer, K., Singer, Y.: On the algorithmic implementation of multiclass kernel-based vector machines. J. Mach. Learn. Res. **2**, 265–292 (2001)
8. Dietterich, T., Bakiri, G.: Solving multiclass learning problems via error-correcting output codes. J. Artif. Intell. Res. **2**, 263–286 (1995)
9. Fung, G., Mangasarian, O.: Proximal support vector machine classifiers. In: Provost, F., Srikant, R. (eds.) Proceedings KDD-2001: Knowledge Discovery and Data Mining, August 26–29, 2001, San Francisco, CA, pp. 77–86. Asscociation for Computing Machinery, New York (2001)
10. Fung, G., Mangasarian, O.: Multicategory proximal support vector machine classifiers. Mach. Learn. **59**(1–2), 77–97 (2005)
11. Ganapathiraju, A., Hamaker, J., Picone, J.: Applications of support vector machines to speech recognition. IEEE Trans. Signal Process. **52**(8), 2348–2355 (2004)
12. Guermeur, Y.: Combining discriminant models with new multi-class svms. Pattern Anal. Appl. **5**(2), 168–179 (2002)
13. He, X., Wang, Z., Jin, C., Zheng, Y., Xue, X.Y.: A simplified multi-class support vector machine with reduced dual optimization. Pattern Recognit. Lett. **33**, 71–82 (2012)
14. Hsu, C., Lin, C.: A comparison of methods for multi-class support vector machines. IEEE Trans. Neural Netw. **13**, 415–425 (2002)
15. Hsu, C., Lin, C.: A simple decomposition method for support vector machines. Mach. Learn. **46**, 291–314 (2002)
16. Hsu, C., Lin, C.: Bsvm. http://mloss.org/software/view/62/ (2008)
17. Hull, J.: A database for handwritten text recognition research. IEEE Trans. Pattern Anal. Mach. Intell. **16**(5), 550–554 (1994)
18. Khan, L., Awad, M., Thuraisingham, B.: A new intrusion detection system using support vector machines and hierarchical clustering. VLDB J. **16**(4), 507–521 (2007)
19. King, R., Feng, C., Sutherland, A.: Statlog: comparison of classification algorithms on large real-world problems. Appl. Artif. Intell. **9**(3), 289–333 (1995)
20. Knerr, S., Personnaz., L., Dreyfus, G.: Single-layer learning revisited: a stepwise procedure for building and training neural network. In: Fogelman, J. (ed.) Neurocomputing: Algorithms, Architectures and Applications. Springer, Berlin (1990)
21. Kreßel, U.: Pairwise classification and support vector machines. In: Schölkopf, B., Burges, C., Smola, A. (eds.) Advances in Kernel Methods: Support Vector Learning, pp. 255–268. MIT Press, Cambridge (1999)
22. Lee, Y., Lin, Y., Wahba, G.: Multicategory support vector machines. In: Wegman, E., Braverman, A., Goodman, A., Smyth, P. (eds.) Computing Science and Statistics, vol. 33, pp. 498–512. Interface Foundation of North America, Inc., Fairfax Station, VA, USA (2002)
23. Lee, Y., Lin, Y., Wahba, G.: Multicategory support vector machines: theory and application to the classification of microarray data and satellite radiance data. J. Am. Stat. Assoc. **99**, 67–81 (2004)

24. Liu, Y.: Fisher consistency of multicategory support vector machines. In: Proceedings of the Eleventh International Conference on Artificial Intelligence and Statistics (AISTATS-07) (2007)
25. Mangasarian, O., Musicant, D.: Successive overrelaxation for support vector machines. IEEE Trans. Neural Netw. **10**(5), 1032–1037 (1999)
26. Mangasarian, O., Musicant, D.: Lagrangian support vector machines. J. Mach. Learn. Res. **1**, 161–177 (2001)
27. Mori, S., Suen, C., Yamamoto, K.: Historical review of OCR research and development, pp. 244–273. IEEE Computer Society Press, Los Alamitos (1995)
28. Platt, J., Cristianini, N., Shawe-Taylor, J.: Large margin dags for multiclass classification. In: Advances in Neural Information Processing Systems, vol. 12, pp. 547–553. MIT Press, Cambridge (2000)
29. Rifkin, R., Klautau, A.: In defense of one-vs-all classification. J. Mach. Learn. Res. **5**, 101–141 (2004)
30. Schölkopf, B., Platt, J., Shawe-Taylor, J., Smola, A., Williamson, R.: Estimating the support of a high-dimensional distribution. Neural Comput. **13**(7), 1443–1471 (2001)
31. Suykens, J., Vandewalle, J.: Least squares support vector machine classifiers. Neural Process. Lett. **9**(3), 293–300 (1999)
32. Suykens, J., Vandewalle, J.: Multiclass least squares support vector machines. In: Proceedings of the International Joint Conference on Neural Networks (IJCNN99). World Scientific, Washington, DC (1999)
33. Szedmak, S., Shawe-Taylor, J., Saunders, C., Hardoon, D.: Multiclass classification by l1 norm support vector machine. In: Pattern Recognition and Machine Learning in Computer Vision Workshop (2004)
34. Tang, Y., Zhang, Y., Chawla, N., Krasser, S.: Svms modeling for highly imbalanced classification. IEEE Trans. Syst. Man Cybern. Part B **39**(1), 281–288 (2009)
35. Tax, D., Duin, R.: Data domain description using support vectors. In: Proceedings of the European Symposium on Artificial Neural Networks, pp. 251–256 (1999)
36. Vapnik, V.: Statistical Learning Theory. Wiley, New York (1998)
37. Wang, L., Shen, X.: On l1-norm multiclass support vector machines: methodology and theory. J. Am. Stat. Assoc. **102**, 583–594 (2007)
38. Weston, J., Watkins, C.: Multi-class support vector machines. In: Proceedings of ESANN99 (1999)
39. Xia, X., Li, K.: A sparse multi-class least-squares support vector machine. In: IEEE International Symposium on Industrial Electronics, 2008 (ISIE 2008), pp. 1230–1235 (2008)

Chapter 3
Novel Inductive and Transductive Transfer Learning Approaches Based on Support Vector Learning

Zhaohong Deng and Shitong Wang

Abstract In this chapter, two novel transfer learning approaches based on support vector learning are involved. For inductive transfer learning, the knowledge-leverage-based TSK fuzzy system (KL-TSK-FS) is proposed, which demonstrates the good privacy-protection abilities and strong adaptability for the situations where the data are only partially available from the target domain while some useful knowledge of the source domains is available. For transductive transfer learning, domain adaptation kernelized support vector machine (DAKSVM) and its two extensions are proposed, which can reduce the distribution gap between different domains in an RKHS as much as possible by integrating the large margin learner with the proposed generalized projected maximum distribution distance (GPMDD) metric.

3.1 Introduction

3.1.1 Background

Recently, transfer learning has been studied extensively for different applications [1], such as text classification and indoor WiFi location estimation. Referring to Fig. 3.1 and the explanations given in Table 3.1, transfer learning is an approach to obtain an effective model of data from the target domain by effectively leveraging the useful information from source domains in the learning procedure.

Situations requiring transfer learning are becoming common in real-world applications. The modeling of fermentation process [2] is one example where the transfer learning is required. In the target domain of a microbiological fermentation process,

Z. Deng • S. Wang (✉)
School of Digital Media, Jiangnan University, Wuxi, Jiangsu, P.R. China
e-mail: wxwangst@yahoo.com.cn

Y. Ma and G. Guo (eds.), *Support Vector Machines Applications*,
DOI 10.1007/978-3-319-02300-7_3,
© Springer International Publishing Switzerland 2014

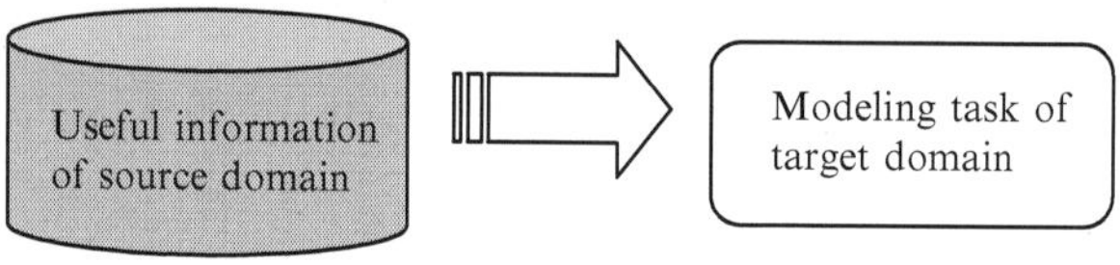

Fig. 3.1 An illustration of transfer learning for regression

Table 3.1 Some terms used for transfer leaning in the text

Terms	Explanations
Domain	A domain is a situation where a modeling task is to be accomplished. It is usually characterized by: (1) the *data* collected in this domain and (2) the *learning task* to be performed in this domain
Target domain	In transfer learning, it is referred to as the domain with insufficient data for proper modeling while a modeling task is required to be effectively implemented
Source domain	It is the domain related to the target domain, with similar data distribution and learning task. There may be differences between the source domain and the target domain, but it is assumed that the source domain can provide some useful information for the modeling task of the target domain

the data collected may be insufficient or some of the data may be missing due to the deficiency of the sensor setup. Thus, we cannot effectively model the fermentation process for this domain with the collected data. However, data available from other similar microbiological fermentation process could be sufficient and considered as source domains for the target domain. Hence, transfer learning can be exploited to make use of the information from the source domain to improve the modeling effect of the target domain, thereby resulting in a model with better generalization capability. In this case, transfer learning is an effective solution to the corresponding modeling task because it can enhance the model by leveraging the information available from the source domains, such as the data collected in other time frames or with other setups.

A comprehensive survey about transfer learning can be discovered in [1]. In general, the existing work about transfer learning can be categorized into three types: (1) transfer learning for classification [3–15]; (2) transfer learning for unsupervised learning (clustering [16, 17] and dimensionality reduction [18, 19]); and (3) transfer learning for regression [20–24]. According the setting whether there are the labeled data in the target domain available, all the transfer learning methods can also be classified as inductive transfer learning methods and transductive learning methods. While there are a few labeled data for the supervised learning in the inductive transfer learning methods, all the data are unlabeled in the target domain for the transductive learning methods and the unsupervised learning is implemented accordingly While there are a few labeled data for the inductive transfer learning, all the data are unlabeled in the target domain for the transductive learning method. Among the existing transfer methods, a lot of them are based on the support vector learning. In this chapter, we mainly focus on the novel support vector learning-based methods for the inductive and transductive learning.

3.1.2 *Support Vector Learning-Based Inductive Transfer Learning*

In the inductive transfer learning setting, the target task is different from the source task. In this case, some labeled data in the target domain are required to induce an objective predictive model for use in the target domain. The representative inductive transfer learning algorithms are reviewed below. Dai et al. [25] proposed a boosting algorithm with the support vector machine (SVM) as the learner, TrAdaBoost, which is an extension of the AdaBoost algorithm, to address the inductive transfer learning problems. TrAdaBoost attempts to iteratively reweight the source domain data to reduce the effect of the "bad" source data while encouraging the "good" source data to contribute more for the target domain. Wu and Dietterich [26] integrated the source domain (auxiliary) data and SVM framework for improving the classification performance. Evgeniou and Pontil [27] borrowed the idea of hierarchical Bayesian to SVMs for multitask learning. The proposed method assumed that the parameter, $\mathbf{w}$, in SVMs for each task can be separated into two terms. One is a common term over tasks and the other is a task-specific term.

3.1.3 *Support Vector Learning-Based Transductive Transfer Learning*

For support vector learning-based transductive transfer learning, a major computational problem is how to reduce the difference between the distributions of the source and target domains. There have existed several works describing how to measure the distance between distributions [28, 29]. Intuitively, discovering a good feature representation across domains is crucial [13, 30]. A good feature representation should be able to reduce the distribution discrepancy between two domains as much as possible, while at the same time preserving the underlying geometric structures (or scatter information) of both source and target domain data as much as possible. Ben-David et al. [31] used an example of hyperplane classifiers to show that the performance of the hyperplane classifier that could best separate the data could provide a good method for measuring the distribution distance for different data representations. Along these same lines, Gretton et al. [32] showed that for a given class of functions, the measure could be simplified by computing the discrepancy between two means of the distributions in a reproducing kernel Hilbert space (RKHS), thus resulting in the maximum mean discrepancy (MMD) measure. Inspired by the ideas of both transductive SVM (TSVM) and MMD, Brian et al. [28] proposed a so-called large margin kernel projected (LMPROJ) TSVM paradigm for domain adaptation problems based on the projected distance measure in an RKHS. The basic idea of LMPROJ is to minimize the distribution mean distance between source and target domain data by finding a feature translation in an RKHS. By the same way of LMPROJ, based on multiple kernel learning framework, Duan et al.

also proposed a domain transfer SVM (DTSVM) for domain adaptation learning (DAL) problem such as video concept detection. Further details about DTSVM can be found in [29].

3.1.4 *Main Work in This Study*

In this study, one support vector learning-based inductive learning approach and one support vector learning-based transductive learning approach are proposed, respectively.

3.1.4.1 Support Vector Learning-Based Inductive Transfer Learning with Knowledge-Leveraged Fuzzy Logic Systems

As support vector learning-based fuzzy system modeling is a type of important modeling methods [2, 33], it is promising to incorporate transfer learning with the fuzzy model. To the best of our knowledge, however, the study of transfer learning for support vector learning-based fuzzy system modeling has not yet been reported before. For support vector learning-based fuzzy system modeling, transfer learning is very useful in real-world modeling tasks where traditional fuzzy modeling methods may not work very well. For example, the trained fuzzy systems are much weaker in generalization capability when the training data are insufficient or only partially available [34, 35]. The situation is common in real-world applications in which the sensors and setups for data sampling are not steady due to noisy environment or other malfunctions that lead to insufficiency of data for the modeling task.

In order to tackle the problems with traditional support vector learning-based fuzzy system modeling as described above, a feasible remedy strategy is to boost up the performance by taking advantage of the useful information from source domains (or related domains), which can be the data in the domains, or the relevant knowledge like the density distribution and/or fuzzy rules. The simplest way to obtain the information from source domains is to directly use the data, collected from the source domains, but this approach leads to two major challenges. First, due to the necessity of privacy protection in some proprietary applications, such as the aforementioned fermentation process, the data of the source domains cannot always be obtained. Under this situation, the knowledge about the source domains, e.g. the density distribution and model parameters, can be obtained more easily to enhance the modeling of the target domain. Second, drifting phenomenon may exist between the source domain and the target domain, which makes it inappropriate to directly use the data from the former in the latter, or negative effect on the modeling task will be produced. These two issues should be properly addressed in order to develop an effective transfer learning modeling strategy for fuzzy systems.

In this study, a support vector learning-based fuzzy system modeling approach with knowledge-leverage capability from source domains is exploited for the inductive transfer learning. In view of its popularity, the Takagi–Sugeno–Kang-type fuzzy system (TSK-FS) is chosen to incorporate with a knowledge-leverage mechanism and hence the knowledge-leveraged TSK-type fuzzy system (KL-TSK-FS) is proposed. A novel objective criterion is proposed to integrate the model knowledge of the source domains and the data of the target domain, and the induced fuzzy rules of the model are learned accordingly. The knowledge of the source domain will effectively make up the deficiency in learning due to the lack of data in the target domain. Hence, the proposed system—KL-TSK-FS is more adaptive to the situations where the data are only partially available from the target domain while some useful knowledge of source domains is available. Besides, the proposed method is distinctive in preserving data privacy as only the knowledge (e.g., the corresponding model parameters) rather than the data of the source domain is used.

3.1.4.2 Support Vector Learning-Based Transductive Transfer Learning with DAL

As we may know well, mean (or expectation) and variance (or scatter) are two main features characterizing the distribution of samples which measure order one and order two statistics, respectively. However, most existing DAL methods for support vector learning-based transductive learning focus only on the first-order statistics matching which attempts to make the empirical means of the training and testing instances from source and target domain to be closer in an RKHS [36]. Intuitively, it is not enough to measure the distribution distance discrepancy between two domains to some extent only by considering the mean of the distribution of samples [13, 29, 36]. Hence, the state-of-the-art DAL MMD-based methods [28, 29, 37], which are only focused on the first-order statistics of the data distributions still have considerable limitation in the generalization capacity for specific domain adaptation transfer learning problems. What is more, since LMPROJ or DTSVM only focuses on the consistency of domain distributions in an RKHS, they sometimes project the data onto some noisy directions of separation which are completely irrelevant to the target learning task [13], and even result in poor performance.

In this study, we claim that it is indispensable to consider both mean and variance (or scatter) of data distribution in order to efficiently measure the distribution discrepancy between source and target domains. This motivates us to definitely utilize both MMD and scatter information of both domains to sufficiently evaluate their distribution discrepancy. In order to overcome the drawbacks of the MMD-based methods, we proposed a novel domain adaptation kernelized SVM (DAKSVM) using GPMDD discrepancy metric on RKHS embedding domain distributions, which can simultaneously consider both the distribution mean and scatter discrepancies between source and target domains. The idea is to find an RKHS for which the means and variances of the training and test data distributions are brought to be consistent, so that the labeled training data can be used to learn a model for the test data. Particularly, we aim to obtain a linear kernel classifier based on the Representer Theorem [32], in an RKHS, such that it achieves a trade-off between the maximal margin between classes and the minimal discrepancy between the training and test distributions.

Compared with the existing state-of-the-art DAL methods, our main contributions include the following aspects: (1) the proposed methods inherit the potential advantages of classical TSVMs and MMD-based methods described as above, and further extend them to DAL; (2) as a novel large margin domain adaptation classifier, the proposed methods can reduce the distribution gap between different domains in an RKHS as much as possible, since they effectively integrate the large margin learner with the proposed GPMDD metric; (3) in addition, we propose two extensions to the standard formulation of DAKSVM based on both ν-SVM and least-square SVM (LS-SVM), respectively.

The rest of this chapter is organized as follows. In Sect. 3.2, inductive transfer learning with support vector learning-based fuzzy systems is proposed; in Sect 3.3, transductive transfer learning with support vector learning-based domain adaptation transfer learning SVM by using the GPMDD metric is proposed; in Sect. 3.4, the experimental results about the proposed inductive transfer learning approach are reported; in Sect. 3.5, the experimental results about the proposed transductive transfer learning approach are reported; and The conclusions are given in the final section.

3.2 Inductive Transfer Learning with Support Vector Learning-Based Fuzzy Systems

3.2.1 Support Vector Learning-Based Fuzzy Systems

Support vector learning has been extensively used in the machine learning methods, such as kernel methods and other intelligence modeling methods. In this section, the support vector learning-based fuzzy systems, which have strong learning abilities and nicer interpretation properties, are introduced to develop the inductive transfer learning method.

3.2.1.1 Concept and Principle of TSK-FS

Classical fuzzy logic system models include the TSK model [38], Mamdani–Larsen (ML) model [39], and generalized fuzzy model [40]. Among them, the TSK model is the most popular one due to its effectiveness. In this study, the TSK model is adopted to develop the KL-TSK-FS for implementing the inductive transfer learning.

For TSK fuzzy logic systems, the most commonly used fuzzy inference rules are defined as follows:

TSK Fuzzy Rule R^k:

$$\begin{aligned} &\text{IF } x_1 \text{ is } A_1^k \wedge x_2 \text{ is } A_2^k \wedge \cdots \wedge x_d \text{ is } A_d^k \\ &\text{Then } f^k(\mathbf{x}) = p_0^k + p_1^k x_1 + \cdots + p_d^k x_d \quad k = 1, \cdots, K \end{aligned} \tag{3.1}$$

In Eq. (3.1) A_i^{kK} is a fuzzy subset subscribed by the input variable x_i for the kth rule; K is the number of fuzzy rules, and $\wedge$ is a fuzzy conjunction operator. Each rule is premised on the input vector $\mathbf{x} = [x_1, x_2, \cdots, x_d]^T$, and maps the fuzzy sets in the input space $A^k \subset R^d$ to a varying singleton denoted by $f^k(\mathbf{x})$. When *multiplicative conjunction* is employed as the conjunction operator, *multiplicative implication* as the implication operator, and *additive disjunction* as the disjunction operator, the output of the TSK fuzzy model can be formulated as

$$y^0 = \sum_{k=1}^{K} \frac{\mu^k(\mathbf{x})}{\sum_{k'=1}^{K} \mu^{k'}(\mathbf{x})} \cdot f^k(\mathbf{x}) = \sum_{k=1}^{K} \tilde{\mu}^k(\mathbf{x}) \cdot f^k(\mathbf{x}), \tag{3.2a}$$

where $\mu^k(\mathbf{x})$ and $\tilde{\mu}^k(\mathbf{x})$ denote the fuzzy membership function and the normalized fuzzy membership associated with the fuzzy set A^k. These two functions can be calculated by using

$$\mu^k(\mathbf{x}) = \prod_{i=1}^{d} \mu_{A_i^k}(x_i) \tag{3.2b}$$

and

$$\tilde{\mu}^k(\mathbf{x}) = \mu^k(\mathbf{x}) / \sum_{k'=1}^{K} \mu^{k'}(\mathbf{x}). \tag{3.2c}$$

A commonly used fuzzy membership function is the Gaussian membership function which can be expressed by

$$\mu_{A_i^k}(x_i) = \exp\left(\frac{-\left(x_i - c_i^k\right)^2}{2\delta_i^k}\right), \tag{3.2d}$$

where the parameters c_i^k, δ_i^k can be estimated by clustering techniques or other partition methods. For example, with fuzzy c-means (FCM) clustering, c_i^k, δ_i^k can be estimated as follows:

$$c_i^k = \sum_{j=1}^{N} u_{jk} x_{ji} / \sum_{j=1}^{N} u_{jk}, \tag{3.2e}$$

$$\delta_i^k = h \cdot \sum_{j=1}^{N} u_{jk} \left(x_{ji} - c_i^k\right)^2 / \sum_{j=1}^{N} u_{jk}, \tag{3.2f}$$

where u_{jk} denotes the fuzzy membership of the jth input data $\mathbf{x}_j = (x_{j1}, \cdots, x_{jd})^T$, belonging to the kth cluster obtained by FCM clustering [41]. Here, h is a scale parameter and can be adjusted manually.

When the antecedents of the TSK fuzzy model are determined, let

$$\mathbf{x}_e = \left(1, \mathbf{x}^T\right)^T, \tag{3.3a}$$

$$\tilde{\mathbf{x}}^k = \tilde{\mu}^k(\mathbf{x})\,\mathbf{x}_e, \tag{3.3b}$$

$$\mathbf{x}_g = \left(\left(\tilde{\mathbf{x}}^1\right)^T, \left(\tilde{\mathbf{x}}^2\right)^T, \cdots, \left(\tilde{\mathbf{x}}^K\right)^T\right)^T, \tag{3.3c}$$

$$\mathbf{p}^k = \left(p_0^k, p_1^k, \cdots, p_d^k\right)^T \tag{3.3d}$$

and

$$\mathbf{p}_g = \left(\left(\mathbf{p}^1\right)^T, \left(\mathbf{p}^2\right)^T, \cdots, \left(\mathbf{p}^K\right)^T\right)^T, \tag{3.3e}$$

then Eq. (3.2a) can be formulated as the following linear regression problem [33]

$$y^o = \mathbf{p}_g^T \mathbf{x}_g. \tag{3.3f}$$

Thus, the problem of TSK fuzzy model training can be transformed into the learning of the parameters in the corresponding linear regression model [2, 33].

3.2.1.2 Support Vector Learning-Based TSK-FS Training

Given a training dataset $D_{tr} = \{\mathbf{x}_i, y_i | \mathbf{x}_i \in R^d, y_i \in R, i = 1, \cdots, N\}$, for fixed antecedents obtained via clustering of the input space (or by other partition techniques), the least-square (LS) solution to the consequents is to minimize the following LS criterion function [29], that is,

$$\min_{\mathbf{p}_g} E = \sum_{i=1}^{N} (y_i^o - y_i)^2 = \sum_{i=1}^{N} \left(\mathbf{p}_g{}^T \mathbf{x}_{g_i} - y_i\right)^2 = (\mathbf{y} - \mathbf{X}_g \mathbf{p}_g)^T (\mathbf{y} - \mathbf{X}_g \mathbf{p}_g), \tag{3.4}$$

where $\mathbf{X}_g = [\mathbf{x}_{g1}, \cdots, \mathbf{x}_{gN}]^T \in R^{N \times K \cdot (d+1)}$ and $\mathbf{y} = [y_1, \cdots, y_N]^T \in R^N$.

The most popular LS criterion-based TSK-FS training algorithm is the one used in the adaptive-network-based fuzzy inference systems [42]. For LS criterion-based algorithms, a main shortcoming is that they usually have weak robustness for modeling tasks involving noisy and/or small datasets. Besides the LS criterion-based

TSK-FS training methods, the more promising TSK-FS training methods are the support vector learning-based training algorithms, which are reviewed as follows.

Support Vector Learning-Based TSK-FS Training with L1-Norm Penalty

In addition to the LS criterion, another important criterion for TSK-FS training is the ε-insensitive criterion [33]. Given a scalar g and a vector $\mathbf{g}=[g_1,\cdots,g_g]^T$, the corresponding ε-insensitive loss functions take the following forms, respectively [33]: $|g|_\varepsilon = g-\varepsilon\ (g>\varepsilon)$, $|g|_\varepsilon = 0\ (g\le 0)$ and $|\mathbf{g}|_\varepsilon = \sum_{i=1}^{d}|g_i|_\varepsilon$. For the linear regression problem of the TSK-FS in Eq. (3.3f), the corresponding ε-insensitive loss-based criterion function [33] is defined as

$$\min_{\mathbf{p}_g} E = \sum_{i=1}^{N}|y_i^o - y_i|_\varepsilon = \sum_{i=1}^{N}\left|\mathbf{p}_g^T\mathbf{x}_{g_i} - y_i\right|_\varepsilon \tag{3.5a}$$

In general, the inequalities $y_i - \mathbf{p}_g^T\mathbf{x}_{gi} < \varepsilon$ and $\mathbf{p}_g^T\mathbf{x}_{gi} - y_i < \varepsilon$ are not satisfied for all data pairs $(\mathbf{x}_{gi},y_i)$. By introducing the slack variables $\xi_i^+ \ge 0$ and $\xi_i^- \ge 0$, Eq. (3.5a) can be equivalently written as

$$\min_{\mathbf{p}_g,\xi_i^+,\xi} E = \sum_{j=1}^{N}\left(\xi_i^+ + \xi_i^-\right) \tag{3.5b}$$

$$\text{s.t.}\quad \begin{cases} y_i - \mathbf{p}_g^T\mathbf{x}_{gi} < \varepsilon + \xi_i^+ \\ \mathbf{p}_g^T\mathbf{x}_{gi} - y_i < \varepsilon + \xi_i^- \end{cases},\qquad \xi_i^+ \ge 0, \xi_i^- \ge 0 \quad \forall i.$$

Further, by introducing the regularization term [30], Eq. (3.5b) is modified to become

$$\min_{\mathbf{p}_g,\xi^+,\xi^+} \quad E = \frac{1}{\tau}\sum_{j=1}^{N}\left(\xi_i^+ + \xi_i^-\right) + \frac{1}{2}\mathbf{p}_g{}^T\mathbf{p}_g \tag{3.5c}$$

$$\text{s.t.}\quad \begin{cases} y_i - \mathbf{p}_g^T\mathbf{x}_{gi} < \varepsilon + \xi_i^+ \\ \mathbf{p}_g^T\mathbf{x}_{gi} - y_i < \varepsilon + \xi_i^- \end{cases},\qquad \xi_i^+ \ge 0, \xi_i^- \ge 0 \quad \forall i,$$

where $\tau > 0$ controls the trade-off between the complexity of the regression model and the tolerance of the errors. Here, ξ_i^+ and ξ_i^- can be taken as the L1-norm penalty terms and thus Eq. (3.5c) is an objective function based on L1-norm penalty terms. TSK training algorithm of this type is referred to as support vector learning-based L1-TSK-FS, which has the similar learning way as the classical SVM. The dual optimization in Eq. (3.5c) is a quadratic programming (QP) problem, which can be expressed as

$$\max_{\alpha^+,\alpha^-} -\frac{1}{2}\sum_{i=1}^{N}\sum_{j=1}^{N}\left(\alpha_i^+ - \alpha_i^-\right)\left(\alpha_j^+ - \alpha_j^-\right)\mathbf{x}_{gi}^T\mathbf{x}_{gi} - \sum_{i=1}^{N}\varepsilon\left(\alpha_j^+ + \alpha_j^-\right) + \sum_{i=1}^{N} y_i\left(\alpha_j^+ - \alpha_j^-\right) \tag{3.5d}$$

$$\text{s.t.} \quad \sum_{i=1}^{N}\left(\alpha_j^+ - \alpha_j^-\right) = 0, \quad \alpha_j^+, \alpha_j^- \in [0,\tau] \quad \forall i.$$

Compared with the LS-criterion-based algorithms, the support vector learning-based L1-TSK-FS with the ε-insensitive criterion has been shown to be more robust when dealing with noisy and small datasets.

Support Vector Learning-Based TSK-FLS Training with L2-Norm Penalty

Instead of the L1-norm penalty terms in Eq. (3.5c), another representative support vector learning-based TSK-FS learning method is the one developed by employing the L2-norm penalty terms [3]. The insensitive parameter ε is also added as a penalty term in the objective function. This is similar to the approaches used in other existing L2-norm penalty-based methods, e.g. L2-norm support vector regression (L2-SVR) [43]. For TSK fuzzy model training, the ε-insensitive objective function based on L2-norm penalty terms is then given by

$$\min_{\mathbf{p}_g,\xi^+,\xi^+,\varepsilon} \; g\left(\mathbf{p}_g,\xi^+,\xi^+,\varepsilon\right) = \frac{1}{\tau}\cdot\frac{1}{N}\sum_{j=1}^{N}\left(\left(\xi_i^+\right)^2 + \left(\xi_i^-\right)^2\right) + \frac{1}{2}\mathbf{p}_g^T\mathbf{p}_g + \frac{2}{\tau}\cdot\varepsilon \tag{3.6a}$$

$$\text{s.t.} \quad \begin{cases} y_i - \mathbf{p}_g^T\mathbf{x}_{gi} < \varepsilon + \xi_i^+ \\ \mathbf{p}_g^T\mathbf{x}_{gi} - y_i < \varepsilon + \xi_i^- \end{cases} \quad \forall i.$$

Compared with the L1-norm penalty-based ε-insensitive criterion, the L2-norm penalty-based criterion is advantageous because of the following characteristics: (1) the constraints $\xi_i^+ \geq 0$ and $\xi_i^- \geq 0$ in Eq. (3.5c) are not needed for the optimization; (2) the insensitive parameter ε can be obtained automatically by optimization without the need of manual setting. Similar properties can also be found in other L2-norm penalty-based machine learning algorithms, such as L2-SVR [43]. For convenience, the L2-norm penalty-based ε-insensitive TSK fuzzy model training is referred to as L2-TSK-FS in this chapter. Based on the optimization theory, the dual problem in Eq. (3.6a) can be formulated as the following QP problem.

$$\begin{aligned}\max_{\alpha^+,\alpha^-} \quad & -\sum_{i=1}^{N}\sum_{j=1}^{N}\left(\alpha_i^+-\alpha_i^-\right)\left(\alpha_i^+-\alpha_i^-\right)\cdot \mathbf{x}_{gi}^T\mathbf{x}_{gj}-\sum_{i=1}^{N}\frac{N\tau}{2}\left(\alpha_i^+\right)^2-\sum_{i=1}^{N}\frac{N\tau}{2}\left(\alpha_i^-\right)^2\\ & +\sum_{i=1}^{N}\alpha_i^+\cdot y_i\cdot\tau-\sum_{i=1}^{N}\alpha_i^-\cdot y_i\cdot\tau\end{aligned} \tag{3.6b}$$

$$\text{s.t.} \quad \sum_{i=1}^{N}\left(\alpha_j^+ + \alpha_j^-\right)=1, \quad \alpha_j^+,\alpha_j^- \geq 0 \quad \forall i$$

Notably, the characteristic of the QP problem in Eq. (3.6b) enables the use of core-set-based minimal enclosing ball (MEB) approximation technique to solve problems involving very large datasets [43]. The scalable L2-TSK-FS learning algorithm (STSK) has thus been proposed in this regard [3].

3.2.2 *Inductive Transfer Learning with Support Vector Learning-Based TSK-FS*

3.2.2.1 Framework of Knowledge-Leveraged Inductive Transfer Learning with TSK-FS

Most inductive transfer learning algorithms are developed to learn from the data in the source domain directly with some strategies. Recently, the transfer learning from the knowledge in the source domain rather than the original data is investigated with the knowledge-leveraged transfer learning framework [44], by observing the characteristics of two types of the learning ways below from the source domain, i.e., from the original data and from the induced knowledge.

1. For the data in the source domains, it is the original information and is also the most commonly used information for transfer learning. However, the data are not always available in some situations. For example, many data samples cannot be made open due to the necessity of privacy protection in the real world. Moreover, even if the data of source domains are available, it may not be always appropriate to directly adopt these data for the modeling task in the target domain due to the following issues: first, it is difficult to control and balance the similarity and difference of distributions of the source and target domains by using the data directly; secondly, there possibly exists a drifting between the distributions of different domains and thus some data from the source domain may result in an obvious negative influence on the modeling effect of the target domain.

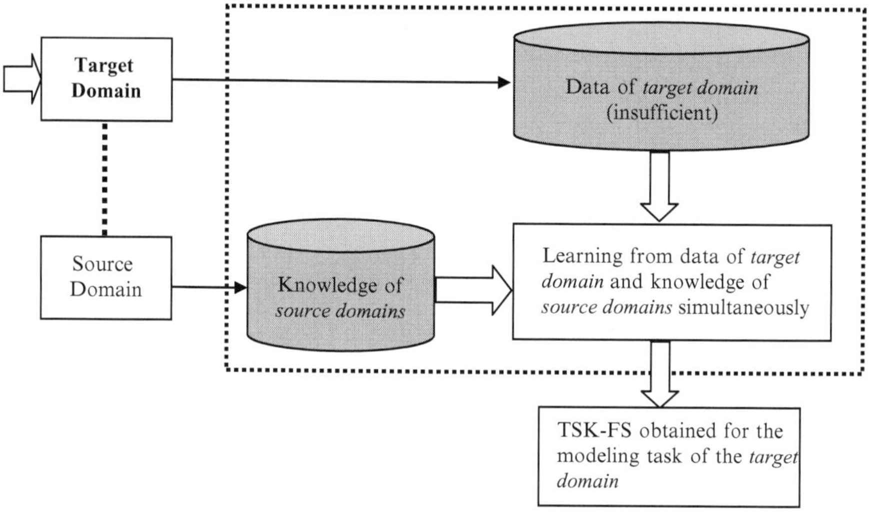

Fig. 3.2 Framework of knowledge-leveraged TSK-FS learning

2. For the knowledge in the source domains, it is another kind of important information. The types of knowledge are diverse, such as density distribution and model parameters. Most of them can be obtained by some learning procedures in the past. For example, the model parameters for the source domain can be learned by a certain modeling algorithm based on the data collected from that domain in a certain historical modeling task. Despite the fact that most of the knowledge obtained cannot be inversely mapped to the original data, it is a good property from a privacy preservation point of view and the important information from the source domains to improve the modeling effect of the target domain.

Thus, the characteristics above show that it should be more appropriate to exploit the use of knowledge rather than data from the source domains to enhance the modeling/learning performance of the models obtained in the target domain. As shown in Fig. 3.2, a generalized learning framework was proposed in [44] for knowledge-leveraged transfer learning. Under this framework, the model in the target domain can be learned from the data in the target domain and the knowledge in the source domain simultaneously. In this study, the knowledge-leveraged inductive transfer learning for the support vector learning-based TSK-FS will be studied accordingly.

3.2.2.2 Inductive Transfer Learning with Support Vector Learning-Based TSK-FS

To take advantage of knowledge-leveraged learning mechanism for TSK-FS, KL-TSK-FS is proposed by using support vector learning and the L2-norm penalty-based TSK-FS learning strategy with the corresponding knowledge-leverage mechanism. The goal is to effectively use the knowledge of the source domains to remedy the deficiency caused by data insufficiency in the target domain and develop an efficient learning algorithm for TSK-FS.

Objective Criterion Integrating the Knowledge of Source Domain

For a TSK-FS constructed by the support vector learning-based technique, the corresponding model parameters obtained in the source domains can be regarded as the knowledge. To develop an effective KL-TSK-FS for model learning of the target domain, we propose an optimization criterion which is integrated with the knowledge of the source domain as follows:

$$\min_{\mathbf{p}_g} \sum_{i=1}^{N} \left| \mathbf{p}_g{}^T \mathbf{x}_{gi} - y_i \right|_{\varepsilon} + \lambda (\mathbf{p}_g - \mathbf{p}_{g0})^T (\mathbf{p}_g - \mathbf{p}_{g0}). \tag{3.7}$$

The optimization criterion in Eq. (3.7) contains two terms. The first term refers to the learning from the data of the target domain for the desired TSK-FS. This term is included so that the desired TSK-FS will fit the sampled training data of the target domain as accurate as possible. The second term refers to the knowledge-leverage of the source domain, with $\mathbf{p}_{g0}$ denoting model parameters learned from the source domains. The purpose is to estimate the desired parameters by approximating the model obtained from the source domains. The parameter λ in Eq. (3.7) is used to balance the influence of these two terms and the optimal value can be determined by using the commonly used cross-validation strategy in machine learning. As in L2-TSK-FS [20], we introduce the terms structure risk and ε-insensitive penalty into Eq. (3.7) to obtain the following objective criterion

$$\begin{aligned} \min_{\mathbf{p}_g, \xi^+, \xi^-, \varepsilon} \quad & \frac{1}{\tau} \cdot \frac{1}{N} \sum_{i=1}^{N} \left((\xi_i^+)^2 + (\xi_i^-)^2 \right) + \frac{1}{2} (\mathbf{p}_g^T \mathbf{p}_g) \\ & + \frac{2}{\tau} \cdot \varepsilon + \lambda (\mathbf{p}_g - \mathbf{p}_{g0})^T (\mathbf{p}_g - \mathbf{p}_{g0}) \end{aligned} \tag{3.8}$$

$$\text{s.t.} \begin{cases} y_i - \mathbf{p}_g^T x_{gi} < \varepsilon + \xi_i^+ \\ \mathbf{p}_g^T x_{gi} - y_i < \varepsilon + \xi_i^- \end{cases}, \forall i$$

In fact, the former three terms in Eq. (3.8) are directly inherited from the L2-TSK-FS [20] and the last term is referred to as the knowledge-leverage term which is used to learn the knowledge from the source domains. Based on the objective criterion in Eq. (3.8), we can derive the corresponding learning rules for the proposed KL-TSK-FS.

Parameter Solution for KL-TSK-FS

Given the optimization problem in Eq. (3.8), Theorem 1 below is proposed for parameter solution.

Theorem 1 *The dual problem of Eq.* (3.8) *is a QP problem as shown in Eq.* (3.9).

$$\begin{aligned}\max_{\boldsymbol{\alpha}^-,\boldsymbol{\alpha}^+} & -\tfrac{1}{2(1+2\lambda)}\sum_{i=1}^{N}\sum_{i=1}^{N}\left(\alpha_i^- - \alpha_i^+\right)\left(\alpha_j^- - \alpha_j^+\right)\mathbf{x}_{gi}^T\mathbf{x}_{gj} - \tfrac{N\tau}{4}\sum_{i=1}^{N}\left(\left(\alpha_i^+\right)^2 + \left(\alpha_i^-\right)^2\right) \\ & -\tfrac{2\lambda}{1+2\lambda}\sum_{i=1}^{N}\left(\alpha_i^- - \alpha_i^+\right)\left(\mathbf{p}_{g0}^T\mathbf{x}_{gi} + y_i\right) + \tfrac{\lambda}{1+2\lambda}\mathbf{p}_{g0}^T\mathbf{p}_{g0} \\ \text{s.t.} & \quad \sum_{i=1}^{N}\alpha_i^- + \sum_{i=1}^{N}\alpha_i^+ = \frac{2}{\tau}, \qquad \alpha_i^- \geq 0, \alpha_i^+ \geq 0.\end{aligned} \tag{3.9}$$

Proof By using the Lagrangian optimization theorem, we can obtain the following Lagrangian function for Eq. (3.8)

$$\begin{aligned} L\left(\mathbf{p}_g, \boldsymbol{\xi}^+, \boldsymbol{\xi}^-, \varepsilon, \boldsymbol{\alpha}^+, \boldsymbol{\alpha}^-\right) &= \frac{1}{\tau}\cdot\frac{1}{N}\sum_{i=1}^{N}\left(\left(\xi_i^+\right)^2 + \left(\xi_i^-\right)^2\right) \\ &+ \frac{1}{2}\left(\mathbf{p}_g^T\mathbf{p}_g\right) + \frac{2}{\tau}\cdot\varepsilon + \lambda\left(\mathbf{p}_g - \mathbf{p}_{g0}\right)^T\left(\mathbf{p}_g - \mathbf{p}_{g0}\right) \\ &+ \sum_{i=1}^{N}\alpha_i^+\left(y_i - \mathbf{p}_g^T\mathbf{x}_{gi} - \varepsilon - \xi_i^+\right) + \sum_{i=1}^{N}\alpha_i^-\left(\mathbf{p}_g^T\mathbf{x}_{gi} - y_i - \varepsilon - \xi_i^-\right). \end{aligned} \tag{3.10}$$

According to the dual theorem, the minimum of the Lagrangian function in Eq. (3.10) with respect to $\mathbf{p}_g, \boldsymbol{\xi}^+, \boldsymbol{\xi}^-, \varepsilon$ is equal to the maximum of the function with respect to $\boldsymbol{\alpha}^+, \boldsymbol{\alpha}^-$. Then the following equations can be considered as the necessary conditions of the optimal solution:

$$\frac{\partial L}{\partial \mathbf{p}_g} = \mathbf{p}_g + 2\lambda\left(\mathbf{p}_g - \mathbf{p}_{g0}\right) - \sum_{i=1}^{N}\left(\alpha_i^+ - \alpha_i^-\right)\mathbf{x}_{gi} = 0, \tag{3.11a}$$

$$\frac{\partial L}{\partial \xi_i^+} = \frac{2}{N\tau}\xi_i^+ - \alpha_i^+ = 0, \tag{3.11b}$$

$$\frac{\partial L}{\partial \xi_i^-} = \frac{2}{N\tau}\xi_i^- - \alpha_i^- = 0, \tag{3.11c}$$

$$\frac{\partial L}{\partial \varepsilon} = \frac{2}{\tau} - \sum_{i=1}^{N}\alpha_i^- - \sum_{i=1}^{N}\alpha_i^+ = 0. \tag{3.11d}$$

From Eqs. (3.11a)–(3.11d), we have

$$\mathbf{p}_g = \frac{2\lambda \mathbf{p}_{g0} + \sum_{i=1}^{N} (\alpha_i^+ - \alpha_i^-) \mathbf{x}_{gi}}{1+2\lambda}, \tag{3.12a}$$

$$\xi_i^+ = \frac{N\tau \cdot \alpha_i^+}{2}, \tag{3.12b}$$

$$\xi_i^- = \frac{N\tau \cdot \alpha_i^-}{2}, \tag{3.12c}$$

$$\sum_{i=1}^{N} \alpha_i^- + \sum_{i=1}^{N} \alpha_i^+ = \frac{2}{\tau}. \tag{3.12d}$$

Substituting Eqs. (3.12a)–(3.12d) into Eq. (3.10), we obtain the dual problem for Eq. (3.8), i.e.,

$$\begin{aligned} \max_{\boldsymbol{\alpha}^-, \boldsymbol{\alpha}^+} \ & \tfrac{-1}{2(1+2\lambda)} \cdot \sum_{i=1}^{N} \sum_{i=1}^{N} (\alpha_i^+ - \alpha_i^-) \left(\alpha_j^+ - \alpha_j^-\right) \mathbf{x}_{gi}^T \mathbf{x}_{gj} - \tfrac{N\tau}{4} \cdot \sum_{i=1}^{N} \left((\alpha_i^+)^2 + (\alpha_i^-)^2 \right) \\ & - \tfrac{2\lambda}{1+2\lambda} \cdot \sum_{i=1}^{N} (\alpha_i^+ - \alpha_i^-) \mathbf{p}_{g0}^T \mathbf{x}_{gi} + \sum_{i=1}^{N} (\alpha_i^+ - \alpha_i^-) y_i + \tfrac{\lambda}{1+2\lambda} \cdot \mathbf{p}_{g0}^T \mathbf{p}_{g0} \\ \text{s.t.} \ & \sum_{i=1}^{N} \alpha_i^- + \sum_{i=1}^{N} \alpha_i^+ = \frac{2}{\tau}, \quad \alpha_i^- \geq 0, \alpha_i^+ \geq 0, \ \forall i. \end{aligned} \tag{3.12e}$$

Since the optimal solution of the dual problem, i.e., $(\boldsymbol{\alpha}^+)^*, (\boldsymbol{\alpha}^-)^*$, is independent of $\frac{\lambda}{1+2\lambda} \cdot \mathbf{p}_{g0}^T \mathbf{p}_{g0}$, Eq. (3.12e) is equivalent to the following equation:

$$\begin{aligned} \max_{\boldsymbol{\alpha}^-, \boldsymbol{\alpha}^+} \ & \tfrac{-1}{2(1+2\lambda)} \cdot \sum_{i=1}^{N} \sum_{i=1}^{N} (\alpha_i^+ - \alpha_i^-)(\alpha_j^+ - \alpha_j^-) \mathbf{x}_{gi}^T x_{gj} - \frac{N\tau}{4} \cdot \sum_{i=1}^{N} \left((\alpha_i^+)^2 + (\alpha_i^-)^2 \right) \\ & - \tfrac{2\lambda}{1+2\lambda} \cdot \sum_{i=1}^{N} (\alpha_i^+ - \alpha_i^-) (p_{g0}^T x_{gi} + y_i) + \sum_{i=1}^{N} (\alpha_i^+ - \alpha_i^-) y_i \\ \text{s.t.} \ & \sum_{i=1}^{N} \alpha_i^- + \sum_{i=1}^{N} \alpha_i^+ = \frac{2}{\tau}, \alpha_i^- \geq 0, \alpha_i^+ \geq 0. \end{aligned} \tag{3.12f}$$

Thus, Theorem 1 is hold.

It is clear from the above results that the optimization problem in Eq. (3.9) for TSK-FS training can be transformed into a QP problem that can be directly solved by the traditional QP solutions [45].

With the optimal solution $(\boldsymbol{\alpha}^{+})^{*}, (\boldsymbol{\alpha}^{-})^{*}$ of the dual problem in Eq. (3.9), we can get the optimal solution of the primal problem in Eq. (3.8) based on the relations presented in Eqs. (3.12a)–(3.12d). The optimal model parameters of trained TSK-FS, i.e., $(\mathbf{p}_g)^{*}$, is then given by

$$(\mathbf{p}_g)^* = \frac{2\lambda \mathbf{p}_{g0} + \sum_{i=1}^{N} \left((\alpha_i^+)^* - (\alpha_i^-)^* \right) \mathbf{x}_{gi}}{1+2\lambda}, \tag{3.13a}$$

which can be further expressed as

$$(\mathbf{p}_g)^* = \gamma \mathbf{p}_{g0} + (1\text{-}\gamma)\, \mathbf{p}_{gc}, \tag{3.13b}$$

with $\gamma = \frac{2\lambda}{1+2\lambda}$, $\mathbf{p}_{gc} = \sum_{i=1}^{N} \left((\alpha_i^+)^* - (\alpha_i^-)^* \right) \mathbf{x}_{gi}$.

From Eq. (3.13b), we can see that the final optimal parameter $(\mathbf{p}_g)^{*}$ obtained for the desired TSK-FS contains two parts, i.e. $\gamma \cdot \mathbf{p}_{g0}$ and $(1-\gamma) \cdot \mathbf{p}_{gc}$. While $(1-\gamma) \cdot \mathbf{p}_{gc}$ can denote the knowledge learned from the data of the target domain, $\gamma \cdot \mathbf{p}_{g0}$ can be taken as the knowledge inherited from the source domains. Thus, the final model parameter $(\mathbf{p}_g)^{*}$ is a balance between these two kinds of knowledge.

Learning Algorithm For KL-TSK-FS

Based on the findings in the previous section, the learning algorithm of the proposed KL-TSK-FS is developed and described as follows:

Algorithm KL-TSK-FS	
Step 1	Introduce the knowledge of the source domains, i.e., the model parameter.
Step 2	Set the balance parameters τ, λ in Eq. (3.8).
Step 3	Use the antecedent parameters of the fuzzy model obtained from the source domains and Eqs. (3.2d) and (3.3e) to construct the dataset $\mathbf{x}_{gi}$ for the corresponding model task, i.e., the linear regression model in Eq. (3.3f), associated with the fuzzy system to be constructed for the target domain.
Step 4	Use Eqs. (3.9) and (3.13a) to obtain the final consequent parameters $(\mathbf{p}_g)^{*}$ of the desired TSK-FS in the target domain.
Step 5	Use the antecedent parameters inherited from the source domains and the consequent parameters obtained in step 4 to generate the fuzzy system for the target domain.

Computational Complexity Analysis

The computational complexity of the above algorithm is analyzed as follows. The whole algorithm includes two main parts: (1) acquisition of the antecedent parameters of the fuzzy system and (2) learning of the consequent parameters. For the first part, since the antecedent parameters are inherited directly from the reference scene as the available knowledge, the computational complexity is O(1). For the second, the consequent parameters are obtained by solving the QP problem in Eq. (3.9) and the complexity is usually $O(N^2)$ for typical QP problems. However, it can be further reduced to $O(N)$ with some sophisticated algorithms, such as the working set-based algorithm [33]. Therefore, the computational complexity of the proposed algorithm is between $O(N)$ and $O(N^2)$, depending on the QP solutions used. In this study, we adopt the working set-based QP solution [33] for solving the QP problem concerned.

3.3 Transductive Transfer Learning with DAKSVM

3.3.1 Concepts and Problem Formulation

In this subsection, we introduce several definitions to clarify our terminology and propose our algorithm and analysis on the domain adaptation transfer learning problems.

Definition 1 (Domain) A domain D is composed of both feature space χ and marginal probabilistic distribution $P(\mathbf{X})$, i.e., $D=\{\chi, P(\mathbf{X})\}$, where $\mathbf{X}=\{\mathbf{x}_i\}_{i=1}^N \in \chi$.

Definition 2 (Task) Given a specific domain $D=\{\chi, P(\mathbf{X})\}$, a task is composed of both tag space $\mathbf{Y}$ and target prediction function $f(\cdot)$, i.e., $T=\{\mathbf{Y}, f(\cdot)\}$, where $f(\cdot)$ learned from the training dataset $\{\mathbf{x}_i, y_i\}$, where $\mathbf{x}_i \in \mathbf{X}$, $y_i \in \mathbf{Y}$. The function $f(\cdot)$ can be used to make prediction for the tag $f(\mathbf{x})$ corresponding with $\mathbf{X}$. From a probabilistic point of view $f(\mathbf{x})=P(y|\mathbf{x})$.

Definition 3 (Domain Adaptation Learning, DAL) Given a source domain D_s with its learning task T_s and target domain D_t with its learning task T_t, respectively, we refer to domain adaptation learning (DAL) as the following problem: given a set of labeled training dataset $\mathbf{X}_s=\{(\mathbf{x}_i, y_i)\}_i \in D_s \times \{\pm 1\}$, where $y_i \in \mathbf{Y}_s \subset \mathbf{Y}$ is the class label corresponding to $\mathbf{x}_i$, from source domain D_s. Thus, we need to make prediction $f_t(\cdot)$ for some unlabeled test dataset $X_t=\{\mathbf{x}_j\}_j \in D_t$ from target domain D_t. D_s with its task T_s and D_t with its task T_t are different, respectively, in the same feature space. When $D_s=D_t$ and $T_s=T_t$, DAL will be degenerated into classical machine learning problems.

Given an input space $\mathbf{X}$ and a label set $\mathbf{Y}$ of classes, a classifier is a function as $f(\mathbf{x}): \mathbf{X} \rightarrow \mathbf{Y}$ which maps data $\mathbf{x} \in \mathbf{X}$ to label set $\mathbf{Y}$. In the context, let us consider two datasets $\mathbf{X}_s = \{(\mathbf{x}_{s1},y_{s1}), \ldots,(\mathbf{x}_{sn},y_{sn})\}$ drawn from $\mathbf{X} \times \mathbf{Y}$ with probabilistic distribution $P_s(\mathbf{x}_s,y_s)$ and $\mathbf{X}_t = \{\mathbf{x}_{t1}, \ldots,\mathbf{x}_{tm}\}$ drawn from $\mathbf{X}$ with probabilistic distribution $P_t(\mathbf{x}_t,y_t)$ where y_t needs to be predicted, which are composed of n source domain and m target domain patterns, respectively, and usually $0 \leq m << n$. $\mathbf{x}_s$ and $\mathbf{x}_t$ are denoted by d-dimensional feature vectors with respect to $\mathbf{X}_s$ and $\mathbf{X}_t$, respectively. The classical large margin learning machines (such as SVMs) work well under such hypothesis as $P_s(\mathbf{x}_s,y_s) = P_t(\mathbf{x}_t,y_t)$. However, DAL can make accurate prediction for the unlabeled target data to some extent by learning a classifier under even such hypothesis as $P_s(\mathbf{x}_s,y_s) \neq P_t(\mathbf{x}_t,y_t)$. The performance of DAL depends on both the complexity of the investigated problems and the correlation between $P_s(\mathbf{x}_s,y_s)$ and $P_t(\mathbf{x}_t,y_t)$ [6]. In this chapter, the proposed method is formulated under the following hypothesis:

1. There are only one source domain and one target domain sharing the same feature space in DAL problems, which is the most popular hypothesis used by the state-of-the-art methods.
2. A training dataset $\mathbf{X}_s = \{(\mathbf{x}_{si},y_{si})\}_i$ is available for D_s while a testing dataset $\mathbf{X}_t = \{(\mathbf{x}_{tj},y_{tj})\}_j$ is available for D_t with y_{tj} which is unknown.
3. $P_s(\mathbf{x}_s,y_s) \neq P_t(\mathbf{x}_t,y_t)$ and $P_s(y_s|\mathbf{x}_s) \neq P_t(y_t|\mathbf{x}_t)$.

3.3.2 Distribution Discrepancy Metrics on RKHS Embedding Domain Distributions

Kernel methods are broadly used as an effective way of constructing nonlinear algorithms from linear ones by embedding datasets into some higher dimensional RKHSs [46]. A generalization of this idea is to embed probabilistic distributions into RKHS, giving us a linear method for dealing with higher order statistics [47, 48]. Let a complete inner product space H of functions F, and for $g \in F$, $g : \mathbf{X} \to \mathbf{R}$, where $\mathbf{X}$ is a nonempty compact set, if the linear dot function mapping $g \to g(x)$ exists for all $\mathbf{x} \in \mathbf{X}$, we call H as an RKHS. Under the aforementioned conditions, $g(\boldsymbol{x})$ can be denoted as an inner product: $g(\mathbf{x}) = < g, \phi(\mathbf{x}) >_H$, where $\phi : \mathbf{X} \to H$ denotes the feature space projection from $\mathbf{x}$ to H. And the inner product of the images of any points $\mathbf{x}$ and $\mathbf{x}'$ in feature space is called kernel $k(\mathbf{x},\mathbf{x}') = < \phi(\mathbf{x}), \phi(\mathbf{x}') >_H$. It is pointed out in [48] that RKHS with Gaussian kernel is universal.

Definition 4 (Integral Probability Metric on RKHS Embedding Distributions [2]) Given the set Θ of all Borel probabilistic measures defined on the topological space M, and the RKHS (H, k) of functions on M with k as its reproducing kernel. For any $P \in \Theta$, denotes by $Pk := \int_M k(.,\boldsymbol{x})dP(\boldsymbol{x})$. If k is measurable and bounded, then we may define the embedding of P in H as $Pk \in H$. Then, the RKHS embedding distributions distance between two such mappings associated with $P, Q \in \Theta$ is defined as follows:

$$\gamma_k(P,Q) = ||Pk - Qk||_H \tag{3.14}$$

We may say k is a characteristic kernel (CK) if the mapping $P \mapsto Pk$ is injective [48], in which case $\gamma_k(P,Q)=0$ if and only if $P=Q$ [49]. Hence γ_k is viewed as the distance metric on Θ. The RKHS embedding distributions cannot be distinguished when k is not a CK, thus leading to the failure of RKHS embedding distribution measure. Hence, it is a key factor for the success of RKHS embedding distribution measure that whether k is a CK or not. Fortunately, many popular kernel functions, such as polynomial kernel function, Gaussian kernel function, and Laplace kernel function, are all CK and universal ones [48]. Particularly, it is worth noting that Gaussian kernel mapping can provide us an effective RKHS embedding skill for the consistency estimation of the probability distribution distance between different domains [48]. Hence, in the sequel, we adopt the Gaussian kernel function $k_\sigma(\mathbf{x},\mathbf{z})=\exp\left(-\frac{1}{2\sigma^2}\|\mathbf{x}-\mathbf{z}\|^2\right)$, where $\mathbf{x},\mathbf{z}\in\mathbf{X}$, and σ denotes the kernel bandwidth, as the reproducing kernel in Hilbert space in this work. It is worthy to note that instead of using a fixed and parameterized kernel, one can also use a finite linear combination of kernels to compute γ_k.

For domain adaptation transfer learning problems, let D_s and D_t denote source and target domain, respectively, and $\mathbf{X}_s\in D_s$, $\mathbf{X}_t\in D_t$ denote sample from D_s and D_t, respectively, with probability measures P_s and P_t, respectively. Let P_{x_s,x_t} denote the joint probability measure of $\mathbf{X}_s\times\mathbf{X}_t$. Assume all measures are Borel ones and $\mathbf{X}_s$, $\mathbf{X}_t$ are two compact sets. Besides, let an RKHS H of a class of functions F with kernel k, then for $g\in F$, $g:\mathbf{X}\to\mathbf{R}$, where $\mathbf{X}$ is a nonempty compact set, there exists the reproducing property as follows: $<g(\cdot),k(\mathbf{x},\cdot)>=g(\mathbf{x})$, $<k(\mathbf{x},\cdot),k(\mathbf{x}',\cdot)>=k(\mathbf{x},\mathbf{x}')$, where $<,>$ denotes inner product operator. Thus, by Definition 1, the unbiased empirical estimator of maximum mean distance (MMD) on RKHS embedding domain distributions is defined as [50]:

$$MMD(F,\mathbf{X}_s,\mathbf{X}_t)=\left\|\frac{1}{n}\sum_{i=1}^{n}\varphi(\mathbf{x}_i)-\frac{1}{m}\sum_{j=1}^{m}\varphi(\mathbf{z}_j)\right\|^2, \tag{3.15}$$

where $\mathbf{x}_i\in\mathbf{X}_s$, $\mathbf{z}_j\in\mathbf{X}_t$.

Specifically, by Definition 1, we can have the following definitions on RKHS embedding distribution distance metric.

Definition 5 (Projected Maximum Mean Distance Metric on RKHS Embedding Domain Distributions) Let linear function f: $f(\mathbf{x})=\langle\mathbf{w},\phi(\mathbf{x})\rangle$, where $\mathbf{w}$ is a projection vector. Then the projected maximum mean distance metric on RKHS embedding domain distributions is defined as follows:

$$\begin{aligned}\gamma_{KM}(f,\mathbf{X}_s,\mathbf{X}_t)&=\left\|\tfrac{1}{n}\sum_{i=1}^{n}\mathbf{w}^T\varphi(\mathbf{x}_i)-\tfrac{1}{m}\sum_{j=1}^{m}\mathbf{w}^T\varphi(\mathbf{z}_j)\right\|^2\\&=\mathbf{w}^T\left(\tfrac{1}{n}\sum_{i=1}^{n}\varphi(\mathbf{x}_i)-\tfrac{1}{m}\sum_{j=1}^{m}\varphi(\mathbf{z}_j)\right)\left(\tfrac{1}{n}\sum_{i=1}^{n}\varphi(\mathbf{x}_i)-\tfrac{1}{m}\sum_{j=1}^{m}\varphi(\mathbf{z}_j)\right)^T\mathbf{w},\end{aligned} \tag{3.16}$$

where $\boldsymbol{x}_i\in\mathbf{X}_s$, $z_j\in\mathbf{X}_t$.

Definition 6 (Projected Maximum Scatter Distance Metric on RKHS Embedding Domain Distributions) Let linear function f: $f(\mathbf{x}) = \langle \mathbf{w}, \phi(\mathbf{x}) \rangle$, where $\mathbf{w}$ is a projection vector. Then, along the same line of Definition 2, the projected maximum scatter distance metric on RKHS embedding domain distributions is defined as

$$\gamma_{KS}(f, \mathbf{X}_s, \mathbf{X}_t) = \mathbf{w}^T \left| \frac{1}{n}\sum_{i=1}^{n} \varphi(\mathbf{x}_i)\,[\varphi(\mathbf{x}_i)]^T - \frac{1}{m}\sum_{j=1}^{m} \varphi(\mathbf{z}_j)\,[\varphi(\mathbf{z}_j)]^T \right| \mathbf{w}, \tag{3.17}$$

where $\mathbf{x} \in \mathbf{X}_s, \mathbf{z} \in \mathbf{X}_t$.

Definition 7 (GPMDD Metric on RKHS Embedding Domain Distributions) By Definitions 2 and 3, generalized projected maximum distribution distance metric on RKHS embedding domain distributions with probabilistic distribution $p, q \in P$ is defined as

$$\gamma_{KMS}(f, \mathbf{X}_s, \mathbf{X}_t) = (1-\lambda)\,\gamma_{KM}(f, \mathbf{X}_s, \mathbf{X}_t) + \lambda\,\gamma_{KS}(f, \mathbf{X}_s, \mathbf{X}_t), \tag{3.18}$$

where $\lambda \in [0,1]$ and when $\lambda = 0$, $\gamma_{KMS} = \gamma_{KM}$. The parameter λ is treated as a balance between probabilistic distribution mean and scatter (or variance). When λ increases, γ_{KMS} is biased in favor of preserving the distribution scatter consistency between both domains. When λ decreases, γ_{KMS} is biased in favor of preserving the distribution mean consistency between both domains. Hence, the proposed method can preserve both the distributions consistency between domains and the discriminative information in both domains.

It can be guaranteed by the following theorem that the GPMDD between both domains can be measured sufficiently.

Theorem 1 [51] Let F is a unit ball defined in some universal RKHS H with a kernel $k(\cdot,\cdot)$, which are all defined in a compact metric space. And let $\mathbf{X}_s$, $\mathbf{X}_t$ are two compact sets generated from Borel probability metrics p and q, respectively, in the metric space with p and q. Then $\gamma_{KMS}(F, \mathbf{X}_s, \mathbf{X}_t) = 0$ if and only if $p = q$.

3.3.3 Domain Adaptation Kernelized Support Vector Machine

Inspired by the idea of manifold regularization, MMD-based methods for transductive transfer learning (e.g., LMPROJ [28] and DTSVM [29], etc.) can be formulated as follows:

$$\begin{aligned} f = &\min_{\mathbf{w} \in H_K} C\sum_{i=1}^{n} \xi_i + \tfrac{1}{2}\|\mathbf{w}\|_K^2 + \lambda\,\gamma_{KM}(f, \mathbf{X}_s, \mathbf{X}_t) \\ s.t.\ & y_i\left(\mathbf{w}^T \phi(\mathbf{x}_i) + b\right) \geq 1 - \xi_i, \\ & \xi_i \geq 0, i = 1, \ldots, n \end{aligned} \tag{3.19}$$

where $\mathbf{w}$ is a normal projection vector, k is a kernel with ϕ a kernel mapping, H_K is a set of functions in the kernel space, λ is a balance parameter, and $\gamma_{KM}(f,\mathbf{X}_s,\mathbf{X}_t)$ is the projected distribution mean distance metric between source and target domains, where $\mathbf{x}_i \in \mathbf{X}_s$.

However, Eq. (3.6) discloses a key limitation of MMD-based methods to some extent, i.e., they ignore considering sufficiently the potential scatter statistics, which may include underlying discriminative information in both domains for DAL, such that they may lead to "overfitting" phenomenon in some specific pattern recognition applications. Therefore, in this chapter, we propose a robust DAKSVM regularized by GPMDD metric on RKHS embedding domain distributions, which partially extends the ideas of classical SVMs and MMD. The key goals of our methods are to find a feature transform such that the mean and variance distances between the distributions of the testing and training data are minimized sufficiently, while at the same time maximizing the class margin or certain classification performance criterion for the training data, thus learning a robust model to effectively make prediction for target domain.

3.3.3.1 Objective Function of DAKSVM

For simplicity, firstly we only consider binary pattern classification problems, and secondly we propose a so-called least-square DAKSVM (LSDAKSVM) based on the classical LS-SVM [52] as an extension to the standard DAKSVM method for multi-class pattern classification problems.

For DAL problems, DAKSVM aims to find a linear transform $f(\mathbf{x}) = \mathbf{w}^T \varphi(\mathbf{x})$ in a universal RKHS with Gaussian kernel mapping, where $\mathbf{w}$ is a linear projection vector, in order to minimize the distribution discrepancy between-domain as well as to reduce the empirical risk of the classification decision function as much as possible, thus implementing transfer learning in cross-domains. DAKSVM can be formulated as

$$\min f = C\sum_{i=1}^{n} V(\mathbf{x}_i, y_i, f) + \gamma_{KMS}(f, \mathbf{X}_s, \mathbf{X}_t), \tag{3.20}$$

where $\mathbf{x}_i \in \mathbf{X}_s$ is a set of training data and matrix $\phi(\mathbf{X}_s) = (\phi(\mathbf{x}_{s1}), \phi(\mathbf{x}_{s2}), \ldots, \phi(\mathbf{x}_{sn}))$, $y_i \in \mathbf{Y}_s$ is the class label corresponding to $\mathbf{x}_i$, $C > 0$ is a regularization coefficient, and V measures the fitness of the function in terms of predicting the class labels for the training data and is called the risk function. The hinge loss function is a commonly used risk function in the form of $V = (1 - y_i f(x_i))_+$ [53] in which $(x)_+ = x$ if $x \geq 0$ and zero otherwise.

Therefore, the linear function f in Eq. (3.20) represented by a vector $\mathbf{w}$ can be represented as

$$\begin{aligned} &\arg\min_{\mathbf{w},b,\xi} f = C\sum_{i=1}^{n} \xi_i + \gamma_{KMS}(f, \mathbf{X}_s, \mathbf{X}_t) \\ &\text{s.t. } y_i((w, \varphi(\mathbf{x}_i)) + b) \geq 1 - \zeta_i, i = 1, 2, \ldots, n \end{aligned} \tag{3.21}$$

In order to solve the primal in Eq. (3.21) effectively, we introduce the following revised Representer Theorem for DAL problems as follows:

Theorem 2 (Representer Theorem [54] for DAL) For a DAL problem, let $\psi:[0,\infty)\rightarrow\mathbf{R}$ denote a strictly monotonic increasing function, $\mathbf{X}=\mathbf{X}_s\cup\mathbf{X}_t$ be a dataset, and $c:(\mathbf{X}\times\mathbf{R}^2)^n\rightarrow\mathbf{R}\cup\{\infty\}$ be any loss function. Then the regularized risk function is defined as

$$R(f)=c\left((x_i,y_i,f(x_i))_{i=1}^{n}\right)+\psi\left(\|f\|_H^2\right), \tag{3.22}$$

where $f\in H$ is represented as

$$f(x)=\sum_{i=1}^{m}\beta_i k(x_i,x)+\sum_{j=1}^{n}\beta_j k(z_j,x), \tag{3.23}$$

where k is a kernel, $\mathbf{x}_i\in\mathbf{X}_s$, $\mathbf{y}_i\in\mathbf{Y}_s$, $\mathbf{z}_j\in\mathbf{X}_t$ and β_i is a coefficient.

By Theorem 2, we can have the following theorem.

Theorem 3 The primal of DAKSVM can be formulated as

$$\min_{\beta,\xi,b} f=\tfrac{1}{2}\beta^T\Omega\beta+C\sum_{i=1}^{N}\xi_i, \tag{3.24a}$$

$$\text{s.t. } y_i\left(\sum_{j=1}^{n+m}\beta_j k_\sigma(\mathbf{x}_i,\mathbf{x}_j)+b\right)\geq 1-\xi_i,\quad i=1,\ldots,n,$$

where $\mathbf{x}_i\in\mathbf{X}_s$, $\mathbf{x}_j\in\mathbf{X}_s\cup\mathbf{X}_t$, Ω is a positive semi-definite kernel matrix with

$$\Omega=(1-\lambda)\Omega_1+\lambda\Omega_2 \tag{3.24b}$$

where Ω_1 is a $(n+m)\times(n+m)$ symmetrical positive semi-definite kernel matrix defined as

$$\Omega_1=\frac{1}{n^2}\mathbf{K}_s[1]^{n\times n}\mathbf{K}_s^T+\frac{1}{m^2}\mathbf{K}_t[1]^{m\times m}\mathbf{K}_t^T-\frac{1}{nm}\left(\mathbf{K}_s[1]^{n\times m}\mathbf{K}_t^T+\mathbf{K}_t[1]^{m\times n}\mathbf{K}_s^T\right) \tag{3.24c}$$

and Ω_2 is a $(n+m)\times(n+m)$ symmetrical positive semi-definite kernel matrix defined as

$$\Omega_2=\left|\frac{1}{n}\mathbf{K}_s\mathbf{K}_s^T-\frac{1}{m}\mathbf{K}_t\mathbf{K}_t^T\right| \tag{3.24d}$$

where $\mathbf{K}_s$ is a $(n+m)\times n$ kernel matrix for the training data, $\mathbf{K}_t$ is a $(n+m)\times m$ kernel matrix for test data, and $[\mathbf{1}]^{k\times l}$ is a $k\times l$ matrix of all ones.

Theorem 4 The dual of the primal in Eq. (3.24) can be formulated as

$$\min_{\alpha}\frac{1}{2}\alpha^T\mathbf{H}^{\phi}\alpha - 1^T\alpha \tag{3.25}$$

$$\text{s.t. } 0 \le \alpha_i \le C, \ i = 1,\ldots,n,$$

$$\sum_{i=1}^{n}\alpha_i y_i = 0,$$

where $\mathbf{H}^{\phi} = \tilde{\mathbf{Y}}\mathbf{K}_s^T(\Omega)^{-1}\mathbf{K}_s\tilde{\mathbf{Y}}$, and $\tilde{\mathbf{Y}} = diag\,(y_1, y_2, \ldots, y_n)$, $y_i \in \mathbf{Y}_s$.

By the same way of the classical SVM, the biased variable b^{φ} in the kernel space can be formulated as

$$b^{\phi} = -\frac{1}{2}\left(\frac{1}{|\mathbf{X}_{s+}|}\sum_{\mathbf{x}\in\mathbf{X}_{s+}}\sum_{j=1}^{n+m}\beta_j k_{\sigma}(\mathbf{x}_j,\mathbf{x}) + \frac{1}{|\mathbf{X}_{s-}|}\sum_{\mathbf{x}\in\mathbf{X}_{s-}}\sum_{j=1}^{n+m}\beta_j k_{\sigma}\left(\mathbf{x}_j,\mathbf{x}\right)\right) \tag{3.26a}$$

Meanwhile, we can get the solution of β with dual theory as follows:

$$\beta = (\Omega)^{-1}\mathbf{K}_s\tilde{\mathbf{Y}}\alpha \tag{3.26b}$$

3.3.3.2 Learning Algorithm of DAKSVM

The proposed DAKSVM algorithm can be summarized as follows.

Algorithm DAKSVM

Input: Dataset matrix $\mathbf{X} = (\{\mathbf{x}_i, y_i\}_{i=1}^n, \{\mathbf{z}_j\}_{j=1}^m)$, $\mathbf{x}_i \in \mathbf{X}_s$, $y_i \in \mathbf{Y}_s$, $\mathbf{z}_j \in \mathbf{X}_t$, set Gaussian kernel bandwidths σ, σ/γ, respectively, in γ_{KM} and γ_{KS} of GPMDD.

Output: Decision function $f(\mathbf{x})$.

Step 1: Determine the parameter γ in γ_{KS} of GPMDD such that the scatter consistency between source and target domains is maximized.

Step 2: Compute the matrices Ω_1 and Ω_2, respectively, by Eqs. (3.24a) and (3.24b). In terms of λ given by users to construct matrix $\Omega = (1-\lambda)\Omega_1 + \lambda\Omega_2$.

Step 3: For the given C, find out the optimal vector $\boldsymbol{\beta}$ by applying Theorem 4 to solve the corresponding dual. And then recover the optimal normal vector $\mathbf{w}$ and bias b^{ϕ} by $\boldsymbol{\beta}$;

Step 4: Output the decision function $f(\mathbf{x}) = \mathbf{w}^T\phi(\mathbf{x}) + b^{\phi}$.

3.3.4 Variants and Extensions

3.3.4.1 Least-Square DAKSVM

One variant of DAKSVM is the LSDAKSVM which is also based on the idea of LS-SVM [52], which can be formulated as:

$$\begin{aligned}
&\arg\min_{\mathbf{w},b,\xi} f = \frac{C}{2}\sum_{i=1}^{n} \xi_i^2 + \gamma_{KMS}(p,q) \\
&\text{s.t.}\ \ (\mathbf{w}, \phi(\mathbf{x}_i)) + b = y_i - \xi_i, i = 1,2,\ldots,n.
\end{aligned} \tag{3.27}$$

Along the same line of DAKSVM, the primal of Eq. (3.17) is defined as

$$\min_{\beta,\xi,b} f = \frac{1}{2}\beta^T \Omega \beta + \frac{C}{2}\sum_{i=1}^{n} \xi_i^2, \tag{3.28}$$

$$\text{s.t.}\ \ \sum_{j=1}^{n+m} \beta_j k_{\sigma/\gamma}(\mathbf{x}_i, \mathbf{x}_j) + b = y_i - \xi_i, \quad \xi_i \geq 0, i = 1,\ldots,n.$$

Theorem 5 (Analytic Solution to Binary Class Case) Given the parameter $\lambda \in [0,1]$, for a binary classification problem, the optimal solution of Eqs. (3.27) and (3.28) is equivalent to the linear system of equations with respect to variable α as follows:

$$\begin{bmatrix} 0 & \mathbf{1}_n^T \\ \mathbf{1}_n & \tilde{\Omega} \end{bmatrix} \begin{bmatrix} b \\ \alpha \end{bmatrix} = \begin{bmatrix} 0 \\ \mathbf{Y}_s \end{bmatrix}, \tag{3.29}$$

where $\mathbf{1}_n = [1, \ldots, 1]^T$, $\alpha = [\alpha_1, \ldots, \alpha_n]^T$, $\mathbf{Y}_s = [y_1, \ldots, y_n]^T$, $\tilde{\Omega} = \mathbf{K}_s^T(\Omega)^{-1}\mathbf{K}_s + \frac{\mathbf{I}_n}{C}$, $\mathbf{I}_n$ is an n-dimensional identity matrix.

As for multi-class classification problems, the traditional skills are to separate a multi-class classification problem into several binary classification problems in one-against-one (OAO) or one-against-all (OAA) way. However, the main drawbacks of these skills deal with high computational complexity and imbalance between classes. Hence, here we introduce the vector labeled outputs into the solution of LSDAKSVM, which can make the corresponding computational complexity independent of the number of classes and require no more computations than a single binary classifier [55]. Furthermore, Szedmak and Shawe-Taylor [55] pointed out that this technique does not reduce the classification performance of a learning model but in some cases can improve it, with respect to OAO and OAA. Therefore, we represent the class labels according to the one-of-c rule, namely, if training sample $\boldsymbol{x}_i$ $(i = 1, \ldots, n)$ belongs to the kth class, then the class label of $\boldsymbol{x}_i$ is

$Y_i = \left[\underbrace{0,\ldots,1}_{k},\ldots,0\right]^T \in \mathbf{R}^c$, where the kth element is one and all the other elements are zero. Hence, for some multi-class classification problems, the optimal problem of LSDAKSVM can be formulated as

$$\min_{\beta,\xi,b} f = \tfrac{1}{2}\tilde{\beta}^T\Omega\tilde{\beta} + \tfrac{C}{2}\sum_{i=1}^{n}\xi_i^2, \tag{3.30}$$

$$\text{s.t.}\ \ \tilde{\beta}^T\mathbf{K}_s + b = Y_i - \xi_i, i = 1,\ldots,n,$$

where $\tilde{\beta} \in R^{n\times c}$, $b \in \mathbf{R}^c$.

Theorem 6 (Analytic Solution to Multi-Class Case) Given the parameter $\lambda \in [0,1]$, for a multi-class classification problem, the optimal solution of Eq. (3.30) is equivalent to the linear system of the following equation.

$$\begin{bmatrix} b & \alpha \end{bmatrix} \begin{bmatrix} 0 & 1_n^T \\ 1_n & \tilde{\Omega} \end{bmatrix} = \begin{bmatrix} 0_c & \tilde{\mathbf{Y}}_s \end{bmatrix}, \tag{3.31}$$

where $\mathbf{0}_c = [0, \ldots, 0]^T$, $\alpha = [\alpha_1, \ldots, \alpha_n]^T$, $\tilde{\mathbf{Y}}_s = [Y_1, Y_2, \ldots, Y_n]^T$, $\widetilde{\Omega}$ is the same as in Theorem 6.

Theorems 5 and 6 actually provide us the LSDAKSVM versions for both binary and multi-class classification problems, respectively. It is clearly shown from Eqs. (3.20) and (3.23) that LSDAKSVM keeps the same solution framework for both binary and multi-class cases.

3.3.4.2 μ-Domain Adaptation Kernelized Support Vector Machine

The ν-SVM [56] is a typical extension of SVM for classification in which Schölkopf et al. introduced a new parameter ν instead of C in SVM to control the number of support vectors and the training errors. More details about ν-SVM can be found in [56]. Hence, as the second variant of DAKSVM based on ν-SVM, μ-DAKSVM can be formulated as:

$$\min_{\beta,\xi,b} f = \tfrac{1}{2}\beta^T\Omega\beta - \mu\rho + \tfrac{1}{N}\sum_{i=1}^{n}\xi_i, \tag{3.32}$$

$$\text{s.t.}\ \ y_i\left(\sum_{j=1}^{N}\beta_j k_{\sigma/\gamma}(\mathbf{x}_i,\mathbf{x}_j) + b\right) \geq \rho - \xi_i,\ \ i = 1,\ldots,n,$$

where the variables $N = n + m$, $\rho \geq 0$, $\mu > 0$ and $\xi_i \geq 0$ have the same meaning as in ν-SVM. Similar to ν-SVM, the dual of the primal in Eq. (3.32) can be formulated as:

$$\min_{\alpha} \frac{1}{2} \alpha^T \mathbf{H}^{\phi} \alpha \tag{3.33}$$

$$\text{s.t.} \ \ 0 \leq \alpha_i \leq \frac{1}{N}, \ \ i = 1, \ldots, n,$$

$$\sum_{i=1}^{n} \alpha_i y_i = 0,$$

$$\sum_{i=1}^{n} \alpha_i \geq \mu,$$

where $\mathbf{H}^{\varphi} = \widetilde{\mathbf{Y}} \mathbf{K}_s^T \left(\widetilde{\Omega}\right)^{-1} \mathbf{K}_s \widetilde{\mathbf{Y}}$, and $\widetilde{\mathbf{Y}} = diag\left(y_1, y_2, \ldots, y_n\right)$, $y_i \in \mathbf{Y}_s$.

3.3.5 Computational Complexity Analysis

In terms of Algorithm 1, DAKSVM and its variants can be implemented by using standard SVM solver (e.g., LibSVM [57]) with the quadratic form induced by matrix Ω aforementioned above, and using the optimal solution to obtain the expansion coefficients by Eqs. (3.35) and (3.13)–(3.15) respectively. It is worth noting that our algorithms compute the inverse of a dense Gram matrix Ω which leads to $O((n+m)^3)$ training complexity comparable to SVM. This seems to be impractical for large datasets. However, for highly sparse datasets, for example, in text categorization problems, effective conjugate gradient schemes can be used in a large-scale implementation [58]. For the nonlinear case, one may obtain approximate solutions (e.g., using greedy, matching pursuit techniques) where the optimization problem is solved over the span of a small set of basis functions instead of using the full representation in $f(x) = \mathbf{w}^T \phi(\boldsymbol{x})$. Besides, CVM [59] may be an alternative choice in addressing scalability issues occurring in SVM learning. The testing complexity of DAKSVM depends on the number of support vector learned from the training stage. In fact, the proposed method DAKSVM and its variants take less than half a minute to finish the whole prediction for test samples from target domain in most of the following experiments.

3.4 Experimental Results of KL-TSK-FS

3.4.1 Experimental Settings

The proposed inductive transfer learning method KL-TSK-FS is evaluated by using both synthetic and real-world datasets. Details about the evaluation will be described in detail in Sects. 3.4.2 and 3.4.3, respectively. For clarity, the notations for the datasets and their definitions are listed in Table 3.2. Here, datasets generated from the source domain and the target domain are denoted by D1 and D2, respectively. The proposed support learning-based KL-TSK-FS algorithm is evaluated from the following two aspects.

1. *Comparison with traditional support vector learning-based L2-TSK-FS.* The performance of KL-TSK-FS is compared comprehensively with that of three L2-TSK-FS methods implemented under different conditions. That is, four TSK-FS systems are constructed by (a) L2-TSK-FS based on the data in the source domain, (b) L2-TSK-FS based on the data in the target domain, (c) L2-TSK-FS based on the data in both the target domain and the source domain, and (d) the proposed KL-TSK-FS. They are denoted by L2-TSK-FS(D1), L2-TSK-FS (D2), L2-TSK-FS (D1 + D2), and KL-TSK-FS(D2 + Knowledge), respectively. With these four fuzzy systems, the testing data, i.e. D2_test, of the target domain are used to evaluate their generalization capability.
2. *Comparison with regression methods designed for datasets with missing or noisy data.* Three related regression methods are employed to compare with the proposed KL-TSK-FS for performance evaluation. The three methods include: (a) TS-fuzzy system-based support vector regression (TSFS-SVR) [60]; (b) fuzzy system learned through fuzzy clustering and SVM (FS-FCSVM) [61]; and (c) Bayesian task-level transfer learning for nonlinear regression method (HiRBF) [20].

The methods adopted for performance comparison from these two aspects are summarized in Table 3.3. The following generalization performance index J is used in the experiments [2],

Table 3.2 Notations of the adopted datasets and their definitions

Notation	Definitions
D1	Dataset generated from the source domain
D2	Dataset generated from the target domain for training
D2_test	Dataset generated from the target domain for testing
r	Relation parameter between the source domain and the target domain, which is used to construct the synthetic datasets

Table 3.3 The methods adopted for performance comparison

Support vector learning and L2-norm penalty-based TSK-FS modeling methods		Four methods designed for noisy/ missing data	
(1) The proposed KL-TSK-FS (D2 + Knowledge)	(2) L2-TSK-FS (D1) [2]	(1) The proposed KL-TSK-FS	(5) TSFS-SVR [60]
	(3) L2-TSK-FS (D2) [2]		(6) FS-FCSVM [61]
	(4) L2-TSK-FS (D1 + D2) [2]		(7) HiRBF [20]

$$J = \sqrt{\frac{\frac{1}{N}\sum_{i=1}^{N}\left(y_i' - y_i\right)^2}{\frac{1}{N}\sum_{i=1}^{N}\left(y_i - \overline{y}\right)^2}}, \tag{3.34}$$

where N is the number of test datasets, y_i is the output for the ith test input, y_i' is the fuzzy model output for the ith test input, and $\overline{y} = \frac{1}{N}\sum_{i=1}^{N} y_i$. The smaller the value of J, the better the generalization performance.

In the experiments, the hyperparameters of all the methods adopted are determined by using the fivefold cross-validation strategy with the training datasets. All the algorithms are implemented using MATLAB on a computer with Intel Core 2 Duo P8600 2.4 GHz CPU and 2GB RAM.

3.4.2 Synthetic Datasets

3.4.2.1 Generation of Synthetic Datasets

Synthetic datasets are generated to simulate the domains in the study and the following requirements need to be satisfied: (1) the source domain should be related to the target domain, i.e., the source and target domains are different but related; (2) some of the data of the target domain are not available or missing. In other words, the data available from the target domain are insufficient.

Based on the above requirements, the function $Y = f(x) = \sin(x) * x, x \in [-10, 10]$ is used to describe the source domain and to generate the dataset D1. On the other hand, the function $y = r * f(x) = r * \sin(x) * x, x \in [-10, 10]$ is used to describe the target domain and to generate the dataset D2 and testing dataset D2_test for the target domain. Here, r is a relation parameter between the source domain and the target domain. The parameter is used to control the degree of similarity/difference between these two domains. When $r = 1$, the two domains are identical. On the other hand,

Table 3.4 Details of the synthetic datasets

Source domain	Target domain		
Dataset	Training set		Testing set
Size of dataset	Interval with missing data	Size of dataset	Size of dataset
400	[−6, −3] and [0, 4]	144	200
Relation parameter between the two domains: $r = 0.9, 0.85, 0.8, 0.75$ and 0.7			

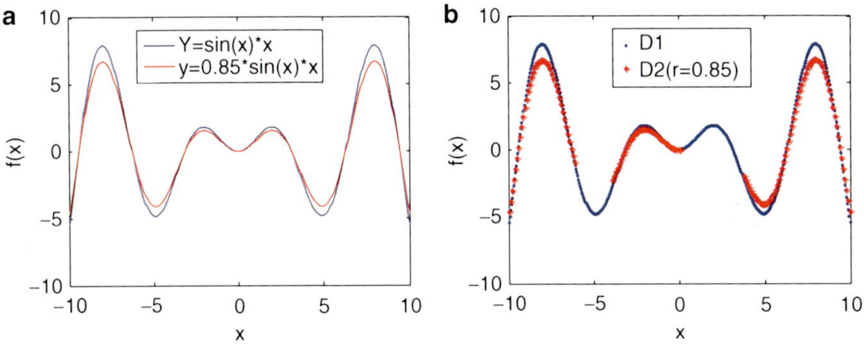

Fig. 3.3 Functions representing two different domains with the relation parameter $r = 0.85$ and the corresponding sampled data from these domains: (**a**) the functions representing the source domain (Y) and the target domain (y); (**b**) the data of the source domain and the training data of the target domain with missing data in the intervals [−6, −3] and [0, 4]

the lack of information (data insufficiency) is simulated by introducing intervals with missing data into the training set generated for the target domain. The settings for generating the synthetic datasets are described in Table 3.4. For example, the two functions used to simulate the two related domains, with the relation parameter $r = 0.85$, are shown in Fig. 3.3a. The datasets of the source domain and the training sets of the target domain, generated with the same relation parameter (i.e. $r = 0.85$), are shown in Fig. 3.3b. The figures also show the two data-missing intervals [−6, −3] and [0, 4] introduced into the dataset.

3.4.2.2 Comparing with the Traditional Support Vector Learning-Based L2-TSK-FS Modeling Methods

The performance of the proposed KL-TSK-FS and the three traditional L2-norm penalty-based TSK-FS modeling methods is evaluated and compared using the synthetic datasets. The experimental results are shown in Table 3.5 and Fig. 3.4. In Table 3.5 and other tables in this paper, the bold values denote the best results obtained among all the methods. The following observations can be made from the results:

Table 3.5 Generalization performance (J) of the proposed KL-TSK-FS method and the traditional L2-TSK-FS methods on the synthetic datasets

Interval with data missing	Relation parameter (r)	L2-TSK-FS (D1)	L2-TSK-FS (D2)	L2-TSK-FS (D1 + D2)	KL-TSK-FS (D2 + Knowledge)
$[-6, -3]$	0.9	0.1343	0.2858	0.1012	**0.0501**
and $[0, 4]$	0.85	0.1908	0.2813	0.1434	**0.0516**
	0.8	0.2574	0.2864	0.1983	**0.1094**
	0.75	0.3525	0.2841	0.2627	**0.1534**
	0.7	0.4406	0.2821	0.3432	**0.2388**

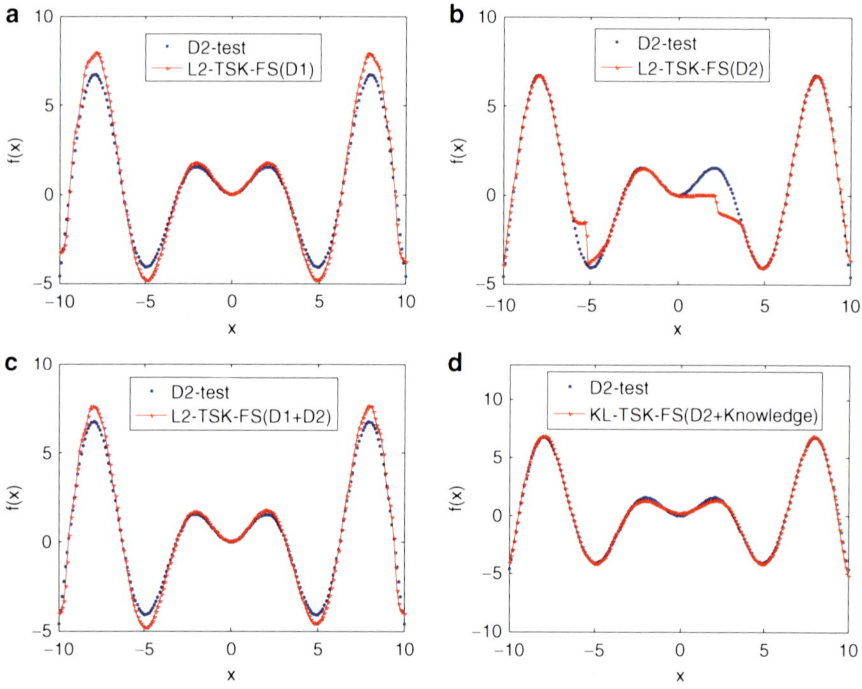

Fig. 3.4 Modeling results of the proposed KL-TSK-FS method and three traditional L2-TSK-FS methods by using the synthetic datasets shown in Fig. 3.5b: (**a**) L2-TSK-FS based on the data of the source domain (D1); (**b**) L2-TSK-FS based on the data of the target domain (D2); (**c**) L2-TSK-FS based on the data of both the reference and target domains (D1 + D2); (**d**) the proposed KL-TSK-FS (D2 + Knowledge)

1. It can be seen from Table 3.4 that the generalization performance of the knowledge-leverage-based fuzzy system KL-TSK-FS is better than that of the traditional L2-TSK-FS methods.
2. Figure 3.4a shows the modeling results of the L2-TSK-FS obtained by using the data of the source domain only. The results indicate that drifting exists between the source domain and the target domain, as evident from the discrepancies

between the two curves in the figure. Hence, the generalization performance of the TSK-FS obtained by L2-TSK-FS from the source domain is weak for the target domain. The findings show that the use of the data of the source domain alone is not appropriate for the modeling of the target domain.

3. Figure 3.4b shows the modeling results of the L2-TSK-FS obtained by using the data of the target domain only. The results indicate that the generalization performance of the TSK-FS obtained by L2-TSK-FS is even much weaker for the target domain. An obvious reason is that the data in the training set is insufficient, which degrades the generalization capability of the obtained TSK-FS. The prediction performance is especially poor in the intervals with missing data in the training dataset.
4. Figure 3.4c shows the modeling results of the L2-TSK-FS obtained by using the data of both the target domain and the source domain. Although the data of both domains have been used for training, the generalization performance of the obtained TSK-FS is still not good enough for the target domain. This can be explained by two reasons. First, drifting occurs between the reference and target domains, i.e., not all data in the source domain are useful for the modeling task of the target domain. Some of them may even have negative influence. Second, the size of the source domain is larger than that of the target domain, which makes the obtained TSK-FS more apt to approximate the source domain rather than the target domain.
5. Figure 3.4d shows the modeling results of the proposed KL-TSK-FS. The following observations can be made by comparing its results with the results of the three L2-TSK-FS methods, respectively. First, by inspecting Fig. 3.4a, d, we see that the KL-TSK-FS demonstrates better prediction results than the L2-TSK-FS which only uses the data of source domain. Second, it is evident from Fig. 3.4b, d that, by introducing the knowledge-leverage mechanism, the proposed KL-TSK-FS has effectively remedied the deficiency of the L2-TSK-FS obtained by the data of the target domain. By comparing Fig. 3.4c, d, we also find that the KL-TSK-FS has demonstrated better generalization performance than the L2-TSK-FS which employs the data of both the reference and target domains. It is noteworthy to point out that the KL-TSK-FS also has better privacy-protection capability than the methods that use the data of source domains directly. When the data in the source domains are not available due to the necessity of privacy protection, or in situations where knowledge are only partially revealed, methods requiring the data of all domains are no longer feasible. Therefore, the proposed KL-TSK-FS is particularly suitable for these situations attributed to its distinctiveness in privacy protection.

3.4.2.3 Comparing with Regression Methods Designed for Missing or Noisy Data

The performance of the proposed KL-TSK-FS method is evaluated by comparing its performance with that of three regression methods designed for handling

Table 3.6 Generalization performance (J) of the proposed KL-TSK-FS method and three related regression methods on the synthetic datasets

Interval with missing data	Relation parameter (r)	TSFS-SVR	FS-FCSVM	HiRBF	KL-TSK-FS
$[-6, -3]$ and $[0, 4]$	0.9	0.2972	0.3161	0.2621	**0.0501**
	0.85	0.2989	0.3179	0.2619	**0.0516**
	0.8	0.2983	0.3170	0.2687	**0.1094**
	0.75	0.2933	0.3167	0.2639	**0.1534**
	0.7	0.2970	0.3185	0.2611	**0.2388**

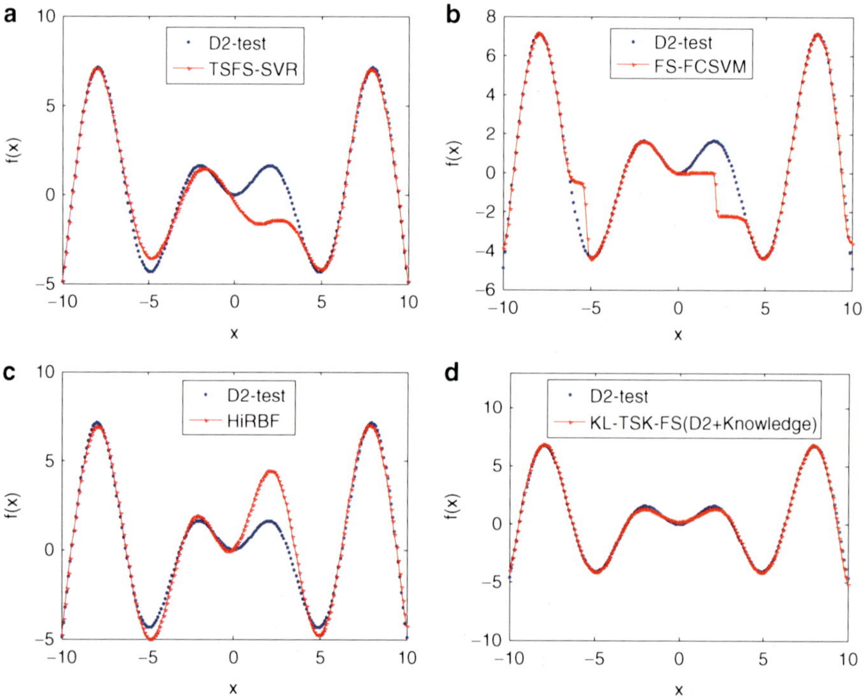

Fig. 3.5 Modeling results of the proposed KL-TSK-FS method and three related regression methods using the synthetic datasets in Fig. 3.5b: **a** TSFS-SVR, **b** FS-FCSVM, **c** HiRBF, and **d** KL-TSK-FS

noisy/missing data, i.e., TSFS-SVR, FS-FCSVM, and HiRBF. The evaluation is performed on the synthetic datasets. The experimental results are shown in Table 3.6 and Fig. 3.5, and the following observations can be obtained:

1. KL-TSK-FS has demonstrated better generalization performance than the other three related methods.
2. The results in Fig. 3.5a, b show that the support vector learning-based fuzzy modeling methods TSFS-SVR and FS-FCSVM are able to give better generalization

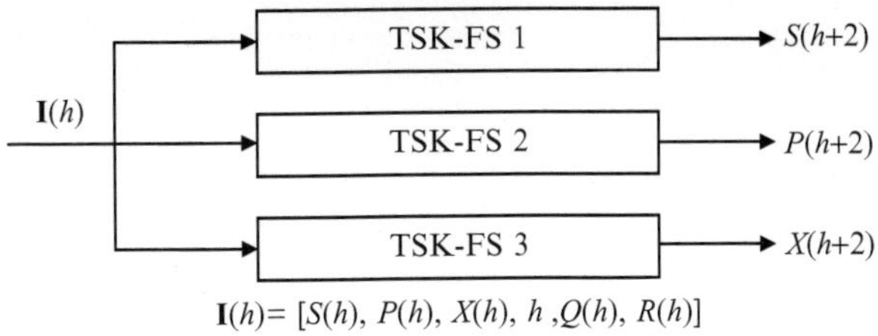

Fig. 3.6 Illustration of the glutamic acid fermentation process prediction model based on the TSK-FS

performance to a certain extent. For example, although the data in the interval $[-6, -3]$ are missing, these two methods still demonstrate promising generalization capability at this interval. However, the generalization abilities of these two methods in the other data-missing interval $[0, 4]$ are not satisfactory.

3. Although the transfer learning-based method HiRBF has used the data in both the target domain and the source domain in the training, it is evident form Fig. 3.5c that this method cannot effectively cope with the problem caused by the missing data, still exhibiting poor generalization ability in the two data-missing intervals.
4. Figure 3.5d shows that the proposed method KL-TSK-FS is able to give acceptable generalization capability in the two data-missing intervals, indicating that the method has effectively leveraged the useful knowledge from the source domain and remedy the generalization abilities in the training procedure.

3.4.3 *Real-World Datasets*

3.4.3.1 The Glutamic Acid Fermentation Process Modeling

To further evaluate the performance of the proposed method, an experiment is conducted to apply the method to model a biochemical process with real-world datasets [2]. The datasets adopted originates from the glutamic acid fermentation process, which is a multiple-input–multiple-output system. The input variables of the dataset include the fermentation time h, glucose concentration $S(h)$, thalli concentration $X(h)$, glutamic acid concentration $P(h)$, stirring speed $R(h)$, and ventilation $Q(h)$, where $h = 0, 2, \cdots, 28$. The output variables are glucose concentration $S(h+2)$, thalli concentration $X(h+2)$, and glutamic acid concentration $P(h+2)$ at a future time $h+2$. The TSK-FS-based biochemical process prediction model is illustrated in Figs. 3.6. The data in this experiment were collected from 21 batches of fermentation processes, with each batch containing 14 effective data samples. In this experiment, in order to match the situation discussed in this study, the data are divided into two domains, i.e., the source domain and the target domain, as described in Table 3.7.

Table 3.7 The fermentation process modeling datasets

	Data of source domain (D1)	Data of target domain	
		Training set (D2)[a]	Testing set (D2_test)
Batches	1–16	17–19	20–21
Size of dataset	224	30	28

[a]For training set of the target domain, information is missing at time $h = 6, 8, 10, 12$

Table 3.8 Generalization performance (J) of the proposed KL-TSK-FS method and the traditional L2-TSK-FS methods in fermentation process modeling

Output	L2-TSK-FS (D1)	L2-TSK-FS (D2)	L2-TSK-FS (D1 + D2)	KL-TSK-FS (D2 + Knowledge)
$S(h+2)$	0.2792	0.3944	0.2804	**0.1239**
$X(h+2)$	0.8342	1.1203	1.0642	**0.4548**
$P(h+2)$	0.2842	0.3255	0.2533	**0.1482**

3.4.3.2 Comparing with the Traditional L2-TSK-FS Modeling Methods

The experimental results of fermentation process modeling using the proposed inductive transfer learning method KL-TSK-FS and the traditional L2-TSK-FS are given in Table 3.8 and Fig. 3.7. The findings are similar to those presented in section IV-B for the experiments performed on the synthetic datasets. The modeling results of the KL-TSK-FS are better than that of the three traditional L2-TSK-FS methods. As the proposed method can effectively exploit not only the data of the target domain but also the useful knowledge of the source domains, the obtained TSK-FS has demonstrated better adaptive abilities. It can be seen from the experimental results that, even if the data in the training data of the target domain are missing, the generalization capability of the TSK-FS obtained by the proposed KL-TSK-FS does not degrade significantly. This remarkable feature is very valuable for the task of biochemical process modeling since the lack of sampled data is common due to poor sensitivity of sensors in the noisy environment.

3.4.3.3 Comparing with the Regression Methods Designed for Missing or Noisy Data

The experimental results of fermentation process modeling using the proposed inductive transfer learning method KL-TSK-FS and three regression methods (i.e., TSFS-SVR, FS-FCSVM, and HiRBF) are shown in Table 3.9 and Fig. 3.8. Similar to the findings presented in Sect. 3.4.2.3 for the experiments conducted with the synthetic datasets, in general, the proposed KL-TSK-FS has demonstrated better generalization performance than the other three regression methods in fermentation

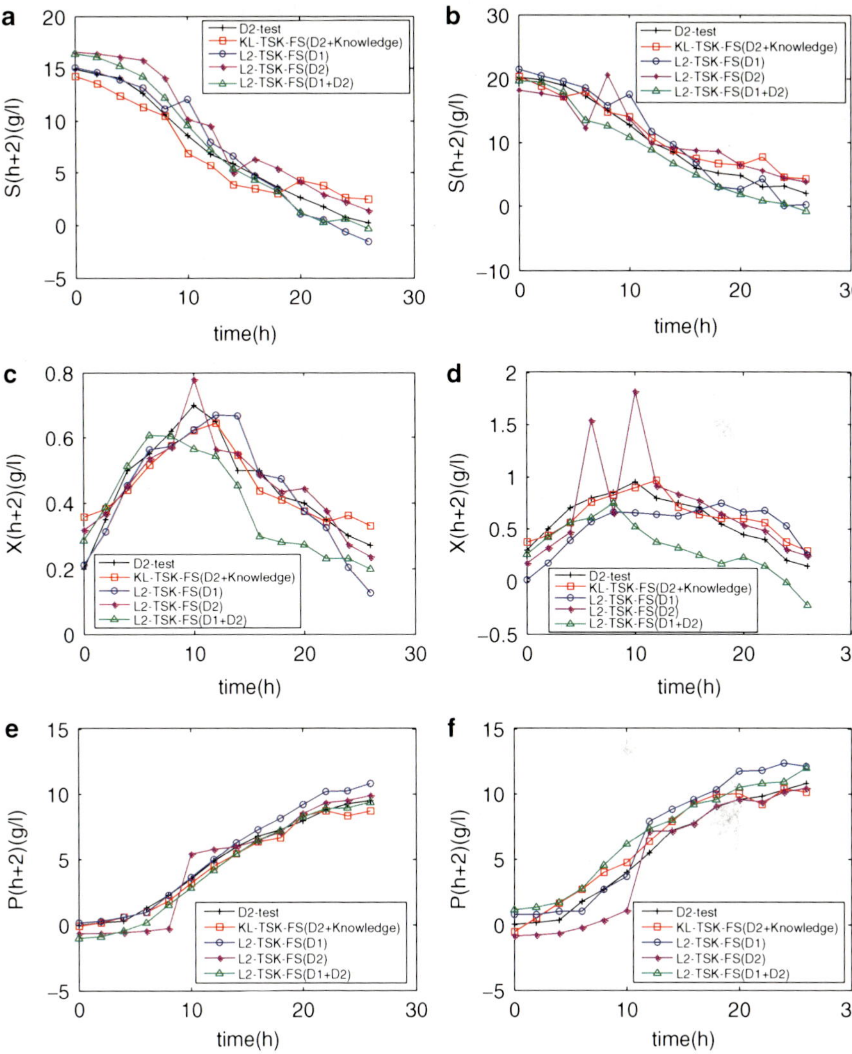

Fig. 3.7 Performance comparison between the proposed KL-TSK-FS method and three traditional L2-TSK-FS methods in fermentation process modeling: the prediction results of **a** S(h + 2) for the 20th batch; **b** S(h + 2) for the 21st batch; **c** X(h + 2) for the 20th batch; **d** X(h + 2) for the 21st batch; **e** P(h + 2) for the 20th batch; and **f** P(h + 2) for the 21st batch

process modeling. This can be explained again by the fact that the proposed KL-TSK-FS has effectively leveraged the useful knowledge from the source domain in the training procedure such that the influence of the missing data can be properly reduced.

Table 3.9 Generalization performance (J) of the proposed KL-TSK-FS method and several related regression methods in fermentation process modeling

Output	TSFS-SVR	FS-FCSVM	HiRBF	KL-TSK-FS
$S(h+2)$	0.3452	0.3750	0.3510	**0.1239**
$X(h+2)$	0.7295	0.6118	0.7026	**0.4548**
$P(h+2)$	0.3574	0.4144	0.4117	**0.1482**

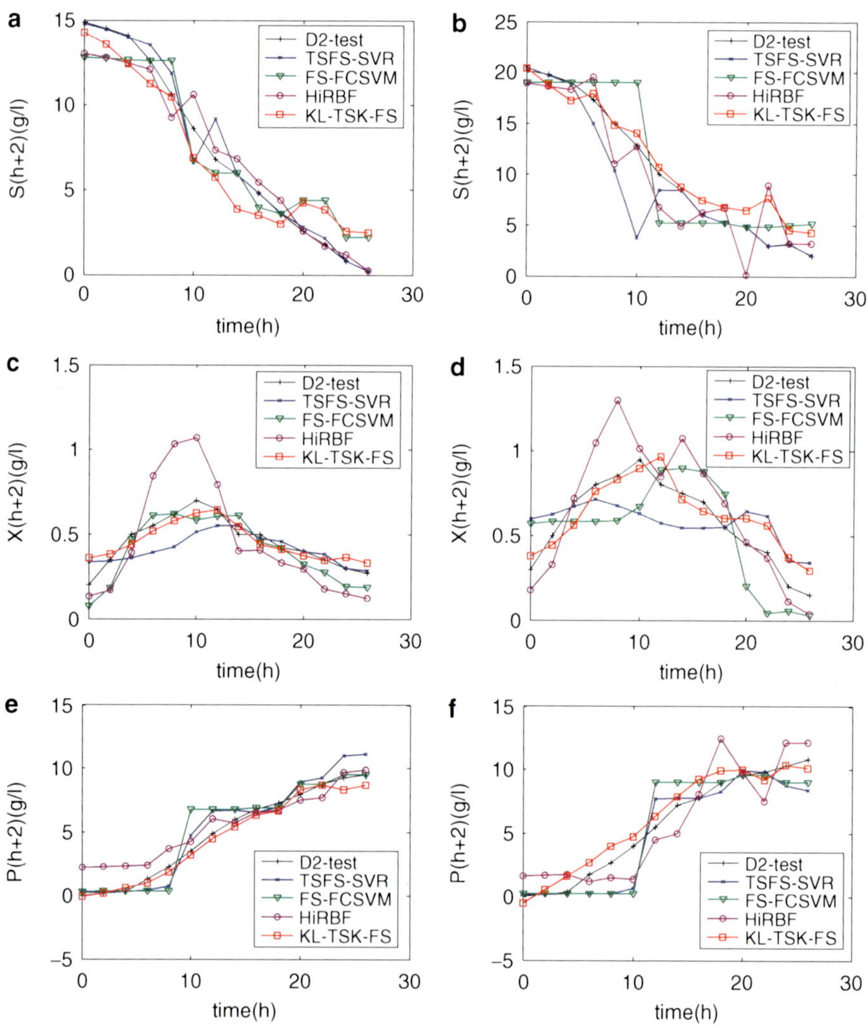

Fig. 3.8 Performance comparison between the proposed KL-TSK-FS method and three regression methods in fermentation process modeling: the prediction results of **a** S(h + 2) for the 20th batch; **b** S(h + 2) for the 21st batch; **c** X(h + 2) for the 20th batch; **d** X(h + 2) for the 21st batch; **e** P(h + 2) for the 20th batch; and **f** P(h + 2) for the 21st batch

3.5 Experimental Results of DAKSVM

3.5.1 *Experiment Settings*

To evaluate the effectiveness of the proposed transductive learning method DAKSVM and its extensions for DAL problems, we systematically compare them with several state-of-the-art algorithms on different datasets. We investigate three classes of domain adaptation problems: (1) a series of two-dimensional synthetic problems having different complexities with a two-moon dataset, (2) several real-world cross-domain text classification problems with different domain adaptation datasets such as 20Newsgroups, Reuters, Email Spam Filtering, web query set, and Amazon sentiment reviews set, and (3) a real problem in the context of multi-class classification in intra-domain on face recognition with Yale and ORL datasets. For all these datasets, true labels are available for both source and target-domain instances. However, prior information related to the target domain D_t is considered only for an objective and quantitative assessment of the performances of the proposed algorithms.

We construct synthetic datasets (two-moon) to exhibit the performance of the proposed method and choose real-world datasets to show the classification performance of the proposed method DAKSVM and its extension μ-DAKSVM. We also carry out a multi-class classification experiment to show the performance of the proposed method LSDAKSVM in multi-class classification problems.

In the sequel, we will first describe the whole experimental details. Throughout this experimental part, we use standard Gaussian kernel function as for several related kernel methods such as SVM, TSVM, KMM, TCA, LMPROJ, and DTSVM. For multiple kernel learning in DTSVM, according to the setting in [29], we use four Gaussian base kernels with the bandwidth $1.2^{\delta}\sigma$, where δ is set as $\{0,0.5,1,1.5\}$. For our methods, we use the parameterized Gaussian kernel as $k_{\sigma/\gamma}(\mathbf{x},\mathbf{x}_i) = \exp(-\|\mathbf{x}-\mathbf{x}_i\|^2/2(\sigma/\gamma)^2)$ in γ_{KS} of GPMDD, where the kernel parameter σ can be obtained by minimizing MMD with the most conservative test, which follows the setting in [46]. Empirically, we first select σ as the square root of the mean norm of the training data for binary classification and $\sigma\sqrt{c}$(where c is the number of classes) for multi-class classification. The tunable parameter γ can be set by minimizing GPMDD with the most optimal target test.

Presently, how to choose the algorithm parameters for the kernel methods still keeps an open and hot topic. In general, the algorithm parameters are manfully set. In order to evaluate the performance of the algorithm, a strategy, as is pointed out in [62], is that a set of the prior parameters is first given and then the best cross-validation mean rate among the set is used to estimate the generalized accuracy. In this work we adopted this strategy. The fivefold cross validation is used on the training set for parameter selection. Finally, the mean of experimental results on the test data is used for the performance evaluation. We chose the percentage overall accuracy AC% (i.e., the percentage of correctly labeled samples over the number of the whole samples) as the classification accuracy measure.

In the context, SVMs (such as SVM or ν-SVM, TSVM) is implemented by the state-of-the-art software package such as LIBSVM [57]. As the experiments in Sect. 3.4, all the algorithms are implemented using MATLAB on a computer with Intel Core 2 Duo P8600 2.4 GHz CPU and 2GB RAM.

3.5.2 Synthetic Datasets

3.5.2.1 Generation of Synthetic Datasets

In this subsection, we construct a serial of trials on two-moon datasets to justify our method DAKSVM. In this toy problem, a serial of two-moon datasets with different complexities are used to exhibit the generalization capability of the proposed method DAKSVM on domain adaptation transfer learning. We compare the proposed method DAKSVM with SVM and LMPROJ on this toy data.

A synthetic dataset containing 600 samples generated according to a bi-dimensional pattern of two intertwining moons associated with two specific information classes (300 samples each) is taken as the source domain data, as shown in Fig. 3.9a. Target data were generated by rotating anticlockwise the original source dataset 11 times by 10°, 15°, 20°, 25°, 30°, 35°, 40°, 45°, 50°, 55°, and 60°, respectively. Due to rotation, source and target-domain data exhibit different distributions. Particularly, the greater the rotation angle, the more complex the resultant domain adaptation problem, as confirmed by the values for Jensen–Shannon scatter (D_{JS}) [6] shown in Fig. 3.10a. The proposed DAKSVM algorithm is proved to be particularly effective for solving this kind of problems with high accuracy. Figure 3.9b, c shows the target domain data with the rotation angle 30° and 60°, respectively.

3.5.2.2 Comparing with the Related Methods

Figure 3.9d–i shows the learning accuracy rates of different methods on the datasets shown in Fig. 3.9b, c. And Fig. 3.10b shows the performance comparison among different methods on 11 target datasets aforementioned above. From Figs. 3.9b–d and 3.10b, we can observe that with appropriate learning parameters, the proposed method can obtain perfect separation between classes even if the rotation angles range from 10° to 50°. Besides, we can also observe several results as follows:

1. From Fig. 3.9d–i, we can observe that the accuracies of DAKSVM and LMPROJ are always higher than those by SVM according to a fivefold cross-validation on source domain data. This result shows that it is unsuitable for SVM on cross-domain learning. With Figs. 3.9 and 3.10, in some angles range (i.e., from 10° to 50°), the proposed method and LMPROJ can preserve the solution consistency

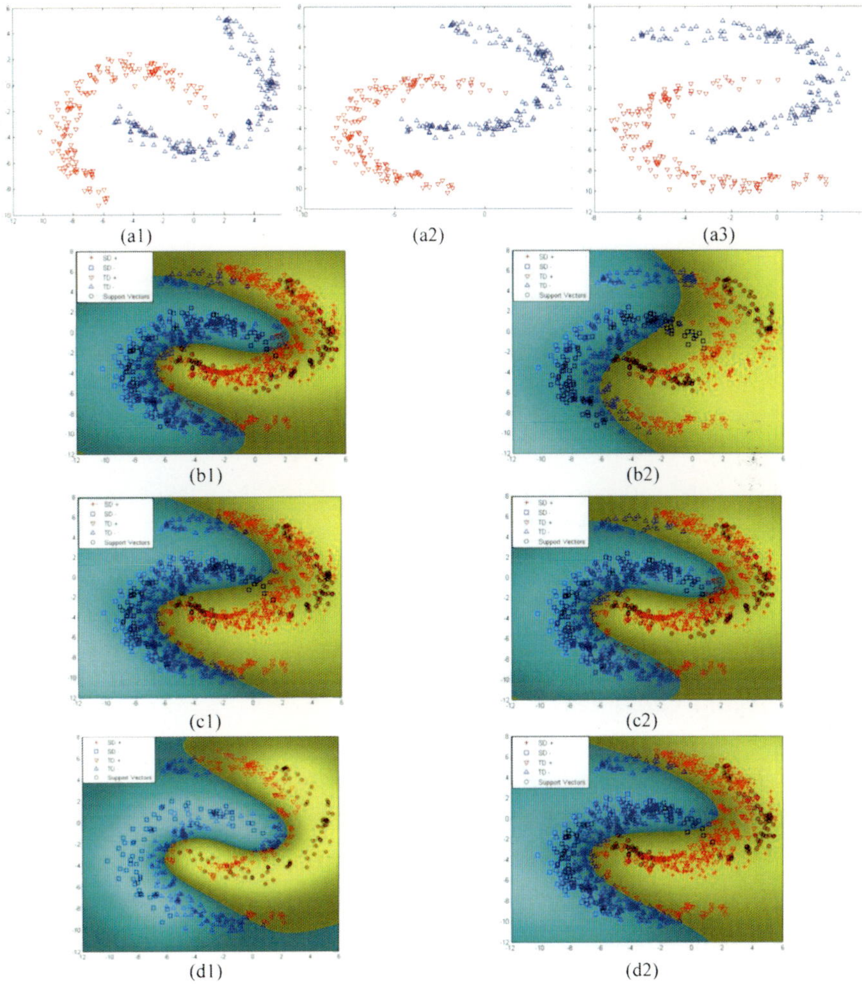

Fig. 3.9 Performance of different classifiers on two two-moon datasets with different complexities. **a** The original two-moon dataset; **b** rotation angle 30°; **c** rotation angle 60°; **d** classification accuracy for SVM: 95.4 %; **e** classification accuracy for SVM: 65 %; **f** classification accuracy for LMPROJ: 97.3 %; **g** classification accuracy for LMPROJ: 78.7 %; **h** classification accuracy for DAKSVM: 98.7 %; **i** classification accuracy for DAKSVM: 87.5 %

well with target domain to some extent, which shows that the proposed method is better than or at least comparable to LMPROJ in this experiment.

2. Figure 3.10b shows that for greater values of rotation angles (i.e., from 50° to 60°), the classification accuracy rates of all methods descend dramatically, which seems reasonable due to the increase of the complexity of the corresponding domain adaptation problems; however, the descendant rate of the proposed method is slower than others due to preserving the distribution consistency of

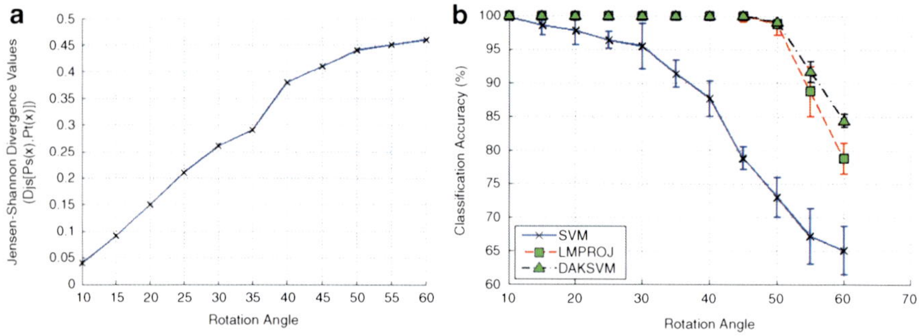

Fig. 3.10 Jenson–Shannon divergence values and classification accuracies on target domain data for different rotation angles: **a** Jenson–Shannon divergence values for different rotation angles (D_{JS}); **b** accuracies exhibited on target domain data for different rotation angles

both means and variances of different domains. When the rotation angle is big enough, all methods will not be able to keep the solution consistency with target domain. If this case happens, the hypothesis aforementioned above will not be satisfied.

3.5.3 *Binary Class Text Classification Datasets*

In this section, we demonstrate the overall efficiency and effectiveness of the proposed method DAKSVM and its variation μ-DKSVM on five different real-world domain adaptation tasks for text datasets such as 20Newsgroups, Reuters, mail spam filtering, web query classification, and Amazon sentiment reviews classification.

Except for SVM, KMM, DTSVM, LMPROJ, and the TSVM, we still choose for comparison another two algorithms from KDD'08. They are cross-domain spectral classifier [63] and locally weighted ensemble (LWE) classifier [9].

Unlike SVM and TSVM with default parameter values are adopted in most cases, in order to make our comparison fair, we report the best performance for each method over a range of parameter selections.

3.5.3.1 Dataset Settings

A brief description of each dataset and its setup is given in this subsection. Tables 3.10 and 3.11 summarize the datasets and give the indices to some of which we will refer in our experimental results. For example, dataset 6 is a 20Newsgroup dataset about *Rec.* vs. *Sci.* where the number of positive and negative training samples is 1,984 and 1,977, respectively, and the number of positive and negative class testing samples is 1,993 and 1,972, respectively.

Table 3.10 Cross-domain text classification tasks

Task		Datasets	Number of training samples		Number of testing samples	
			Positive class	Negative class	Positive class	Negative class
1	Reuters	Orgs vs. People	Documents from sub-categories		Documents from different sub-categories	
2		Orgs vs. Place				
3		People vs. Place				
4	20Newsgroup	Comp vs. Sci	1,958	1,972	2,923	1,977
5		Rec vs. Talk	1,993	1,568	1,984	1,658
6		Rec vs. Sci	1,984	1,977	1,993	1,972
7		Sci vs. Talk	1,971	1,403	1,978	1,850
8		Comp vs. Rec	2,916	1,993	1,965	1,984
9		Comp vs. Talk	2,914	1,568	1,967	1,685
10	Email spam filtering	User1 vs. User2	User1's emails		User2's emails	
11		User2 vs. User3	User2's emails		User3's emails	
12		User3 vs. User1	User3's emails		User1's emails	

Table 3.11 Web query text and sentiment reviews classification tasks

Task		Categories	Number of training samples	Number of testing samples
13	Web query	Business (B)	1,500	1,200
14		Computers (C)	1,500	1,000
15		Education (E)	2,210	2,500
16		Health (H)	1,180	1,190
17		Sports (S)	1,420	660
18	Amazon sentiment reviews	Books (B)	1,000	1,000
19		DVDs (D)	1,000	1,000
20		Electronics (E)	1,000	1,000
21		Kitchen (H)	1,000	1,000

20Newsgroups and Reuters

Reuters and 20Newsgroups are two cross-domain text classification datasets commonly used by the state-of-the-art DAL classifiers [9, 28–30, 36, 64]. These datasets both represent text categorization tasks, Reuters is made up of news articles with five top-level categories, among which, Orgs, Places, and People are the largest, and the 20Newsgroups dataset contains 20newsgroup categories each with approximately 1,000 documents. For these text categorization data, in each case the goal is to correctly discriminate between articles at the top level, e.g. "sci" articles vs. "talk" articles, using different sets of sub-categories within each top-category for training and testing, e.g. *sci.electronics* and *sci.med* vs. *talk.politics.misc* and *talk.religion.misc* for training and *sci.crypt* and *sci.space* vs. *talk.politics.guns* and *talk.politics.mideast* for testing. For more details about the sub-categories, see [65]. Each set of sub-categories represents a different domain in which different words will be more common. Features are given by converting the documents into bag-of-word representations which are then transformed into feature vectors using the term frequency, details about this procedure can also be found in [65]. Table 3.10 shows the more detailed information about the experimental datasets drawn from the aforementioned above datasets.

Email Spam Filtering

In email spam filtering datasets [66], there are three email subsets (denoted by User1, User2, and User3, respectively) annotated by three different users. In this trial, the task is to classify spam and non-spam emails. Since the spam and non-spam emails in the subsets have been identified by different users, the data distributions of the three subsets are different but related. Each subset has 2,500 emails, in which one half of the emails are non-spam (labeled as 1) and the other half of them are spam (labeled as -1). On this dataset, in terms of [54], we consider three settings: (1) User1 (source domain) and User2 (target domain); (2) User2 (source domain) and User3 (target domain), and 3) User3 (source domain) and User1 (target domain). For each setting, the training dataset contains all labeled samples from the source domain. And the samples in the target domain are used as the unlabeled test ones. We report the experimental results with their means and the standard deviations of all methods. Again, the word-frequency feature is used to represent each document as in [66]. The more detailed information about the experimental datasets drawn from Email Spam Filtering datasets can be found in Table 3.2.

Web Query

We also construct a set of tasks on cross-domain query classification for a search engine, e.g. Google. We use a set of search snippets gathered from Google as our training data and some incoming unlabeled queries as the test data. The detailed descriptions of the procedure can be found in [67]. We use the labeled queries from AOL provided by [68] (http://grepgsadetsky.com/aol-data) for evaluation. We consider queries from five classes: *Business*, *Computer*, *Entertainment*, *Health*, *and Sports* which are shown in both training and test datasets. We form ten binary classification tasks for query classification [64]. The more detailed information can be seen in Table 3.11.

Sentiment Reviews

The data of sentiment domain adaptation [69] consist of Amazon product reviews for four different product types, including books, DVDs, electronics, and kitchen appliances. Each review consists of a rating with scores ranging from 0 to 5, a reviewer name and location, a product name, a review title and date, and the review text. Reviews with ratings higher than three are labeled as positive and reviews with ratings lower than three are labeled as negative, the rest are discarded since the polarity of these reviews is ambiguous. The details of the data in different domains are summarized in Table 3.11. The experimental settings are the same as in [69]. To study the performance of our methods in this task, we construct 12 pairs of cross-domain sentiment classification tasks as shown in Table 3.6, e.g., we use the reviews from domain A as the training data and then predict the sentiment of the reviews in the domain B.

3.5.3.2 Comparing with the Related Methods

Tables 3.12, 3.13, and 3.14 and Fig. 3.11 show the means and standard deviations of classification accuracies of different methods on the above domain adaptation transfer learning tasks, respectively. From these results, we can make several interesting observations as follows:

Table 3.12 Means and standard deviations (%) of classification accuracies (ACC) of all methods on the 20Newsgroups, Reuters datasets, and email spam filtering datasets

Datasets		Methods								
		SVM	TSVM	CDCS	LWE	LMPROJ	DTSVM	KMM	DAKSVM	μ-DAKSVM
Reuters	1	80.20*	81.84*	88.5	83.42*	84.63*	**88.77***	85.43*	87.64	88.13
		(±0.45)	(±1.26)	(±4.32)	(±2.01)	(±2.82)	(±**2.14**)	(±3.02)	(±0.42)	(±0.50)
	2	71.35*	75.80*	73.90*	69.70*	80.20*	81.52	79.38*	79.97	**82.64**
		(±2.18)	(±1.72)	(±2.14)	(±0.08)	(±1.16)	(±0.56)	(±0.01)	(±1.06)	(±**0.00**)
	3	65.36*	69.8	64.00*	68.52*	70.80*	74.12*	72.19*	74.4	**74.9**
		(±1.67)	(±0.46)	(±0.32)	(±1.49)	(±1.38)	(±3.26)	(±1.13)	(±**0.42**)	
20NG	4	72.53*	76.75*	69.80*	**85.24**	82.52*	83.52*	78.11*	84.68	85.02
		(±3.42)	(±0.54)	(±0.48)	(±**1.81**)	(±0.86)	(±2.74)	(±2.80)	(±0.63)	(±1.08)
	5	70.10*	73.40*	**82.92**	78.60*	79.30*	80.5*	79.11	82.36	82.67
		(±2.62)	(±2.02)	(±**1.06**)	(±0.34)	(±2.76)	(±2.82)	(±1.00)	(±0.15)	(±0.30)
	6	75.40*	83.9	64.00*	87.2	86.34*	**90.23**	87.52*	88.81	88.81
		(±0.51)	(±1.14)	(±3.1)	(±2.10)	(±3.10)	(±**0.82**)	(±2.69)	(±1.74)	(±0.6)
	7	78.00*	81.20*	70.84*	75.32*	84.68	84.84	81.47*	85.12	**85.37**
		(±0.04)	(±0.06)	(±1.62)	(±0.47)	(±2.11)	(±1.49)	(±3.27)	(±0.04)	(±**0.80**)
	8	83.80*	85.24*	82.72*	88.3	85.40*	91.76	89.46*	89.58	**91.84**
		(±1.13)	(±0.18)	(±0.34)	(±1.08)	(±1.08)	(±1.68)	(±2.07)	(±0.3)	(±**0.10**)
	9	92.70*	88.74*	90.20*	94.00*	93.43*	94.13	92.50*	96.72	**96.78**
		(±0.40)	(±0.70)	(±1.16)	(±2.06)	(±0.79)	(±0.4)	(±0.78)	(±0.60)	(±**0.20**)
Spam filtering	10	96.08	96.21	83.28*	93.51*	93.21*	96.89	96.21	96.49	**97.19**
		(±0.10)	(±0.22)	(±0.20)	(±0.40)	(±0.52)	(±0.1)	(±0.10)	(±0.03)	(±**0.06**)
	11	96.89	97	92.14*	98.74	94.0*	**97.65**	97.13	97.25	97.25
		(±0.0)	(±0.10)	(±1.02)	(±0.4)	(±0.00)	(±**0.20**)	(±0.05)	(±0.4)	(±0.41)
	12	91.7*	91.80*	90.02*	88.78*	88.79*	94.5	91.8*	93.2	**93.85**
		(±0.6)	(±0.3)	(±0.3)	(±0.01)	(±0.24)	(±0.60)	(±0.00)	(±0.20)	(±**0.10**)

Each result in the table is best among all the results obtained by using different parameters
The performance of μ-DAKSVM is statistically significant compared with other seven classifiers at p-value ≤ 0.05

Table 3.13 Means and standard deviations (%) of classification accuracies (ACC) of all methods on the web query dataset

	Datasets									
	Web query data									
Methods	B–C	B–H	B–H	B–S	C–E	C–H	C–S	E–H	E–S	H–S
SVM	82.52* (±1.14)	85.93* (±0.03)	90.94* (±2.67)	87.25 (±0.44)	87.34 (±4.12)	93.19* (±3.10)	84.68* (±2.01)	92.55* (±0.4)	81.59* (±0.2)	93.45* (±1.04)
TSVM	82.64 (±0.2)	85.42* (±0.2)	91.19* (±1.06)	82.60* (±2.50)	83.35* (±1.52)	89.70* (±2.00)	92.01* (±3.24)	93.11 (±0.03)	77.93* (±2.44)	80.84* (±1.60)
CDCS	84.34 (±2.74)	82.86* (±0.33)	96.44 (±3.16)	85.40* (±2.22)	78.70* (±0.50)	91.28* (±1.60)	89.40* (±0.62)	92.76* (±1.10)	**85.55** (±**0.00**)	91.25* (±0.40)
LWE	83.26* (±0.74)	86.72 (±3.02)	93.20* (±1.27)	82.56* (±0.80)	85.80 (±2.28)	93.66 (±1.40)	84.20* (±4.56)	94.40* (±0.72)	79.46* (±3.31)	94.71* (±0.20)
LMPROJ	84.68 (±0.4)	86.48* (±2.33)	94.82* (±1.08)	85.78* (±0.86)	88.52* (±3.26)	92.00* (±1.09)	86.12* (±4.20)	95.38* (±2.02)	82.70* (±1.17)	95.00* (±2.48)
DTSVM	84.22* (±3.12)	**88.36*** (±**1.32**)	96.61 (±0.08)	86.39* (±2.16)	**90.15** (±**0.26**)	94.08 (±0.40)	95.16 (±0.82)	93.08* (±0.00)	83.29* (±0.42)	97.33* (±1.28)
KMM	83.92* (±0.74)	86.86 (±0.30)	95.12* (±0.6)	81.22* (±3.06)	85.52* (±1.90)	93.00* (±0.01)	84.82* (±2.70)	95.11 (±0.02)	80.10* (±0.2)	96.32* (±1.42)
DAKSVM	86.36 (±0.40)	87.82 (±0.56)	98.23 (±1.02)	90.46 (±0.03)	88.78 (±0.00)	95.10 (±0.6)	96.94 (±0.90)	96.81 (±0.12)	83.20 (±0.62)	**98.27** (±**0.2**)
μ-DAKSVM	**86.76** (±**0.00**)	87.87 (±0.2)	**98.75** (±**0.4**)	**91.02** (±**1.01**)	88.78 (±0.00)	**95.42** (±**0.80**)	**97.54** (±0.05)	**96.81** (±0.3)	84.31 (±0.5)	**98.27** (±**0.4**)

Each result in the table is best among all the results obtained by using different parameters
*The performance of μ-DAKSVM is statistically significant compared with other seven classifiers at p-value ≤ 0.05

Table 3.14 Means and standard deviations (%) of classification accuracies (ACC) of all methods on the sentiment reviews dataset

	Datasets											
	Sentiment reviews data											
Methods	D–B	E–B	K–B	B–D	E–D	K–D	B–E	D–E	K–E	B–K	D–K	E–K
SVM	71.29* (±2.14)	67.89* (±1.32)	67.72* (±0.74)	75.69* (±3.22)	67.81* (±0.50)	71.77* (±4.22)	68.83* (±3.80)	69.31* (±0.62)	78.77* (±0.34)	73.13* (±2.18)	73.68* (±4.18)	80.46 (±0.22)
TSVM	71.52* (±0.72)	68.27 (±0.40)	69.96* ±2.18)	74.84* (±2.50)	64.51* (±3.58)	71.63* (±3.04)	71.13* (±1.48)	70.52* (±1.22)	80.16 (±0.30)	76.14* (±2.54)	76.57* (±0.96)	80.99 (±0.12)
CDCS	69.33* (±0.44)	72.30 (±1.05)	74.34 (±0.60)	82.30* (±0.14)	73.00* (±0.54)	80.66 (±0.82)	76.80 (±0.40)	70.36* (±2.08)	84.07* (±0.41)	79.02 (±0.36)	80.58* (±1.16)	**86.47** (±**3.05**)
LWE	74.54* (±1.40)	69.10 (±0.20)	77.60 (±1.01)	78.70* (±0.82)	69.40* (±2.60)	78.21* (±1.36)	78.90* (±0.08)	75.67 (±1.15)	84.73 (±0.68)	78.79* (±1.06)	83.19* (±0.01)	79.89 (±2.76)
LMPROJ	76.71* (±0.61)	71.29* (±1.34)	75.80* (±1.80)	83.65* (±1.74)	73.20* (±0.20)	81.14* (±1.52)	75.20 (±0.32)	77.30 (±2.26)	81.66* (±0.92)	80.04 (±0.00)	85.33 (±0.44)	81.38* (±0.80)
DTSVM	78.41 (±0.30)	72.58 (±0.40)	**78.88** (±**0.66**)	**86.69** (±**1.27**)	74.62* (±0.25)	82.49* (±0.50)	79.21 (±1.14)	77.54 (±0.05)	83.97 (±0.20)	**81.72** (±**2.02**)	87.07 (±0.84)	82.56* (±1.48)
KMM	74.02* (±1.31)	69.58* (±2.04)	77.18 (±0.56)	81.79* (±0.07)	73.16* (±0.05)	82.06* (±1.12)	70.88* (±2.06)	75.82* (±0.42)	83.50 (±0.20)	79.94* (±1.70)	84.27* (±1.46)	80.07* (±0.40)
DAKSVM	78.54 (±0.40)	73.09 (±0.01)	77.82 (±1.06)	86.34 (±0.6)	78.02 (±0.00)	83.07 (±0.56)	78.79 (±0.02)	79.21 (±0.72)	**84.25** (±**0.2**)	81.33 (±0.43)	88.14 (±0.6)	84.57 (±0.05)
μ-DAKSVM	**78.72** (±**0.25**)	**73.74** (±**0.56**)	78.03 (±0.50)	86.34 (±0.71)	**78.43** (±**0.34**)	**84.21** (±**0.20**)	**79.36** (±**0.18**)	**79.54** (±**0.00**)	**84.25** (±**0.3**)	81.68 (±0.08)	**89.36** (±**0.82**)	84.89 (±0.25)

Each result in the table is best among all the results obtained by using different parameters

*The performance of μ-DAKSVM is statistically significant compared with other seven classifiers at p-value $\leq$0.05

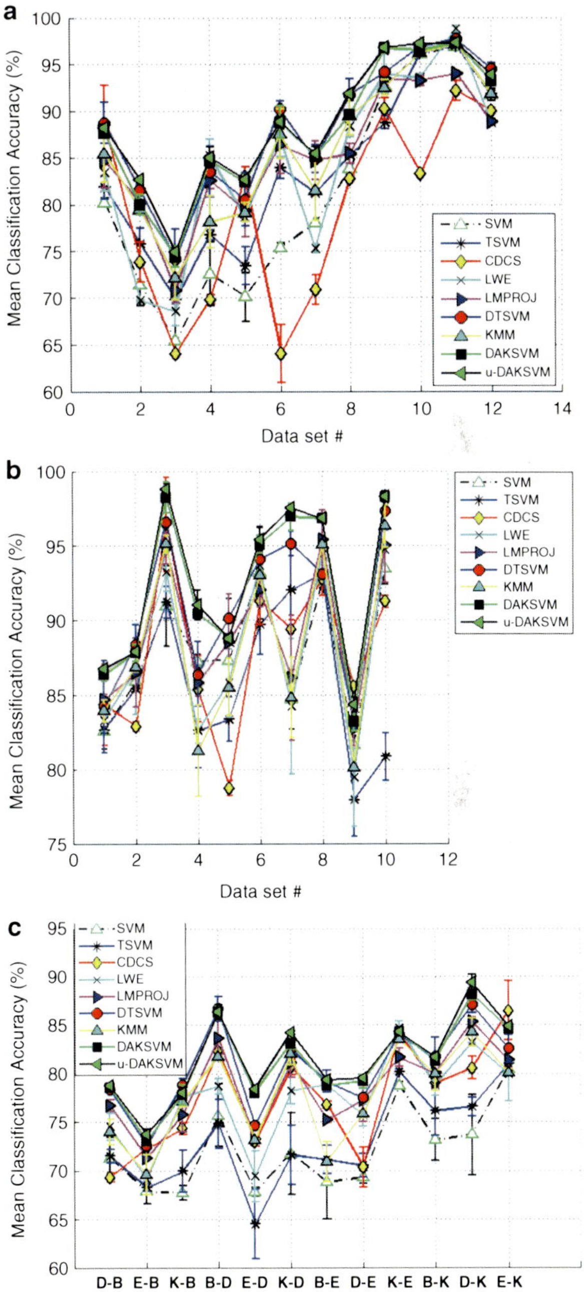

Fig. 3.11 Means and standard deviations (%) of classification accuracies (*ACC*) of all methods on text datasets. **a** Text datasets: Reuters, 20Newsgroups and mail spam filtering; **b** web query dataset; **c** sentiment classification dataset

1. From Tables 3.12, 3.13, and 3.14, we can see that our method achieves very promising result. The major limitation of LMPROJ, DTSVM, and KMM is that they only consider the first-order statistics and thus cannot well generalize their result. However, since our methods definitely consider both the second-order and the first-order statistics between the source and target domains, it yields better generalization capability. It can be observed that our method significantly outperforms other methods. These empirical results again show that considering second-order statistics as well as first-order statistics can help us improve the domain adaptation performance.
2. SVM and TSVM have the worst performance on almost all learning tasks compared to other classifiers, which is consistent with the experimental results of the above toy datasets. Though obtaining better classification on both 20Newsgroup and Reuters datasets, TSVM exhibits its worse classification performance on two web text classification tasks than other methods. It is worth noting that we obtain a little better results for SVM and TSVM than those typically reported in the previous literature on the same datasets used in our trials. This is because in order to make our comparison fair we reported the best results over a set of parameters for SVM and instead of selecting a default parameter on the training data to be performed.
3. In Tables 3.12, 3.13, and 3.14 and Fig. 3.11, we can also observe that although seven methods, i.e., CDCS, LWE, LMPROJ, DTSVM, KMM, DAKSVM, and its variation μ-DAKSVM, exhibit comparable classification capability on all text datasets, the proposed method DAKSVM and its variation μ-DAKSVM always keep significantly high classification accuracy in most cases, which implies that it is more stable than other methods, particularly on two web text classification datasets such as web query and sentiment reviews datasets.
4. The results in Tables 3.12, 3.13, and 3.14 and Fig. 3.11 also show that the proposed method DAKSVM and its variation μ-DAKSVM perform relatively better than MMD-based methods LMPROJ and KMM in almost all datasets, which justifies that the only emphasis on minimizing distribution mean discrepancy between both domains is far from sufficiency for domain adaptation transfer learning. Hence, we should introduce more underlying information, such as distribution scatter discrepancy minimization, into the regularization framework of the classifier to further enhance the classification performance. Besides, it is worth mentioning that DTSVM also obtains fairly robust performance on almost all datasets by adopting multiple kernel learning scheme. A possible explanation is that multiple kernel learning skill can improve learning capability for DAL.
5. μ-DAKSVM keeps obviously superior capability over DAKSVM in classification accuracy for almost all these datasets, which demonstrates that parameter μ can be used to enhance the generalization capability of DAKSVM. Therefore, we use μ-DAKSVM instead of DAKSVM for the performance evaluation hereafter.
6. In order to verify whether the proposed methods are significantly better than the other methods, we also performed the paired two-tailed t-test [70] on the classification results of the 10 runs to calculate the statistical significance of the proposed method μ-DAKSVM. The smaller the p-value, the more significant

the difference of the two average results is, and a p-value of 0.05 is a typical threshold which is considered to be statistically significant. Thus, in Tables 3.12, 3.13, and 3.14, if the p-value of each dataset is less than 0.05, the corresponding results will be denoted "*." Therefore, as shown in Tables 3.12, 3.13, and 3.14, we can clearly find that the proposed method μ-DAKSVM significantly outperforms other methods in most datasets.

3.5.4 Multi-Class Face Recognition Datasets

3.5.4.1 Dataset Settings

In this subsection, in order to evaluate the effectiveness of the proposed methods on multi-class classification problems, we investigate the performance of the proposed algorithms LSDAKSVM and μ-DAKSVM for face recognition on two benchmarking Yale and ORL face databases. The Yale face database was constructed at the Yale Center for Computation Vision and Control. There are 165 images about 15 individuals in this database where each person has 11 images. The images demonstrate face variations under lighting condition (left-right, center-light, right-light) and facial expression (normal, happy, sad, sleepy, surprised and wink) with or without glasses. Each image is cropped to be the size of 32×32 in our experiment. We randomly select 8 images of each individual to construct the source domain dataset; the ORL database contains 400 images grouped into 40 distinct subjects with 10 different images for each. The images are captured at different times, and for some subjects, the images may vary in facial expressions and facial details. All the images are taken against a dark homogeneous background with the tolerance for some side movement of about 20. The original images are all sized 112×92 pixels with 256 gray levels per pixel, which are further down-sampled into 32×32 pixels in our experiment. We randomly select eight images of each individual to construct the source domain training set. Figure 3.12a, c shows the cropped images of one person in Yale and ORL face databases, respectively.

The target datasets are generated by rotating anticlockwise the original source domain dataset three times by 10°, 30°, and 50°, respectively. Due to rotation, source and target-domain data exhibit different distributions. Particularly, the greater the rotation angle is, the more complex the resulting domain adaptation problem becomes. Thus we construct three face domain adaptation transfer learning problems for each face database. Figure 3.12b, d shows the face samples with rotation angle 10°, respectively.

3.5.4.2 Comparing with the Related Methods

We test the performance of LSDAKSVM and μ-DAKSVM in comparison with CDCS, LWE, DTSVM, and LMPROJ. In order to do a comprehensive comparison,

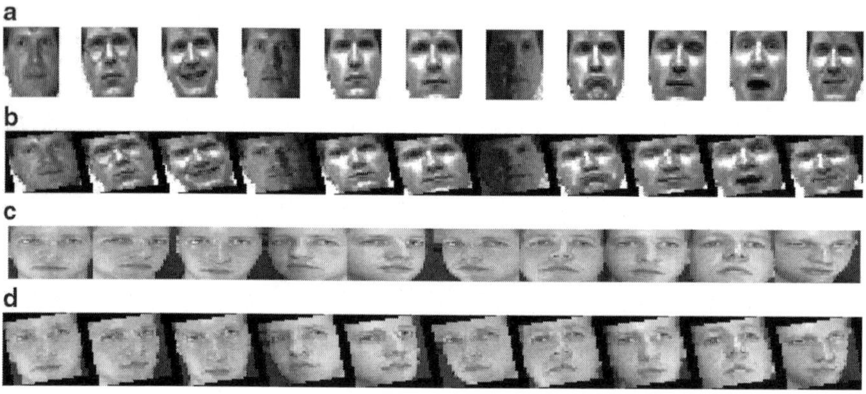

Fig. 3.12 Face examples from the face databases Yale and ORL. **a** Yale faces for an object; **b** Yale faces for an object with rotation angle 10°; **c** ORL faces for an object; **d** ORL faces for an object with rotation angle 10°

we also perform the baseline method LS-SVM for face recognition with different distributions. For the above multi-class classification tasks, μ-DAKSVM, CDCS, LWE, LS-SVM, DTSVM, and LMPROJ adopt OAO multi-class separation strategy to finish the corresponding multi-class classification tasks. For each evaluation, ten rounds of experiments are repeated with randomly selected training data, and the average result is recorded as the final classification accuracy in Table 3.7. Several attractive insights can be obtained from these results as follows:

1. The overall accuracy of LS-SVM is lower than any other classifier on all DAL tasks, which is consistent with SVM.
2. With the increase of rotation angle, the classification performance of all classifiers descends gradually. However, LSDAKSVM seems to decrease more slowly than other methods. Exceptionally, CDCS and DTSVM exhibit competitive performance to some extent compared to other methods, particularly on more complex datasets.
3. As shown in Table 3.15, we can observe that the LSDAKSVM method delivers more stable results across all the datasets and is competitive as the best method for the majority of all the other datasets. It obtains the best classification accuracy more times than any other method. Hence, as discussed in the above section, LSDAKSVM possesses overall DAL advantages over other methods in the sense of both computational complexity and classification accuracy.
4. Table 3.15 also shows that although LSDAKSVM seems to have overall advantage over μ-DAKSVM in classification accuracy, μ-DAKSVM is actually considerably comparable to LSDAKSVM.

Table 3.15 Means and standard deviations (%) of classification accuracies (ACC) of all methods on Yale and ORL with different rotation angles

Faces data		Method LS-SVM	LMPROJ	LWE	CDCS	DTSVM	LSDAKSVM	μ-DAKSVM
Yale	10°	61.78 (±3.45)	68.45 (±0.56)	63.78 (±2.48)	62.47 (±1.10)	68.77 (±0.54)	**70.24** (±0.24)	69.93 (±0.2)
	30°	58.37 (±2.74)	64.13 (±1.07)	61.66 (±1.01)	60.70 (±0.4)	67.28 (±2.34)	**66.47** (±0.5)	66.2 (±0.6)
	50°	52.29 (±2.12)	62.08 (±1.18)	58.78 (±0.41)	60.20 (±0.34)	**65.63** (±1.14)	63.00 (±0.4)	63.7 (±0.34)
ORL	10°	76.30 (±1.00)	85.94 (±1.40)	80.90 (±0.6)	84.64 (±0.2)	84.84 (±0.00)	**86.28** (±0.04)	84.18 (±0.44)
	30°	70.72 (±3.04)	82.00 (±0.76)	79.33 (±1.20)	**83.71** (±2.10)	83.19 (±0.01)	83.10 (±1.06)	83.40 (±0.00)
	50°	65.70 (±0.62)	78.65 (±0.20)	72.22 (±3.54)	79.91 (±1.03)	80.01 (±1.14)	**81.46** (±0.02)	78.10 (±1.68)

The best results among all the results obtained with different parameters are listed in the table

3.6 Conclusions

In this chapter, we propose one inductive learning approach and one transductive learning approach based on support vector learning, respectively. On the one hand, the proposed inductive transfer learning method, i.e., KL-TSK-FS, is more adaptive to the situations where the data are only partially available from the target domain while some useful knowledge of the source domains is available. Besides, the proposed method is distinctive in preserving data privacy as only the knowledge (e.g., the corresponding model parameters) rather than the data of the source domain is adopted. On the other hand, the proposed transductive transfer learning method DAKSVM and its two extensions indeed inherit the potential advantages of classical TSVMs and MMD-based methods and are further extended to DAL. As a novel large margin domain adaptation classifier, the proposed methods can reduce the distribution gap between different domains in an RKHS as much as possible, since they effectively integrate the large margin learner with the proposed GPMDD metric, in which both the distribution mean discrepancy and the distribution scatter discrepancy on RKHS embedding domain distributions are *simultaneously* considered.

References

1. Pan, S.J., Yang, Q.: A survey on transfer learning. IEEE Trans. Knowl. Data Eng. **22**(10), 1345–1359 (2010)
2. Deng, Z.H., Choi, K.S., Chung, F.L., et al.: Scalable TSK fuzzy modeling for very large datasets using minimal-enclosing-ball approximation. IEEE Trans. Fuzzy Syst. **19**(2), 210–226 (2011)
3. Liao, X., Xue, Y., Carin, L.: Logistic regression with an auxiliary data source. In: Proceedings of 21st International Conference Machine Learning, pp. 505–512 (2005)
4. Huang, J., Smola, A., Gretton, A., et al.: Correcting sample selection bias by unlabeled data. In: Proceedings of 19th Annual Conference Neural Information Processing Systems, pp. 601–608 (2007)
5. Bickel, S., Brückner, M., Scheffer, T.: Discriminative learning for differing training and test distributions. In: Proceedings of 24th International Conference Machine Learning, pp. 81–88 (2007)
6. Sugiyama, M., Nakajima, S., Kashima, H., et al.: Direct importance estimation with model selection and its application to covariate shift adaptation. In: Proceedings of 20th Annual Conference Neural Information Processing Systems, Bangkok (December 2008)
7. Lawrence, N.D., Platt, J.C.: Learning to learn with the informative vector machine. In: Proceedings of 21st International Conference Machine Learning (July 2004)
8. Schwaighofer, A., Tresp, V., Yu, K.: Learning Gaussian process kernels via hierarchical Bayes. In: Proceedings 17th Annual Conference Neural Information Processing Systems, pp. 1209–1216 (2005)
9. Gao, J., Fan, W., Jiang, J., et al.: Knowledge transfer via multiple model local structure mapping. In: Proceedings of 14th ACM SIGKDD International Conference Knowledge Discovery and Data Mining, pp. 283–291 (August 2008)

10. Mihalkova, L., Huynh, T., Mooney, R.J.: Mapping and revising Markov logic networks for transfer learning. In: Proceedings of 22nd Association for the Advancement of Artificial Intelligence (AAAI) Conference on Artificial Intelligence, pp. 608–614 (July 2007)
11. Mihalkova, L., Mooney, R.J.: Transfer learning by mapping with minimal target data. In: Proceedings of Association for the Advancement of Artificial Intelligence (AAAI'08) Workshop Transfer Learning for Complex Tasks (July 2008)
12. Davis, J., Domingos, P.: Deep transfer via second-order Markov logic. In: Proceedings of Association for the Advancement of Artificial Intelligence (AAAI'08) Workshop Transfer Learning for Complex Tasks (July 2008)
13. Pan, S.J., Tsang, I.W., Kwok, J.T., et al.: Domain adaptation via transfer component analysis. IEEE Trans. Neural Netw. **22**(2), 199–210 (2011)
14. Duan, L.X., Xu, D., Tsang, I.W.: Domain adaptation from multiple sources: a domain-dependent regularization approach. IEEE Trans. Neural Netw. Learn. Syst. **23**(3), 504–518 (2012)
15. Duan, L.X., Tsang, I.W., Xu, D.: Domain transfer multiple kernel learning. IEEE Trans. Pattern Anal. Mach. Intell. **34**(3), 465–479 (2012)
16. Dai, W., Yang, Q., Xue, G., et al.: Self-taught clustering. In: Proceedings of 25th International Conference Machine Learning, pp. 200–207 (July 2008)
17. Jiang, W.H., Chung, F.L.: Transfer spectral clustering. In: Proceedings of European Conference on Machine Learning and Principles and Practice of Knowledge Discovery in Databases (ECML PKDD), Bristol (24–28 September 2012)
18. Wang, Z., Song, Y., Zhang, C.: Transferred dimensionality reduction. In: Proceedings of European Conference Machine Learning and Knowledge Discovery in Databases (ECML/PKDD'08), pp. 550–565 (September 2008)
19. Gong, B., Shi, Y., Sha, F., et al.: Geodesic flow kernel for unsupervised domain adaptation. In: IEEE Conference on Computer Vision and Pattern Recognition (CVPR), pp. 2066–2073 (2012)
20. Yang, P., Tan, Q., Ding, Y.: Bayesian task-level transfer learning for non-linear regression. In: Proceedings of International Conference on Computer Science and Software Engineering, pp. 62–65 (2008)
21. Borzemski, L., Starczewski, G.: Application of transfer regression to TCP throughput prediction. In: Proceedings of First Asian Conference on Intelligent Information and Database Systems, pp. 28–33 (2009)
22. Mao, W., Yan, G., Bai, J., et al.: Regression transfer learning based on principal curve. Lect. Notes Comput. Sci. **6063**, 365–372 (2010)
23. Liu, J., Chen, Y., Zhang, Y.: Transfer regression model for indoor 3D location estimation. Lect. Notes Comput. Sci. **5916**, 603–613 (2010)
24. Pardoe, D., Stone, P.: Boosting for regression transfer. In: Proceedings of International Conference on Machine Learning, pp. 863–870 (2010)
25. Dai, W., Yang, Q., Xue, G., et al.: Boosting for transfer learning. In: Proceedings of 24th International Conference on Machine Learning, pp. 193–200 (June 2007)
26. Wu, P., Dietterich, T.G.: Improving SVM accuracy by training on auxiliary data sources. In: Proceedings of 21st International Conference on Machine Learning (July 2004)
27. Evgeniou, T., Pontil, M.: Regularized multi-task learning. In: Proceedings of 10th ACM SIGKDD International Conference on Knowledge Discovery and Data Mining, pp. 109–117 (August 2004)
28. Qduanz, B., Huan, J.: Large margin transductive transfer learning. In: Proceedings of 18th ACM conference on Information and knowledge management (CIKM), pp. 1327–1336. ACM, New York (2009)
29. Duan, L., Tsang, I.W., Xu, D., et al.: Domain transfer SVM for video concept detection. In: Proceedings IEEE International Conference on Computer Vision and Pattern Recognition, pp. 1375–1381 (2009)

30. Bruzzone, L., Marconcini, M.: Domain adaptation problems: a DASVM classification technique and a circular validation strategy. IEEE Trans. Pattern Anal. Mach. Intell. **32**(5), 770–787 (2010)
31. Ben-David, S., Blitzer, J., Crammer, K., et al.: Analysis of representations for domain adaptation. In: NIPS (2007)
32. Gretton, A., Harchaoui, Z., Fukumizu, K., Sriperumbudur, B.: A fast, consistent Kernel two-sample test. In: Advances in Neural Information Processing Systems, vol. 22, pp. 673–681. MIT, Cambridge (2010)
33. Leski, J.: TSK-fuzzy modeling based on ε-insensitive learning. IEEE Trans. Fuzzy Syst. **13**(2), 181–193 (2005)
34. Wang, L.X.: Adaptive fuzzy systems and control: design and stability analysis. Prentice-Hall, Upper Saddle River (1994)
35. Jang, J.S.R., Sun, C.T., Mizutani, E.: Neuro-fuzzy and soft-computing. Prentice-Hall, Upper Saddle River (1997)
36. Chen, B., Lam, W., Tsang, I.W., et al.: Location and scatter matching for dataset shift in text mining. In: Proceedings of the IEEE International Conference on Data Mining (IEEE ICDM 2010), Sydney (December 2010)
37. Huang, J., Smola, A., Gretton, A., et al.: Correcting sample selection bias by unlabeled data. In: Proceedings of Twentieth Annual Conference on Neural Information Processing Systems (2006)
38. Takagi, T., Sugeno, M.: Fuzzy identification of systems and its application to modeling and control. IEEE Trans. Syst. Man Cybern. **15**(1), 116–132 (1985)
39. Mamdani, E.H.: Application of fuzzy logic to approximate reasoning using linguistic synthesis. IEEE Trans. Comput. **C-26**(12), 1182–1191 (1977)
40. Azeem, M.F., Hanmandlu, M., Ahmad, N.: Generalization of adaptive neural-fuzzy inference systems. IEEE Trans. Neural Netw. **11**(6), 1332–1346 (2000)
41. Deng, Z.H., Choi, K.S., Chung, F.L., et al.: Enhanced soft subspace clustering integrating within-cluster and between-cluster information. Pattern Recognit. **43**(3), 767–781 (2010)
42. Jang, J.S.R.: ANFIS: adaptive-network-based fuzzy inference systems. IEEE Trans. Syst. Man Cybern. **23**(3), 665–685 (1993)
43. Tsang, I.W., Kwok, J.T., Zurada, J.M.: Generalized core vector machines. IEEE Trans. Neural Netw. **17**(5), 1126–1140 (2006)
44. Deng, Z.H., Jiang, Y.Z., Chung, F.L., et al.: Knowledge-leverage based fuzzy system and its modeling. IEEE Trans. Fuzzy Syst. (2012). doi:10.1109/TFUZZ.2012.2212444
45. Fan, R.E., Chen, P.H., Lin, C.J.: Working Set selection using second order information for training support vector machines. J. Mach. Learn. Res. **6**, 1889–1918 (2005)
46. Hofmann, T., Schölkopf, B., Smola, A.J.: Kernel methods in machine learning. Ann. Stat. **36**, 1171–1220 (2007)
47. Sriperumbadur, B.K., Fukumizu, K., Gretton, A., et al.: Kernel choice and classifiability for RKHS embeddings of probability distributions. In: Advances in Neural Information Processing Systems, vol. 22, pp. 1750–1758. MIT, Cambridge (2010)
48. Smola, A.J., Gretton, A., Song, L., et al.: A Hilbert space embedding for distributions. In: Proceedings of 18th International Conference on Algorithmic Learn. Theory, Sendai, pp. 13–31 (October 2007)
49. Gretton, A., Harchaoui, Z., Fukumizu, K., et al.: A fast, consistent kernel two-sample test. In: Advances in Neural Information Processing Systems, vol. 22, pp. 673–681. MIT, Cambridge (2010)
50. Borgwardt, K.M., Gretton, A., Rasch, M.J., et al.: Integrating structured biological data by kernel maximum mean discrepancy. Bioinformatics (ISMB) **22**(14), e49–e57 (2006)
51. Sriperumbudur, B.K., Gretton, A., Fukumizu, K., et al.: Hilbert space embeddings and metrics on probability measures. J. Mach. Learn. Res. **11**(3), 1517–1561 (2010)
52. Suykens, J.A.K., Vandewalle, J.: Least squares support vector machine classifiers. Neural Process. Lett. **9**(3), 293–300 (1999)

53. Wu, Y., Liu, Y.: Robust truncated hinge loss support vector machines. J. Am. Stat. Assoc. **102**(479), 974–983 (2007)
54. Scholkopf, B., Herbrich, R., Smola, A.J.: Generalized representer theorem. In: Proceedings of COLT'2001, pp. 416–426. Springer, Amsterdam (2001)
55. Szedmak, S., Shawe-Taylor, J.: Multiclass learning at one-class complexity. Technical Report No. 1508, School of Electronics and Computer Science, Southampton (2005)
56. Schölkopf, B., Smola, A.J., Williamson, R., et al.: New support vector algorithms. Neural Comput. **12**(5), 1207–1245 (2000)
57. Chang, C.C., Lin, C.J.: Training ν-support vector classifiers: theory and algorithms. Neural Comput. **13**(9), 2119–2147 (2001)
58. Sindhwani, V., Belkin, M., Niyogi, P.: The geometric basis of semi-supervised learning. In: Chapelle, O., Schölkopf, B., Zien, A. (eds.) Semi-Supervised Learning. MIT, Cambridge (2006)
59. Tsang, I.W., Kwok, J.T.: Very large scale manifold regularization using core vector machines. In: NIPS 2005 Workshop on Large Scale Kernel Machines (2005)
60. Juang, C.F., Chiu, S.H., Shiu, S.J.: Fuzzy system learned through fuzzy clustering and support vector machine for human skin color segmentation. IEEE Trans. Syst. Man Cybern. **37**(6), 1077–1087 (2007)
61. Juang, C.F., Hsieh, C.D.: TS-fuzzy system-based support vector regression. Fuzzy Sets Syst. **60**(17), 2486–2504 (2009)
62. Abril, L.G., Angulo, C., Velasco, F., et al.: A note on the bias in SVMs for multi classification. IEEE Trans. Neural Netw. **19**(4), 723–725 (2008)
63. Ling, X., Dai, W., Xue, G., et al.: Spectral domain transfer learning. In: Proceedings of the 14th ACM SIGKDD International Conference on Knowledge Discovery and Data Mining. ACM, New York (2008)
64. Xiang, E.W., Cao, B., Hu, D.H., et al.: Bridging domains using world wide knowledge for transfer learning. IEEE Trans. Knowl. Data Eng. **22**(6), 770–783 (2010)
65. Dai, W., Xue, G.R., Yu, Y.: Co-clustering based classification for out-of-domain documents. In: Proceedings of the Thirteenth ACM SIGKDD International Conference on Knowledge Discovery and Data Mining, San Jose, pp. 210–219. ACM, New York (August 2007)
66. Bickel, S.: ECML-PKDD discovery challenge 2006 overview. In: Proceedings of ECML/PKDD on Discovery Challenge Workshop (2006)
67. Phan, X.H., Nguyen, M.L., Horiguchi, S.: Learning to classify short and sparse text & web with hidden topics from large-scale data collections. In: Proceedings of 17th International World Wide Web Conference (WWW'08), pp. 91–100 (April 2008)
68. Beitzel, S.M., Jensen, E.C., Frieder, O., et al.: Improving automatic query classification via semi-supervised learning. In: Proceedings of Fifth IEEE International Conference Data Mining (ICDM'05), pp. 42–49 (November 2005)
69. Blitzer, J., Dredze, M., Pereira, F., et al.: Boom-boxes and blenders: domain adaptation for sentiment classification. In: Proceedings of 45th Annual Meeting of the Association for Computational Linguistics (ACL'07), pp. 440–447 (June 2007)
70. Alpaydin, E.: Introduction to machine learning. MIT, Cambridge (2004)

Chapter 4
Security Evaluation of Support Vector Machines in Adversarial Environments

Battista Biggio, Igino Corona, Blaine Nelson, Benjamin I.P. Rubinstein, Davide Maiorca, Giorgio Fumera, Giorgio Giacinto, and Fabio Roli

Abstract Support vector machines (SVMs) are among the most popular classification techniques adopted in security applications like malware detection, intrusion detection, and spam filtering. However, if SVMs are to be incorporated in real-world security systems, they must be able to cope with attack patterns that can either mislead the learning algorithm (poisoning), evade detection (evasion) or gain information about their internal parameters (privacy breaches). The main contributions of this chapter are twofold. First, we introduce a formal general framework for the empirical evaluation of the security of machine-learning systems. Second, according to our framework, we demonstrate the feasibility of evasion, poisoning and privacy attacks against SVMs in real-world security problems. For each attack technique, we evaluate its impact and discuss whether (and how) it can be countered through an adversary-aware design of SVMs. Our experiments are easily reproducible thanks to *open-source* code that we have made available, together with all the employed datasets, on a public repository.

B. Biggio (✉) • I. Corona • D. Maiorca • G. Fumera • G. Giacinto • F. Roli
Department of Electrical and Electronic Engineering, University of Cagliari, Piazza d'Armi 09123, Cagliari, Italy
e-mail: battista.biggio@diee.unica.it; igino.corona@diee.unica.it; davide.maiorca@diee.unica.it; fumera@diee.unica.it; giacinto@diee.unica.it; roli@diee.unica.it

B. Nelson
Institut für Informatik, Universität Potsdam, August-Bebel-Straße 89, 14482 Potsdam, Germany
e-mail: blaine.nelson@gmail.com

B.I.P. Rubinstein
Department of Computing and Information Systems, University of Melbourne, Parkville, 3010 VIC, Australia
e-mail: ben@bipr.net

Y. Ma and G. Guo (eds.), *Support Vector Machines Applications*,
DOI 10.1007/978-3-319-02300-7_4,
© Springer International Publishing Switzerland 2014

4.1 Introduction

Machine-learning and pattern-recognition techniques are increasingly being adopted in security applications like spam filtering, network intrusion detection, and malware detection due to their ability to generalize and to potentially detect novel attacks or variants of known ones. Support vector machines (SVMs) are among the most successful techniques that have been applied for this purpose [28, 55].

However, learning algorithms like SVMs assume *stationarity*: that is, both the data used to train the classifier and the operational data it classifies are sampled from the same (though possibly unknown) distribution. Meanwhile, in adversarial settings such as the above-mentioned ones, intelligent and adaptive adversaries may purposely manipulate data (violating *stationarity*) to exploit existing vulnerabilities of learning algorithms and to impair the entire system. This raises several open issues, related to whether machine-learning techniques can be safely adopted in security-sensitive tasks, or if they must (and can) be redesigned for this purpose. In particular, the main open issues to be addressed include:

1. analyzing the vulnerabilities of learning algorithms;
2. evaluating their security by implementing the corresponding attacks; and
3. eventually, designing suitable countermeasures.

These issues are currently addressed in the emerging research area of *adversarial machine learning*, at the intersection between computer security and machine learning. This field is receiving growing interest from the research community, as witnessed by an increasing number of recent events: the NIPS Workshop on "Machine Learning in Adversarial Environments for Computer Security" [43]; the subsequent Special Issue of the Machine Learning journal titled "Machine Learning in Adversarial Environments" [44]; the 2010 UCLA IPAM workshop on "Statistical and Learning-Theoretic Challenges in Data Privacy"; the ECML-PKDD Workshop on "Privacy and Security issues in Data Mining and Machine Learning" [27]; five consecutive CCS Workshops on "Artificial Intelligence and Security" [2, 3, 19, 22, 34], and the Dagstuhl Perspectives Workshop on "Machine Learning for Computer Security" [37].

In Sect. 4.2, we review the literature of adversarial machine learning, focusing mainly on the issue of *security evaluation*. We discuss both theoretical work and applications, including examples of how learning can be attacked in practical scenarios, either during its training phase (i.e., *poisoning* attacks that contaminate the learner's training data to mislead it) or during its deployment phase (i.e., *evasion* attacks that circumvent the learned classifier).

In Sect. 4.3, we summarize our recently defined framework for the empirical evaluation of classifiers' security [12]. It is based on a *general* model of an adversary that builds on previous models and guidelines proposed in the literature of adversarial machine learning. We expound on the assumptions of the adversary's goal, knowledge, and capabilities that comprise this model, which also easily accommodate application-specific constraints. Having detailed the assumptions of his/her adversary, a security analyst can formalize the adversary's strategy as an optimization problem.

We then demonstrate our framework by applying it to assess the security of SVMs. We discuss our recently devised evasion attacks against SVMs [8] in Sect. 4.4, and review and extend our recent work [14] on poisoning attacks against SVMs in Sect. 4.5. We show that the optimization problems corresponding to the above attack strategies can be solved through simple gradient-descent algorithms. The experimental results for these evasion and poisoning attacks show that the SVM is vulnerable to these threats for both linear and nonlinear kernels in several realistic application domains including handwritten digit classification and malware detection for PDF files. We further explore the threat of privacy-breaching attacks aimed at the SVM's training data in Sect. 4.6 where we apply our framework to precisely describe the setting and threat model.

Our analysis provides useful insights into the potential security threats from the usage of learning algorithms (and, particularly, of SVMs) in real-world applications and sheds light on whether they can be safely adopted for security-sensitive tasks. The presented analysis allows a system designer to *quantify* the security risk entailed by an SVM-based detector so that he/she may weigh it against the benefits provided by the learning. It further suggests guidelines and countermeasures that may mitigate threats and thereby improve overall system security. These aspects are discussed for evasion and poisoning attacks in Sects. 4.4 and 4.5. In Sect. 4.6 we focus on developing countermeasures for privacy attacks that are endowed with strong theoretical guarantees within the framework of *differential privacy*. We conclude with a summary and discussion in Sect. 4.7.

In order to support the reproducibility of our experiments, we published all the code and the data employed for the experimental evaluations described in this paper [24]. In particular, our code is released under open-source license, and carefully documented, with the aim of allowing other researchers to not only reproduce but also customize, extend, and improve our work.

4.2 Background

In this section, we review the main concepts used throughout this chapter. We first introduce our notation and summarize the SVM learning problem. We then motivate the need for the proper assessment of the security of a learning algorithm so that it can be applied to security-sensitive tasks.

Learning can be generally stated as a process by which data is used to form a hypothesis that performs better than an a priori hypothesis formed without the data. For our purposes, the hypotheses will be represented as functions of the form $f : \mathscr{X} \to \mathscr{Y}$, which assign an input sample point $\mathbf{x} \in \mathscr{X}$ to a class $y \in \mathscr{Y}$; that is, given an observation from the input space $\mathscr{X}$, a hypothesis f makes a prediction in the output space $\mathscr{Y}$. For *binary classification*, the output space is binary and we use $\mathscr{Y} = \{-1, +1\}$. In the classical *supervised learning* setting, we are given a paired training dataset $\{(\mathbf{x}_i, y_i) \mid \mathbf{x}_i \in \mathscr{X}, y_i \in \mathscr{Y}\}_{i=1}^n$, we assume each pair is drawn independently from an unknown joint distribution $P(\mathbf{X}, Y)$, and we want to infer a classifier f able to *generalize* well on $P(\mathbf{X}, Y)$; i.e., to accurately predict the label y of an unseen sample $\mathbf{x}$ drawn from that distribution.

4.2.1 Support Vector Machines

In its simplest formulation, an SVM learns a linear classifier for a binary classification problem. Its decision function is thus $f(\mathbf{x}) = \mathrm{sign}(\mathbf{w}^\top \mathbf{x} + b)$, where $\mathrm{sign}(a) = +1\ (-1)$ if $a \geq 0$ $(a < 0)$, and $\mathbf{w}$ and b are learned parameters that specify the position of the decision hyperplane in feature space: the hyperplane's normal $\mathbf{w}$ gives its orientation and b is its displacement. The learning task is thus to find a hyperplane that well separates the two classes. While many hyperplanes may suffice for this task, the SVM hyperplane both separates the training samples of the two classes and provides a maximum distance from itself to the nearest training point (this distance is called the classifier's *margin*), since maximum-margin learning generally reduces *generalization error* [66]. Although originally designed for linearly separable classification tasks (*hard-margin* SVMs), SVMs were extended to nonlinearly separable classification problems by Vapnik [25] (*soft-margin* SVMs), which allow some samples to violate the margin. In particular, a soft-margin SVM is learned by solving the following convex quadratic program (QP):

$$\min_{\mathbf{w}, b, \xi} \quad \frac{1}{2}\mathbf{w}^\top \mathbf{w} + C \sum_{i=1}^{n} \xi_i$$

$$\text{s.t.} \quad \forall\, i = 1, \ldots, n \quad y_i(\mathbf{w}^\top \mathbf{x}_i + b) \geq 1 - \xi_i \quad \text{and} \quad \xi_i \geq 0 \ ,$$

where the margin is maximized by minimizing $\frac{1}{2}\mathbf{w}^\top \mathbf{w}$, and the variables ξ_i (referred to as *slack variables*) represent the extent to which the samples, $\mathbf{x}_i$, violate the margin. The parameter C tunes the trade-off between minimizing the sum of the slack violation errors and maximizing the margin.

While the primal can be optimized directly, it is often solved via its (Lagrangian) dual problem written in terms of Lagrange multipliers, α_i, which are constrained so that $\sum_{i=1}^n \alpha_i y_i = 0$ and $0 \leq \alpha_i \leq C$ for $i = 1, \ldots, n$. Solving the dual has a computational complexity that grows according to the size of the training data as opposed to the feature space's dimensionality. Further, in the dual formulation, both the data and the slack variables become implicitly represented—the data is represented by a *kernel matrix*, $\mathbf{K}$, of all inner products between pairs of data points (i.e., $K_{i,j} = \mathbf{x}_i^\top \mathbf{x}_j$) and each slack variable is associated with a Lagrangian multiplier via the KKT conditions that arise from duality. Using the method of Lagrangian multipliers, the *dual problem* is derived, in matrix form, as

$$\min_{\alpha} \quad \frac{1}{2}\alpha^\top \mathbf{Q}\alpha - \mathbf{1}_n^\top \alpha$$

$$\text{s.t.} \quad \sum_{i=1}^{n} \alpha_i y_i = 0 \quad \text{and} \quad \forall\, i = 1, \ldots, n \quad 0 \leq \alpha_i \leq C \ ,$$

where $\mathbf{Q} = \mathbf{K} \circ yy^\top$ (the Hadamard product of $\mathbf{K}$ and $yy^\top$) and $\mathbf{1}_n$ is a vector of n ones.

Through the kernel matrix, SVMs can be extended to more complex feature spaces (where a linear classifier may perform better) via a *kernel function*—an implicit inner product from the alternative feature space. That is, if some function $\phi : \mathscr{X} \to \Phi$ maps training samples into a higher-dimensional feature space, then K_{ij} is computed via the space's corresponding kernel function, $\kappa(\mathbf{x}_i, \mathbf{x}_j) = \phi(\mathbf{x}_i)^\top \phi(\mathbf{x}_j)$. Thus, one need not explicitly know ϕ, only its corresponding *kernel function*.

Further, the dual problem and its KKT conditions elicit interesting properties of the SVM. First, the optimal primal hyperplane's normal vector, $\mathbf{w}$, is a linear combination of the training samples[1]; i.e., $\mathbf{w} = \sum_{i=1}^{n} \alpha_i y_i \mathbf{x}_i$. Second, the dual solution is *sparse*, and only samples that lie on or within the hyperplane's margin have a nonzero α-value. Thus, if $\alpha_i = 0$, the corresponding sample $\mathbf{x}_i$ is correctly classified, lies beyond the margin (i.e., $y_i(\mathbf{w}^\top \mathbf{x}_i + b) > 1$), and is called a *non-support vector*. If $\alpha_i = C$, the ith sample violates the margin (i.e., $y_i(\mathbf{w}^\top \mathbf{x}_i + b) < 1$) and is an *error vector*. Finally, if $0 < \alpha_i < C$, the ith sample lies exactly on the margin (i.e., $y_i(\mathbf{w}^\top \mathbf{x}_i + b) = 1$) and is a *support vector*. As a consequence, the optimal displacement b can be determined by averaging $y_i - \mathbf{w}^\top \mathbf{x}_i$ over the support vectors.

4.2.2 Machine Learning for Computer Security: Motivation, Trends, and Arms Races

In this section, we motivate the recent adoption of machine-learning techniques in computer security and discuss the novel issues this trend raises. In the last decade, security systems increased in complexity to counter the growing sophistication and variability of attacks; a result of a long-lasting and continuing arms race in security-related applications such as malware detection, intrusion detection, and spam filtering. The main characteristics of this struggle and the typical approaches pursued in security to face it are discussed in Sect. 4.2.3.1. We now discuss some examples that better explain this trend and motivate the use of modern machine-learning techniques for security applications.

In the early years, the attack surface (i.e., the vulnerable points of a system) of most systems was relatively small and most attacks were simple. In this

[1] This is an instance of the Representer Theorem which states that solutions to a large class of regularized ERM problems lie in the span of the training data [61].

era, signature-based detection systems (e.g., rule-based systems based on string-matching techniques) were considered sufficient to provide an acceptable level of security. However, as the complexity and exposure of sensitive systems increased in the Internet Age, more targets emerged and the incentive for attacking them became increasingly attractive, thus providing a means and motivation for developing sophisticated and diverse attacks. Since signature-based detection systems can only detect attacks matching an existing *signature*, attackers used minor variations of their attacks to evade detection (e.g., string-matching techniques can be evaded by slightly changing the attack code). To cope with the increasing variability of attack samples and to detect never-before-seen attacks, machine-learning approaches have been increasingly incorporated into these detection systems to complement traditional signature-based detection. These two approaches can be combined to make accurate and agile detection: signature-based detection offers fast and lightweight filtering of most known attacks, while machine-learning approaches can process the remaining (unfiltered) samples and identify new (or less well-known) attacks.

The Quest of Image Spam. A recent example of the above arms race is *image spam* (see, e.g., [10]). In 2006, to evade the textual-based spam filters, spammers began rendering their messages into images included as attachments, thus producing "image-based spam," or *image spam* for short. Due to the massive volume of image spam sent in 2006 and 2007, researchers and spam-filter designers proposed several different countermeasures. Initially, suspect images were analyzed by OCR tools to extract text for standard spam detection, and then signatures were generated to block the (known) spam images. However, spammers immediately reacted by *randomly* obfuscating images with *adversarial* noise, both to make OCR-based detection ineffective and to evade signature-based detection. The research community responded with (fast) approaches mainly based on machine-learning techniques using visual features extracted from images, which could accurately discriminate between spam images and legitimate ones (e.g., photographs and plots). Although image spam volumes have since declined, the exact cause for this decrease is debatable—these countermeasures may have played a role, but the image spam were also more costly to the spammer as they required more time to generate and more bandwidth to deliver, thus limiting the spammers' ability to send a high volume of messages. Nevertheless, had this arms race continued, spammers could have attempted to evade the countermeasures by *mimicking* the feature values exhibited by legitimate images, which would have, in fact, forced spammers to increase the number of colors and elements in their spam images, thus further increasing the size of such files and the cost of sending them.

Misuse and Anomaly Detection in Computer Networks. Another example of the above arms race can be found in network intrusion detection, where *misuse detection* has been gradually augmented by *anomaly detection*. The former approach relies on detecting attacks on the basis of signatures extracted from (known) intrusive network traffic, while the latter is based upon a statistical model of the *normal profile* of the network traffic and detects *anomalous* traffic that deviates from the assumed model of normality. This model is often constructed using machine-learning techniques, such as one-class classifiers (e.g., one-class SVMs), or, more generally, using density estimators. The underlying assumption of anomaly-detection-based intrusion detection, though, is that *all anomalous* network traffic is, in fact, intrusive. Although intrusive traffic often does exhibit anomalous behavior, the opposite is not necessarily true: some non-intrusive network traffic may also behave anomalously. Thus, accurate anomaly detectors often suffer from high false-alarm rates.

4.2.3 Adversarial Machine Learning

As witnessed by the above examples, the introduction of machine-learning techniques in security-sensitive tasks has many beneficial aspects, and it has been somewhat necessitated by the increased sophistication and variability of recent attacks and zero-day exploits. However, there is good reason to believe that machine-learning techniques themselves will be subject to carefully designed attacks in the near future, as a logical next step in the above-sketched arms race. Since machine-learning techniques were not originally designed to withstand manipulations made by intelligent and adaptive adversaries, it would be reckless to naively trust these learners in a secure system. Instead, one needs to carefully consider whether these techniques can introduce novel vulnerabilities that may degrade the overall system's security, or whether they can be safely adopted. In other words, we need to address the question raised by Barreno et al. [5]: *can machine learning be secure?*

At the center of this question is the effect an adversary can have on a learner by violating the *stationarity assumption* that the *training data* used to train the classifier comes from the same distribution as the *test data* that will be classified by the learned classifier. This is a conventional and natural assumption underlying much of machine learning and is the basis for performance-evaluation-based techniques like cross-validation and bootstrapping as well as for principles like empirical risk minimization (ERM). However, in security-sensitive settings, the adversary may purposely manipulate data to mislead learning. Accordingly, the data distribution is subject to *change*, thereby potentially violating non-stationarity, albeit, in a limited way subject to the adversary's assumed capabilities (as we discuss in Sect. 4.3.1.3). Further, as in most security tasks, predicting how the data distribution will change is difficult, if not impossible [12, 36]. Hence, adversarial learning problems are often addressed as a *proactive* arms race [12], in which the classifier designer tries to anticipate the next adversary's move, by simulating and hypothesizing proper attack scenarios, as discussed in the next section.

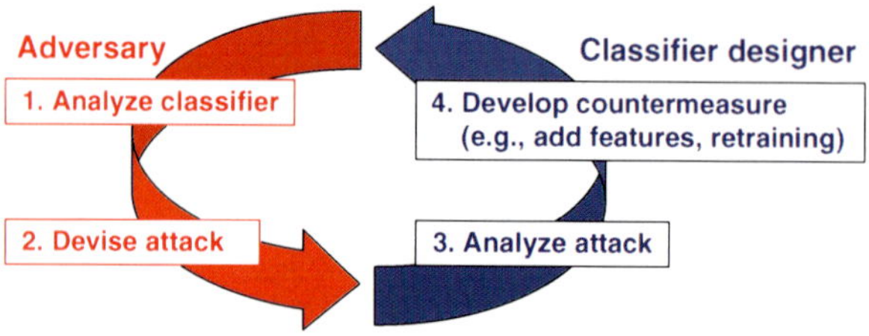

Fig. 4.1 A conceptual representation of the *reactive* arms race [12]

4.2.3.1 Reactive and Proactive Arms Races

As mentioned in the previous sections, and highlighted by the examples in Sect. 4.2.2, security problems are often cast as a long-lasting *reactive* arms race between the classifier designer and the adversary, in which each player attempts to achieve his/her goal by reacting to the changing behavior of his/her opponent. For instance, the adversary typically crafts samples to evade detection (e.g., a spammer's goal is often to create spam emails that will not be detected), while the classifier designer seeks to develop a system that accurately detects most malicious samples while maintaining a very low false-alarm rate; i.e., by not falsely identifying legitimate examples. Under this setting, the arms race can be modeled as the following cycle [12]. First, the adversary analyzes the existing learning algorithm and manipulates his/her data to evade detection (or more generally, to make the learning algorithm ineffective). For instance, a spammer may gather some knowledge of the words used by the targeted spam filter to block spam and then manipulate the textual content of her spam emails accordingly; e.g., words like "cheap" that are indicative of spam can be misspelled as "che4p." Second, the classifier designer reacts by analyzing the novel attack samples and updating his/her classifier. This is typically done by retraining the classifier on the newly collected samples, and/or by adding features that can better detect the novel attacks. In the previous spam example, this amounts to retraining the filter on the newly collected spam and, thus, to adding novel words into the filter's dictionary (e.g., "che4p" may be now learned as a spammy word). This *reactive* arms race continues in perpetuity as illustrated in Fig. 4.1.

However, *reactive* approaches to this arms race do not anticipate the next generation of security vulnerabilities and thus, the system potentially remains vulnerable to new attacks. Instead, computer security guidelines traditionally advocate a *proactive* approach[2]—the classifier designer should *proactively anticipate* the adversary's strategy by (1) identifying the most relevant threats, (2) designing proper countermeasures into his/her classifier, and (3) repeating this process for his/her new

[2]Although in certain abstract models we have shown how regret-minimizing online learning can be used to define reactive approaches that are competitive with proactive security [6].

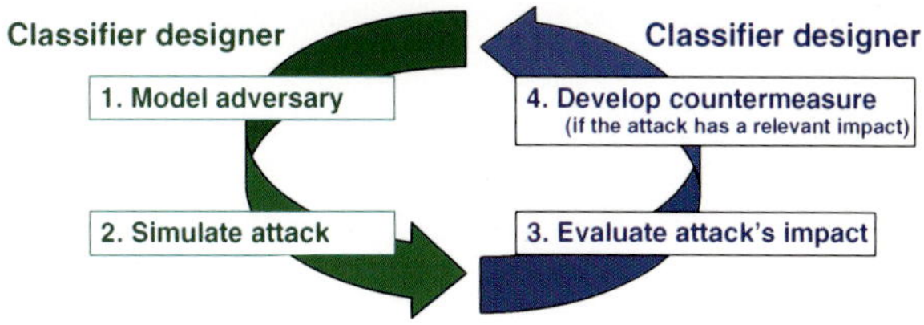

Fig. 4.2 A conceptual representation of the *proactive* arms race [12]

design *before* deploying the classifier. This can be accomplished by modeling the adversary (based on knowledge of the adversary's goals and capabilities) and using this model to simulate attacks, as is depicted in Fig. 4.2 to contrast the reactive arms race. While such an approach does not account for unknown or changing aspects of the adversary, it can indeed lead to an improved level of security by delaying each step of the *reactive* arms race because it should reasonably force the adversary to exert greater effort (in terms of time, skills, and resources) to find new vulnerabilities. Accordingly, proactively designed classifiers should remain useful for a longer time, with less frequent supervision or human intervention and with less severe vulnerabilities.

Although this approach has been implicitly followed in most of the previous work (see Sect. 4.2.3.2), it has only recently been formalized within a more general framework for the empirical evaluation of a classifier's security [12], which we summarize in Sect. 4.3. Finally, although security evaluation may suggest specific countermeasures, designing general-purpose *secure* classifiers remains an open problem.

4.2.3.2 Previous Work on Security Evaluation

Previous work in adversarial learning can be categorized according to the two main steps of the proactive arms race described in the previous section. The first research direction focuses on identifying potential vulnerabilities of learning algorithms and assessing the impact of the corresponding attacks on the targeted classifier; e.g., [4, 5, 18, 36, 40–42, 46]. The second explores the development of proper countermeasures and learning algorithms robust to known attacks; e.g., [26, 41, 58].

Although some prior work does address aspects of the *empirical* evaluation of classifier security, which is often implicitly defined as the performance degradation incurred under a (simulated) attack, to our knowledge a systematic treatment of this process under a unifying perspective was only first described in our recent work [12]. Previously, security evaluation is generally conducted within a specific application domain such as spam filtering and network intrusion detection (e.g., in [26, 31, 41, 47, 67]), in which a different application-dependent criteria is separately defined for each endeavor. Security evaluation is then implicitly undertaken by defining an attack and assessing its impact on the given classifier. For instance, in [31], the authors showed how *camouflage* network packets can mimic legitimate traffic to evade detection; and, similarly, in [26, 41, 47, 67], the content of spam emails was manipulated for evasion. Although such analyses provide indispensable insights into specific problems, their results are difficult to generalize to other domains and provide little guidance for evaluating classifier security in a different application. Thus, in a new application domain, security evaluation often must begin anew and it is difficult to directly compare with prior studies. This shortcoming highlights the need for a more general set of security guidelines and a more systematic definition of classifier security evaluation that we began to address in [12].

Apart from application-specific work, several theoretical models of adversarial learning have been proposed [4, 17, 26, 36, 40, 42, 46, 54]. These models frame the secure learning problem and provide a foundation for a proper security evaluation scheme. In particular, we build upon elements of the models of [4, 5, 36, 38, 40, 42], which were used in defining our framework for security evaluation [12]. Below we summarize these foundations.

4.2.3.3 A Taxonomy of Potential Attacks Against Machine Learning Algorithms

A taxonomy of potential attacks against pattern classifiers was proposed in [4, 5, 36] as a baseline to characterize attacks on learners. The taxonomy is based on three main features: the kind of *influence* of attacks on the classifier, the kind of *security violation* they cause, and the *specificity* of an attack. The attack's influence can be either **causative**, if it aims to undermine learning, or **exploratory**, if it targets the classification phase. Accordingly, a causative attack may manipulate both training and testing data, whereas an exploratory attack only affects testing data. Examples of causative attacks include work in [14, 38, 40, 53, 59], while exploratory attacks can be found in [26, 31, 41, 47, 67]. The security violation can be either an **integrity** violation, if it aims to gain unauthorized access to the system (i.e., to have malicious samples be misclassified as legitimate); an **availability** violation, if the goal is to generate a high number of errors (both false-negatives and false-positives) such that normal system operation is compromised (e.g., legitimate users are denied access to their resources); or a **privacy** violation, if it allows the adversary to obtain confidential information from the classifier (e.g., in biometric recognition, this may amount to recovering a protected biometric template of a system's client). Finally, the attack specificity refers to the samples that are affected by the attack. It ranges continuously from **targeted** attacks (e.g., if the goal of the attack is to have a specific spam email misclassified as legitimate) to **indiscriminate** attacks (e.g., if the goal is to have *any* spam email misclassified as legitimate).

Each portion of the taxonomy specifies a different type of attack as laid out in Barreno et al. [4] and here we outline these with respect to a PDF malware detector. An example of a **causative integrity** attack is an attacker who wants to mislead the malware detector to falsely classify malicious PDFs as benign. The attacker could accomplish this goal by introducing benign PDFs with malicious features into the training set and the attack would be **targeted** if the features corresponded to a particular malware or otherwise an **indiscriminate** attack. Similarly, the attacker could cause a **causative availability** attack by injecting malware training examples that exhibited features common to benign messages; again, these would be **targeted** if the attacker wanted a particular set of benign PDFs to be misclassified. A **causative privacy** attack, however, would require both manipulation of the training and information obtained from the learned classifier. The attacker could inject malicious PDFs with features identifying a particular author and then subsequently test if other PDFs with those features were labeled as malicious; this observed behavior may leak private information about the authors of other PDFs in the training set.

In contrast to the causative attacks, **exploratory** attacks cannot manipulate the learner, but can still exploit the learning mechanism. An example of an **exploratory integrity** attack involves an attacker who crafts a malicious PDF for an existing malware detector. This attacker queries the detector with candidate PDFs to discover which attributes the detector uses to identify malware, thus, allowing his/her to redesign his/her PDF to avoid the detector. This example could be **targeted** to a single PDF exploit or **indiscriminate** if a set of possible exploits are considered. An **exploratory privacy** attack against the malware detector can be conducted in the same way as the **causative privacy** attack described above, but without first injecting PDFs into the training data. Simply by probing the malware detector with crafted PDFs, the attacker may divulge secrets from the detector. Finally, **exploratory availability** attacks are possible in some applications but are not currently considered to be of interest.

4.3 A Framework for Security Evaluation

In Sects. 4.2.3 and 4.2.3.1, we motivated the need for simulating a proactive arms race as a means for improving system security. We further argued that evaluating a classifier's security properties through simulations of different, potential attack scenarios is a crucial step in this arms race for identifying the most relevant vulnerabilities and for suggesting how to potentially counter them. Here, we summarize our recent work [12] that proposes a new framework for designing proactive secure classifiers by addressing the shortcomings of the reactive security cycle raised above. Namely, our approach allows one to empirically evaluate a classifier's security during its design phase by addressing the first three steps of the proactive arms race depicted in Fig. 4.2: (1) identifying potential attack scenarios, (2) devising the corresponding attacks, and (3) systematically evaluating their impact. Although it may also suggest countermeasures to the hypothesized attacks, the final step of the proactive arms race remains unspecified as a unique design step that has to be addressed separately in an application-specific manner.

Under our proposed security evaluation process, the analyst must clearly scrutinize the classifier by considering different attack scenarios to investigate a set of distinct potential vulnerabilities. This amounts to performing a more systematic *what-if analysis* of classifier security [57]. This is an essential step in the design of security systems, as it not only allows the designer to identify the most important and relevant threats, but also it forces him/her to consciously decide whether the classifier can be reasonably deployed, after being made aware of the corresponding risks, or whether it is instead better to adopt additional countermeasure to mitigate the attack's impact *before* deploying the classifier.

Our proposed framework builds on previous work and attempts to systematize and unify their views under a more coherent perspective. The framework defines how an analyst can conduct a security audit of a classifier, which we detail in the remainder of this section. First, in Sect. 4.3.1, we explain how an adversary model is constructed according to the adversary's anticipated goals, knowledge, and capabilities. Based on this model, a simulation of the adversary can be conducted to find the corresponding *optimal* attack strategies and produce simulated attacks, as described in Sect. 4.3.1.4. These simulated attack samples are then used to evaluate the classifier by adding them to either the training or test data, in accordance with the adversary's capabilities from Sect. 4.3.1.3. We conclude this section by discussing how to exploit our framework in specific application domains in Sect. 4.3.2.

4.3.1 Modeling the Adversary

The proposed model of the adversary is based on specific assumptions about his/her goal, knowledge of the system, and capability to modify the underlying data distribution by manipulating individual samples. It allows the classifier designer to model the attacks identified in the attack taxonomy described as in Sect. 4.2.3.3 [4, 5, 36]. However, in our framework, one can also incorporate application-specific constraints into the definition of the adversary's capability. Therefore, it can be exploited to derive practical guidelines for developing optimal attack strategies and to guide the design of adversarially resilient classifiers.

4.3.1.1 Adversary's Goal

According to the taxonomy presented first by Barreno et al. [5] and extended by Huang et al. [36], the adversary's goal should be defined based on the anticipated security violation, which might be an integrity, availability, or privacy violation (see Sect. 4.2.3.3), and also depending on the attack's specificity, which ranges from targeted to indiscriminate. Further, as suggested by Laskov and Kloft [42] and Kloft and Laskov [40], the adversary's goal should be defined in terms of an objective function that the adversary is willing to maximize. This allows for a formal characterization of the *optimal* attack strategy.

For instance, in an indiscriminate integrity attack, the adversary may aim to maximize the number of spam emails that evade detection, while minimally

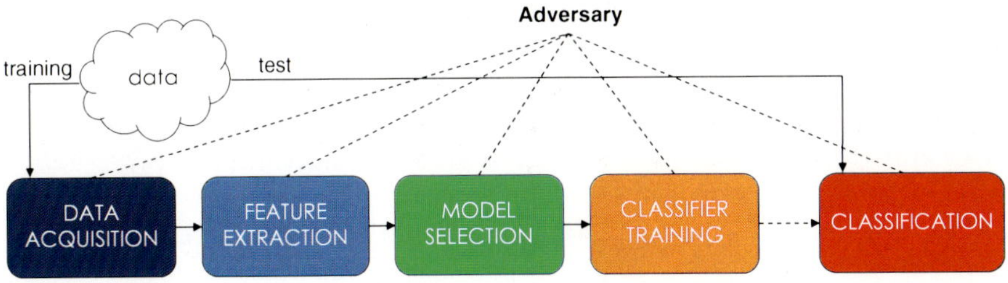

Fig. 4.3 A representation of the design steps of a machine-learning system [29] that may provide sources of knowledge for the adversary

manipulating their content [26, 46, 54], whereas in an indiscriminate availability attack, the adversary may aim to maximize the number of classification errors, thereby causing a general denial-of-service due to an excess of false alarms [14,53].

4.3.1.2 Adversary's Knowledge

The adversary's knowledge of the attacked system can be defined based on the main components involved in the design of a machine-learning system, as described in [29] and depicted in Fig. 4.3.

According to the five design steps depicted in Fig. 4.3, the adversary may have various degrees of knowledge (ranging from no information to complete information) pertaining to the following five components:

($k.i$) the training set (or part of it);
($k.ii$) the feature representation of each sample; i.e., how *real* objects (emails, network packets, etc.) are mapped into the feature space;
($k.iii$) the learning algorithm and its decision function; e.g., that logistic regression is used to learn a linear classifier;
($k.iv$) the learned classifier's parameters; e.g., the actual learned weights of a linear classifier;
($k.v$) feedback from the deployed classifier; e.g., the classification labels assigned to some of the samples by the targeted classifier.

These five elements represent different levels of knowledge about the system being attacked. A typical hypothesized scenario assumes that the adversary has *perfect knowledge* of the targeted classifier ($k.iv$). Although potentially too pessimistic, this worst-case setting allows one to compute a lower bound on the classifier performance when it is under attack [26, 41]. A more realistic setting is that the adversary knows the (untrained) learning algorithm ($k.iii$), and he/she may exploit feedback from the classifier on the labels assigned to some *query* samples ($k.v$), either to directly find optimal or nearly optimal attack instances [46,54] or to learn a surrogate classifier, which can then serve as a template to guide the attack against the actual classifier. We refer to this scenario as a *limited knowledge* setting in Sect. 4.4.

Note that one may also make more restrictive assumptions on the adversary's knowledge, such as considering partial knowledge of the feature representation (*k.ii*), or a complete lack of knowledge of the learning algorithm (*k.iii*). Investigating classifier security against these uninformed adversaries may yield a higher level of security. However, such assumptions would be contingent on *security through obscurity*; that is, the provided security would rely upon *secrets* that must be kept unknown to the adversary even though such a high level of secrecy may not be practical. Reliance on unjustified secrets can potentially lead to catastrophic unforeseen vulnerabilities. Thus, this paradigm should be regarded as being complementary to *security by design*, which instead advocates that systems should be designed from the ground-up to be secure and, if secrets are assumed, they must be well justified. Accordingly, security is often investigated by assuming that the adversary knows at least the learning algorithm and the underlying feature representation.

4.3.1.3 Adversary's Capability

We now give some guidelines on how the attacker may be able to manipulate samples and the corresponding data distribution. As discussed in Sect. 4.2.3.3 [4, 5, 36], the adversary may control both training and test data (causative attacks), or only on test data (exploratory attacks). Further, training and test data may follow different distributions, since they can be manipulated according to different attack strategies by the adversary. Therefore, we should specify:

- (*c.i*) whether the adversary can manipulate training (TR) and/or testing (TS) data; i.e., the attack influence from the taxonomy in [4, 5, 36];
- (*c.ii*) whether and to what extent the attack affects the class priors, for TR and TS;
- (*c.iii*) which and how many samples can be modified in each class, for TR and TS;
- (*c.iv*) which features of each attack sample can be modified and how can these features' values be altered; e.g., correlated feature values cannot be modified independently.

Assuming a generative model $p(\mathbf{X}, Y) = p(Y)p(\mathbf{X}|Y)$ (where we use p_{tr} and p_{ts} for training and test distributions, respectively), assumption (*c.ii*) specifies how an attack can modify the priors $p_{\mathrm{tr}}(Y)$ and $p_{\mathrm{ts}}(Y)$, while assumptions (*c.iii*) and (*c.iv*) specify how it can alter the class-conditional distributions $p_{\mathrm{tr}}(\mathbf{X}|Y)$ and $p_{\mathrm{ts}}(\mathbf{X}|Y)$.

To perform security evaluation according to the hypothesized attack scenario, it is thus clear that the collected data and generated attack samples should be resampled according to the above distributions to produce suitable training and test set pairs. This can be accomplished through existing resampling algorithms like cross-validation or bootstrapping, when the attack samples are independently sampled from an identical distribution (i.i.d.). Otherwise, one may consider different sampling schemes. For instance, in Biggio et al. [14] the attack samples had to be injected into the training data, and each attack sample depended on thecurrent training data, which also included past attack samples. In this case, it was sufficient to add one attack sample at a time, until the desired number of samples was reached.[3]

4.3.1.4 Attack Strategy

Once specific assumptions on the adversary's goal, knowledge, and capability are made, one can compute the optimal attack strategy corresponding to the hypothesized attack scenario; i.e., the adversary model. This amounts to solving the optimization problem defined according to the adversary's goal, under proper constraints defined in accordance with the adversary's assumed knowledge and capabilities. The attack strategy can then be used to produce the desired attack samples, which then have to be merged consistently to the rest of the data to produce suitable training and test sets for the desired security evaluation, as explained in the previous section. Specific examples of how to derive optimal attacks against SVMs and how to resample training and test data to properly include them are discussed in Sects. 4.4 and 4.5.

4.3.2 How to Use Our Framework

We summarize here the steps that can be followed to correctly use our framework in specific application scenarios:

1. hypothesize an attack scenario by identifying a proper adversary's goal, and according to the taxonomy in [4, 5, 36];
2. define the adversary's knowledge according to (*k.i–v*), and capabilities according to (*c.i–iv*);
3. formulate the corresponding optimization problem and devise the corresponding attack strategy;
4. resample the collected (training and test) data accordingly;
5. evaluate classifier's security on the resampled data (including attack samples);
6. repeat the evaluation for different levels of adversary's knowledge and/or capabilities, if necessary; or hypothesize a different attack scenario.

In the next sections we show how our framework can be applied to investigate three security threats to SVMs: evasion, poisoning, and privacy violations. We then discuss how our findings may be used to improve the security of such classifiers to the considered attacks. For instance, we show how careful kernel parameter selection, which trades off between security to attacks and classification accuracy, may complicate the adversary's task of subverting the learning process.

---

[3] See [12] for more details on the definition of the data distribution and the resampling algorithm.

4.4 Evasion Attacks Against SVMs

In this section, we consider the problem of SVM evasion at *test time*; i.e., how to optimally manipulate samples at test time to avoid detection. The problem of evasion at test time has been considered in previous work albeit limited either to simple decision functions such as linear classifiers [26, 46], or to cover any convex-inducing classifiers [54] that partition the feature space into two sets, one of which is convex, but do not include most interesting families of nonlinear classifiers such as neural nets or SVMs. In contrast to this prior work, the methods presented in our recent work [8] and in this section demonstrate that evasion of kernel-based classifiers at test time can be realized with a straightforward gradient-descent-based approach derived from Golland's technique of discriminative directions [33]. As a further simplification of the attacker's effort, we empirically show that, even if the adversary does not precisely know the classifier's decision function, he/she can learn a *surrogate* classifier on a surrogate dataset and reliably evade the targeted classifier.

This section is structured as follows. In Sect. 4.4.1, we define the model of the adversary, including his/her attack strategy, according to our evaluation framework described in Sect. 4.3.1. Then, in Sect. 4.4.2 we derive the attack strategy that will be employed to experimentally evaluate evasion attacks against SVMs. We report our experimental results in Sect. 4.4.3. Finally, we critically discuss and interpret our research findings in Sect. 4.4.4.

4.4.1 Modeling the Adversary

We show here how our framework can be applied to evaluate the security of SVMs against evasion attacks. We first introduce our notation, state our assumptions about attack scenario, and then derive the corresponding optimal attack strategy.

Notation. We consider a classification algorithm $f : \mathscr{X} \mapsto \mathscr{Y}$ that assigns samples represented in some feature space $\mathbf{x} \in \mathscr{X}$ to a label in the set of predefined classes $y \in \mathscr{Y} = \{-1, +1\}$, where -1 $(+1)$ represents the legitimate (malicious) class. The label $f_{\mathbf{x}} = f(\mathbf{x})$ given by a classifier is typically obtained by thresholding a continuous discriminant function $g : \mathscr{X} \mapsto \mathbb{R}$. Without loss of generality, we assume that $f(\mathbf{x}) = -1$ if $g(\mathbf{x}) < 0$, and $+1$ otherwise. Further, note that we use $f_{\mathbf{x}}$ to refer to a label assigned by the classifier for the point $\mathbf{x}$ (rather than the true label y of that point) and the shorthand f_i for the label assigned to the ith training point, $\mathbf{x}_i$.

4.4.1.1 Adversary's Goal

Malicious (positive) samples are manipulated to evade the classifier. The adversary may be satisfied when a sample $\mathbf{x}$ is found such that $g(\mathbf{x}) < -\varepsilon$ where $\varepsilon > 0$ is a small constant. However, as mentioned in Sect. 4.3.1.1, these attacks may be easily defeated by simply adjusting the decision threshold to a slightly more conservative value (e.g., to attain a lower false negative rate at the expense of a higher false positive rate). For this reason, we assume a *smarter* adversary, whose goal is to have his/her attack sample misclassified as legitimate with the largest confidence. Analytically, this statement can be expressed as follows: find an attack sample $\mathbf{x}$ that minimizes the value of the classifier's discriminant function $g(\mathbf{x})$. Indeed, this adversarial setting provides a worst-case bound for the targeted classifier.

4.4.1.2 Adversary's Knowledge

We investigate two adversarial settings. In the first, the adversary has *perfect* knowledge (PK) of the targeted classifier; i.e., he/she knows the feature space (*k.ii*) and function $g(\mathbf{x})$ (*k.iii–iv*). Thus, the labels from the targeted classifier (*k.v*) are not needed. In the second, the adversary is assumed to have *limited* knowledge (LK) of the classifier. We assume he/she knows the feature representation (*k.ii*) and the learning algorithm (*k.iii*), but that he/she does not know the learned classifier $g(\mathbf{x})$ (*k.iv*). In both cases, we assume the attacker does not have knowledge of the training set (*k.i*).

Within the LK scenario, the adversary does not know the true discriminant function $g(\mathbf{x})$ but may approximate it as $\hat{g}(\mathbf{x})$ by learning a *surrogate* classifier on a surrogate training set $\{(\mathbf{x}_i, y_i)\}_{i=1}^{n_q}$ of n_q samples. This data may be collected by the adversary in several ways; e.g., he/she may sniff network traffic or collect legitimate and spam emails from an alternate source. Thus, for LK, there are two sub-cases related to assumption (*k.v*), which depend on whether the adversary can query the classifier. If so, the adversary can build the training set by submitting a set of n_q queries $\mathbf{x}_i$ to the targeted classifier to obtain their classification labels, $y_i = f(\mathbf{x}_i)$. This is indeed the adversary's true learning task, but it requires him/her to have access to classifier feedback; e.g., by having an email account protected by the targeted filter (for public email providers, the adversary can reasonably obtain such accounts). If not, the adversary may use the true class labels for the surrogate data, although this may not correctly approximate the targeted classifier (unless it is very accurate).

4.4.1.3 Adversary's Capability

In the evasion setting, the adversary can only manipulate testing data (*c.i*); i.e., he/she has no way to influence training data. We further assume here that the class priors cannot be modified (*c.ii*), and that all the malicious testing samples are affected by the attack (*c.iii*). In other words, we are interested in simulating an *exploratory*, *indiscriminate* attack. The adversary's capability of manipulating the features of each sample (*c.iv*) should be defined based on application-specific constraints. However, at a more general level we can bound the attack point to lie within some maximum distance from the original attack sample, $d_{\max}$, which then is a parameter of our evaluation. Similarly to previous work, the definition of a suitable distance measure $d : \mathscr{X} \times \mathscr{X} \mapsto \mathbb{R}$ is left to the specific application domain [26, 46, 54]. Note indeed that this distance should reflect the adversary's effort or cost in manipulating samples, by considering factors that can limit the overall attack impact; e.g., the increase in the file size of a malicious PDF, since larger files will lower the infection rate due to increased transmission times. For spam filtering, distance is often given as the number of modified words in each spam [26,46,53,54], since it is assumed that highly modified spam messages are less effectively able to convey the spammer's message.

4.4.1.4 Attack Strategy

Under the attacker's model described in Sects. 4.4.1.1, 4.4.1.2, and 4.4.1.3, for any target malicious sample $\mathbf{x}^0$ (the adversary's true objective), an optimal attack strategy finds a sample $\mathbf{x}^*$ to minimize g or its estimate $\hat{g}$, subject to a bound on its modification distance from $\mathbf{x}^0$:

$$\mathbf{x}^* = \arg\min_{\mathbf{x}} \hat{g}(\mathbf{x}) \quad \text{s.t.} \quad d(\mathbf{x}, \mathbf{x}^0) \leq d_{\max}.$$

For several classifiers, minimizing $g(\mathbf{x})$ is equivalent to maximizing the estimated posterior $p(f_{\mathbf{x}} = -1|\mathbf{x})$; e.g., for neural networks, since they directly output a posterior estimate, and for SVMs, since their posterior can be estimated as a sigmoidal function of the distance of $\mathbf{x}$ to the SVM hyperplane [56].

Generally, this is a nonlinear optimization, which one may optimize with many well-known techniques (e.g., gradient-descent, Newton's method, or BFGS) and below we use a gradient-descent procedure. However, if $\hat{g}(\mathbf{x})$ is not convex, descent approaches may not find a global optima. Instead, the descent path may lead to a flat region (local minimum) outside of the samples' support where $p(\mathbf{x}) \approx 0$ and the classification behavior of g is unspecified and may stymie evasion attempts (see the *upper left* plot in Fig. 4.4).

Unfortunately, our objective does not utilize the evidence we have about the distribution of data $p(\mathbf{x})$, and thus gradient descent may meander into unsupported regions ($p(\mathbf{x}) \approx 0$) where g is relatively unspecified. This problem is further compounded since our estimate $\hat{g}$ is based on a finite (and possibly small) training set making it a poor estimate of g in unsupported regions, which may lead to false evasion points in these regions. To overcome these limitations, we introduce an additional component into the formulation of our attack objective, which estimates $p(\mathbf{x}|f_{\mathbf{x}} = -1)$ using density-estimation techniques. This second component acts as a penalizer for $\mathbf{x}$ in low density regions and is weighted by a parameter $\lambda \geq 0$ yielding the following modified optimization problem:

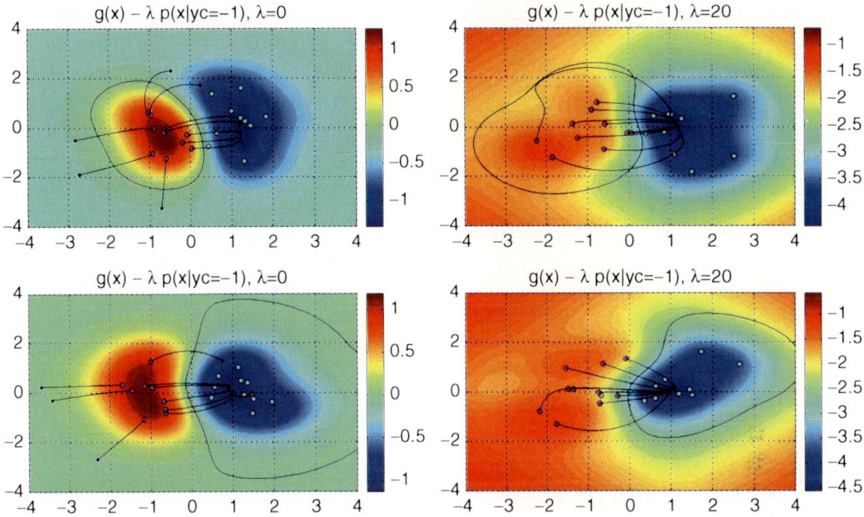

Fig. 4.4 Different scenarios for gradient-descent-based evasion procedures. In each, the function $g(\mathbf{x})$ of the learned classifier is plotted with a color map with high values (*red-orange-yellow*) corresponding to the malicious class, low values (*green-cyan-blue*) corresponding to the benign class, and a black decision boundary for the classifier. For every malicious sample, we plot the path of a simple gradient-descent evasion for a classifier with a closed boundary around the malicious class (*upper left*) or benign class (*bottom left*). Then, we plot the modified objective function of Eq. (4.1) and the paths of the resulting density-augmented gradient descent for a classifier with a closed boundary around the malicious (*upper right*) or benign class (*bottom right*)

$$\arg\min_{\mathbf{x}} \; E(\mathbf{x}) = \hat{g}(\mathbf{x}) - \frac{\lambda}{n} \sum_{i|f_i=-1} k\left(\frac{\mathbf{x}-\mathbf{x}_i}{h}\right) \qquad (4.1)$$
$$\text{s.t.} \quad d(\mathbf{x}, \mathbf{x}^0) \leq d_{\max},$$

where h is a bandwidth parameter for a kernel density estimator (KDE) and n is the number of benign samples ($f_{\mathbf{x}} = -1$) available to the adversary. This alternate objective trades off between minimizing $\hat{g}(\mathbf{x})$ (or $p(f_{\mathbf{x}} = -1|\mathbf{x})$) and maximizing the estimated density $p(\mathbf{x}|f_{\mathbf{x}} = -1)$. The extra component favors attack points to imitate features of known samples classified as legitimate, as in mimicry attacks [31]. In doing so, it reshapes the objective function and thereby biases the resulting *density-augmented gradient descent* towards regions where the negative class is concentrated (see the *bottom-right* plot in Fig. 4.4).

Finally, note that this behavior may lead our technique to disregard attack patterns within unsupported regions ($p(\mathbf{x}) \approx 0$) for which $g(\mathbf{x}) < 0$, when they do exist (see, e.g., the *upper right* plot in Fig. 4.4). This may limit classifier evasion especially when the constraint $d(\mathbf{x}, \mathbf{x}^0) \leq d_{\max}$ is particularly strict. Therefore, the trade-off between the two components of the objective function should be carefully considered.

Algorithm 1 Gradient-descent attack procedure

Input: the initial attack point, $\mathbf{x}^0$; the step size, t; the trade-off parameter, λ; and $\varepsilon > 0$.
Output: $\mathbf{x}^*$, the final attack point.

$k \leftarrow 0$.
repeat
    $k \leftarrow k+1$
    Set $\nabla E(\mathbf{x}^{k-1})$ to a unit vector aligned with $\nabla g(\mathbf{x}^{k-1}) - \lambda \nabla p(\mathbf{x}^{k-1}|f_{\mathbf{x}} = -1)$.
    $\mathbf{x}^k \leftarrow \mathbf{x}^{k-1} - t\nabla E(\mathbf{x}^{k-1})$
    if $d(\mathbf{x}^k, \mathbf{x}^0) > d_{\max}$ **then**
        Project $\mathbf{x}^k$ onto the boundary of the feasible region (enforcing application-specific constraints, if any).
    end if
until $E\left(\mathbf{x}^k\right) - E\left(\mathbf{x}^{k-1}\right) < \varepsilon$
return: $\mathbf{x}^* = \mathbf{x}^k$

4.4.2 *Evasion Attack Algorithm*

Algorithm 1 details a gradient-descent method for optimizing problem of Eq. (4.1). It iteratively modifies the attack point $\mathbf{x}$ in the feature space as $\mathbf{x}' \leftarrow \mathbf{x} - t\nabla E$, where ∇E is a unit vector aligned with the gradient of our objective function, and t is the step size. We assume g to be differentiable almost everywhere (subgradients may be used at discontinuities). When g is non-differentiable or is not smooth enough for a gradient descent to work well, it is also possible to rely upon the mimicry / KDE term in the optimization of Eq. (4.1).

In the next sections, we show how to compute the components of ∇E; namely, the gradient of the discriminant function $g(\mathbf{x})$ of SVMs for different kernels, and the gradient of the mimicking component (density estimation). We finally discuss how to project the gradient ∇E onto the feasible region in *discrete* feature spaces.

4.4.2.1 Gradient of Support Vector Machines

For SVMs, $g(\mathbf{x}) = \sum_i \alpha_i y_i k(\mathbf{x}, \mathbf{x}_i) + b$. The gradient is thus given by $\nabla g(\mathbf{x}) = \sum_i \alpha_i y_i \nabla k(\mathbf{x}, \mathbf{x}_i)$. Accordingly, the feasibility of the approach depends on the computability of this kernel gradient $\nabla k(\mathbf{x}, \mathbf{x}_i)$, which is computable for many numeric kernels. In the following, we report the kernel gradients for three main cases: (a) the linear kernel, (b) the RBF kernel, and (c) the polynomial kernel.

(a) *Linear Kernel.* In this case, the kernel is simply given by $k(\mathbf{x}, \mathbf{x}_i) = \langle \mathbf{x}, \mathbf{x}_i \rangle$. Accordingly, $\nabla k(\mathbf{x}, \mathbf{x}_i) = \mathbf{x}_i$ (we remind the reader that the gradient has to be computed with respect to the current attack sample $\mathbf{x}$), and $\nabla g(\mathbf{x}) = \mathbf{w} = \sum_i \alpha_i y_i \mathbf{x}_i$.

(b) *RBF Kernel*. For this kernel, $k(\mathbf{x},\mathbf{x}_i) = \exp\{-\gamma\|\mathbf{x}-\mathbf{x}_i\|^2\}$. The gradient is thus given by $\nabla k(\mathbf{x},\mathbf{x}_i) = -2\gamma\exp\{-\gamma\|\mathbf{x}-\mathbf{x}_i\|^2\}(\mathbf{x}-\mathbf{x}_i)$.
(c) *Polynomial Kernel*. In this final case, $k(\mathbf{x},\mathbf{x}_i) = (\langle\mathbf{x},\mathbf{x}_i\rangle + c)^p$. The gradient is thus given by $\nabla k(\mathbf{x},\mathbf{x}_i) = p(\langle\mathbf{x},\mathbf{x}_i\rangle + c)^{p-1}\mathbf{x}_i$.

4.4.2.2 Gradients of Kernel Density Estimators

As with SVMs, the gradient of kernel density estimators depends on the gradient of its kernel. We considered generalized RBF kernels of the form

$$k\left(\tfrac{\mathbf{x}-\mathbf{x}_i}{h}\right) = \exp\left(-\tfrac{d(\mathbf{x},\mathbf{x}_i)}{h}\right),$$

where $d(\cdot,\cdot)$ is any suitable distance function. We used here the same distance $d(\cdot,\cdot)$ used in Eq. (4.1), but they can be different, in general. For ℓ_2- and ℓ_1-norms (i.e., RBF and Laplacian kernels), the KDE (sub)gradients are, respectively, given by:

$$-\frac{2}{nh}\sum_{i|f_i=-1}\exp\left(-\frac{\|\mathbf{x}-\mathbf{x}_i\|_2^2}{h}\right)(\mathbf{x}-\mathbf{x}_i) \ ,$$

$$-\frac{1}{nh}\sum_{i|f_i=-1}\exp\left(-\frac{\|\mathbf{x}-\mathbf{x}_i\|_1}{h}\right)(\mathbf{x}-\mathbf{x}_i) \ .$$

Note that the scaling factor here is proportional to $O(\frac{1}{nh})$. Therefore, to influence gradient descent with a significant mimicking effect, the value of λ in the objective function should be chosen such that the value of $\frac{\lambda}{nh}$ is comparable to (or higher than) the range of the discriminant function $\hat{g}(\mathbf{x})$.

4.4.2.3 Gradient-Descent Attack in Discrete Spaces

In discrete spaces, gradient approaches may lead to a path through infeasible portions of the feature space. In such cases, we need to find feasible neighbors $\mathbf{x}$ that yield a steepest descent; i.e., maximally decreasing $E(\mathbf{x})$. A simple approach to this problem is to probe E at every point in a small neighborhood of $\mathbf{x}$: $\mathbf{x}' \leftarrow \arg\min_{\mathbf{z}\in\mathscr{N}(\mathbf{x})} E(\mathbf{z})$. However, this approach requires a large number of queries. For classifiers with a differentiable decision function, we can instead use the neighbor whose difference from $\mathbf{x}$ best aligns with $\nabla E(\mathbf{x})$; i.e., the update becomes

$$\mathbf{x}' \leftarrow \arg\max_{\mathbf{z}\in\mathscr{N}(\mathbf{x})} \frac{(\mathbf{z}-\mathbf{x})}{\|\mathbf{z}-\mathbf{x}\|}^{\top} \nabla E(\mathbf{x}) \ .$$

Thus, the solution to the above alignment is simply to modify a feature that satisfies $\arg\max_i |\nabla E(\mathbf{x})_i|$ for which the corresponding change leads to a feasible state. Note, however that, sometimes, such a step may be relatively quite large and may lead the attack out of a local minimum potentially increasing the objective function. Therefore, one should consider the best alignment that effectively reduces the objective function by disregarding features that lead to states where the objective function is higher.

4.4.3 Experiments

In this section, we first report some experimental results on the MNIST handwritten digit classification task [32, 45] that visually demonstrate how the proposed algorithm modifies digits to mislead classification. This dataset is particularly useful because the visual nature of the handwritten digit data provides a *semantic meaning* for attacks. We then show the effectiveness of the proposed attack on a more realistic and practical scenario: the detection of malware in PDF files.

4.4.3.1 Handwritten Digits

We first focus on a two-class sub-problem of discriminating between two distinct digits from the MNIST dataset [45]. Each digit example is represented as a grayscale image of 28×28 pixels arranged in raster-scan-order to give feature vectors of $d = 28 \times 28 = 784$ values. We normalized each feature (pixel) $x_f \in [0,1]^d$ by dividing its value by 255, and we constrained the attack samples to this range. Accordingly, we optimized Eq. (4.1) subject to $0 \leq x_f \leq 1$ for all f.

For our attacker, we assume the perfect knowledge (PK) attack scenario. We used the *Manhattan* distance (ℓ_1-norm) as the distance function, d, both for the kernel density estimator (i.e., a Laplacian kernel) and for the constraint $d(\mathbf{x}, \mathbf{x}^0) \leq d_{\max}$ of Eq. (4.1), which bounds the total difference between the gray-level values of the original image $\mathbf{x}^0$ and the attack image $\mathbf{x}$. We used an upper bound of $d_{\max} = \frac{5000}{255}$ to limit the total change in the gray-level values to 5000. At each iteration, we increased the ℓ_1-norm value of $\mathbf{x} - \mathbf{x}^0$ by $\frac{10}{255}$, which is equivalent to increasing the difference in the gray-level values by 10. This is effectively the gradient step size.

For the digit discrimination task, we applied an SVM with the linear kernel and $C = 1$. We randomly chose 100 training samples and applied the attacks to a correctly classified positive sample.

In Fig. 4.5 we illustrate gradient attacks in which a “3” is to be misclassified as a “7.” The left image shows the initial attack point, the middle image shows the first attack image misclassified as legitimate, and the right image shows the attack point after 500 iterations. When $\lambda = 0$, the attack images exhibit only a weak resemblance to the target class “7” but are, nevertheless, reliably misclassified. This is the same effect we observed in the left plot of Fig. 4.4: the classifier is evaded by making the attack sample dissimilar to the malicious class. Conversely, when $\lambda = 10$ the attack

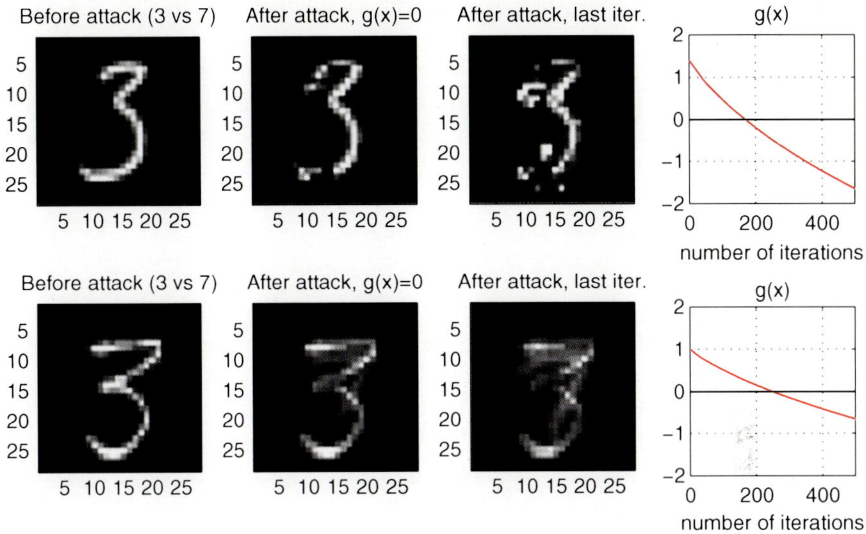

Fig. 4.5 Illustration of the gradient attack on the MNIST digit data, for $\lambda = 0$ (*top row*) and $\lambda = 10$ (*bottom row*). Without a mimicry component ($\lambda = 0$) gradient descent quickly decreases g but the resulting attack image does not resemble a "7." In contrast, the attack minimizes g slower when mimicry is applied ($\lambda = 10$), but the final attack image closely resembles a mixture between "3" and "7," as the term "mimicry" suggests

images strongly resemble the target class because the mimicry term favors samples that are more similar to the target examples. This is the same effect illustrated in the rightmost plot of Fig. 4.4.

4.4.3.2 Malware Detection in PDF Files

We focus now on the problem of discriminating between legitimate and malicious PDF files, a popular medium for disseminating malware [68]. PDF files are excellent vectors for malicious-code, due to their flexible *logical structure*, which can be described by a hierarchy of interconnected objects. As a result, an attack can be easily hidden in a PDF to circumvent file-type filtering. The PDF format further allows a wide variety of resources to be embedded in the document including `JavaScript`, `Flash`, and even binary programs. The type of the embedded object is specified by *keywords*, and its content is in a *data stream*. Several recent works proposed machine-learning techniques for detecting malicious PDFs use the file's logical structure to accurately identify the malware [50, 63, 64]. In this case study, we use the feature representation of Maiorca et al. [50] in which each feature corresponds to the tally of occurrences of a given keyword in the PDF file. Similar feature representations were also exploited in [63, 64].

The PDF application imposes natural constraints on attacks. Although it is difficult to *remove* an embedded object (and its corresponding keywords) without corrupting the PDF's file structure, it is rather easy to *insert* new objects (and, thus, keywords) through the addition of a new *version* to the PDF file [1, 49]. In our feature representation, this is equivalent to allowing only feature increments; i.e., requiring $\mathbf{x}^0 \leq \mathbf{x}$ as an additional constraint in the optimization problem given by Eq. (4.1). Further, the total difference in keyword counts between two samples is their *Manhattan* distance, which we again use for the kernel density estimator and the constraint in Eq. (4.1). Accordingly, $d_{\max}$ is the maximum number of additional keywords that an attacker can add to the original $\mathbf{x}^0$.

Experimental Setup. For experiments, we used a PDF corpus with 500 malicious samples from the *Contagio* dataset[4] and 500 benign samples collected from the web. We randomly split the data into five pairs of training and testing sets with 500 samples each to average the final results. The features (keywords) were extracted from each training set as described in [50]; on average, 100 keywords were found in each run. Further, we also bounded the maximum value of each feature (keyword count) to 100, as this value was found to be close to the 95th percentile for each feature. This limited the influence of outlying samples having very high feature values.

We simulated the *perfect* knowledge (PK) and the *limited* knowledge (LK) scenarios described in Sect. 4.4.1.2. In the LK case, we set the number of samples used to learn the surrogate classifier to $n_q = 100$. The reason is to demonstrate that even with a dataset as small as the 20% of the original training set size, the adversary may be able to evade the targeted classifier with high reliability. Further, we assumed that the adversary uses feedback from the *targeted* classifier f; i.e., the labels $\hat{y}_i = f_i = f(\mathbf{x}_i)$ for each surrogate sample $\mathbf{x}_i$. Similar results were also obtained using the true labels (without relabeling), since the targeted classifiers correctly classified almost all samples in the test set.

As discussed in Sect. 4.4.2.2, the value of λ is chosen according to the scale of the discriminant function $g(\mathbf{x})$, the bandwidth parameter h of the kernel density estimator, and the number of samples n labeled as legitimate in the surrogate training set. For computational reasons, to estimate the value of the KDE at a given point $\mathbf{x}$ in the feature space, we only consider the 50 nearest (legitimate) training samples to $\mathbf{x}$; therefore, $n \leq 50$ in our case. The bandwidth parameter was set to $h = 10$, as this value provided a proper rescaling of the Manhattan distances observed in our dataset for the KDE. We thus set $\lambda = 500$ to be comparable with $O(nh)$.

For each targeted classifier and training/testing pair, we learned five different surrogate classifiers by randomly selecting n_q samples from the test set and averaged their results. For SVMs, we sought a surrogate classifier that would correctly match the labels from the targeted classifier; thus, we used parameters $C = 100$, and $\gamma = 0.1$ (for the RBF kernel) to heavily penalize training errors.

Experimental Results. We report our results in Fig. 4.6, in terms of the false negative (FN) rate attained by the targeted classifiers as a function of the maximum

[4]http://contagiodump.blogspot.it

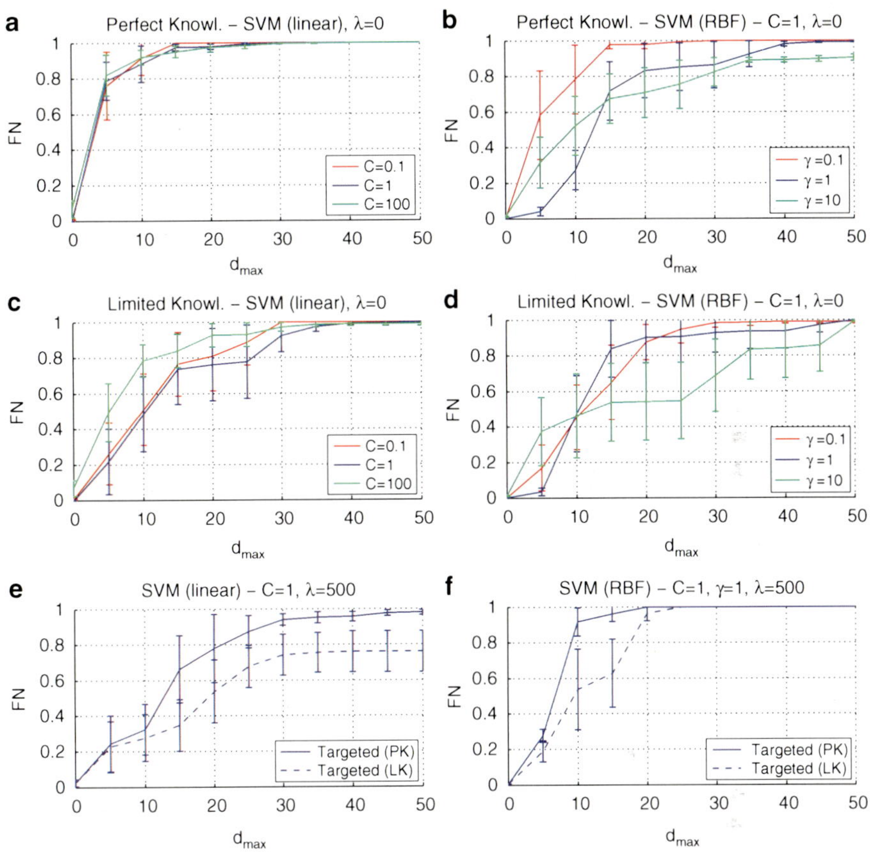

Fig. 4.6 Experimental results for SVMs with the linear and the RBF kernel (first and second columns). We report the FN values (attained at FP=0.5 %) for increasing $d_{\max}$. For the sake of readability, we report the average FN value $\pm$ half standard deviation (shown with error bars). Results for perfect knowledge (PK) attacks with $\lambda = 0$ (without mimicry) are shown in the first row, for different values of the considered classifier parameters, i.e., the regularization parameter C for the linear SVM, and the kernel parameter γ for the SVM with RBF kernel (with $C = 1$). Results for limited knowledge (LK) attacks with $\lambda = 0$ (without mimicry) are shown in the second row for the linear SVM (for varying C), and the SVM with RBF kernel (for varying γ, with $C = 1$). Results for perfect (PK) and limited knowledge (LK) with $\lambda = 500$ (with mimicry) are shown in the third row for the linear SVM (with $C = 1$), and the SVM with RBF kernel (with $\gamma = 1$ and $C = 1$)

allowable number of modifications, $d_{\max} \in [0, 50]$. We compute the FN rate corresponding to a fixed false positive (FP) rate of FP$= 0.5\,\%$. For $d_{\max} = 0$, the FN rate corresponds to a standard performance evaluation using unmodified PDFs. As expected, the FN rate increases with $d_{\max}$ as the PDF is increasingly modified, since the adversary has more flexibility in his/her attack. Accordingly, a more secure classifier will exhibit a more graceful increase of the FN rate.

Results for $\lambda = 0$. We first investigate the effect of the proposed attack in the PK case, without considering the mimicry component (Fig. 4.6, top row), for varying parameters of the considered classifiers. The linear SVM (Fig. 4.6, top-left plot) is almost always evaded with as few as 5–10 modifications, independent of the regularization parameter C. It is worth noting that attacking a linear classifier amounts to always incrementing the value of the same highest-weighted feature (corresponding to the `/Linearized` keyword in the majority of the cases) until it is bounded. This continues with the next highest-weighted non-bounded feature until termination. This occurs simply because the gradient of $g(\mathbf{x})$ does not depend on $\mathbf{x}$ for a linear classifier (see Sect. 4.4.2.1). With the RBF kernel (Fig. 4.6, top-right plot), SVMs exhibit a similar behavior with $C = 1$ and various values of its γ parameter,[5] and the RBF SVM provides a higher degree of security compared to linear SVMs (cf. top-left plot and middle-left plot in Fig. 4.6).

In the LK case, without mimicry (Fig. 4.6, middle row), classifiers are evaded with a probability only *slightly* lower than that found in the PK case, even when only $n_q = 100$ surrogate samples are used to learn the surrogate classifier. This aspect highlights the threat posed by a skilled adversary with incomplete knowledge: only a small set of samples may be required to successfully attack the target classifier using the proposed algorithm.

Results for $\lambda = 500$. When mimicry is used (Fig. 4.6, bottom row), the success of the evasion of linear SVMs (with $C = 1$) decreases both in the PK (e.g., compare the blue curve in the top-left plot with the solid blue curve in the bottom-left plot) and in the LK case (e.g., compare the blue curve in the middle-left plot with the dashed blue curve in the bottom-left plot). The reason is that the computed direction tends to lead to a slower descent; i.e., a less direct path that often requires more modifications to evade the classifier. In the nonlinear case (Fig. 4.6, bottom-right plot), instead, mimicking exhibits some beneficial aspects for the attacker, although the constraint on feature addition may make it difficult to properly mimic legitimate samples. In particular, note how the targeted SVMs with RBF kernel (with $C = 1$ and $\gamma = 1$) in the PK case (e.g., compare the blue curve in the top-right plot with the solid blue curve in the bottom-right plot) are evaded with a significantly higher probability than in the case when $\lambda = 0$. The reason is that a pure descent strategy on $g(\mathbf{x})$ may find local minima (i.e., attack samples) that do not evade detection, while the mimicry component biases the descent towards regions of the feature space more densely populated by legitimate samples, where $g(\mathbf{x})$ eventually attains lower values. In the LK case (e.g., compare the blue curve in the middle-right plot with the dashed blue curve in the bottom-right plot), however, mimicking does not exhibit significant improvements.

Analysis. Our attacks raise questions about the feasibility of detecting malicious PDFs solely based on logical structure. We found that `/Linearized`,

---

[5] We also conducted experiments using $C = 0.1$ and $C = 100$, but did not find significant differences compared to the presented results using $C = 1$.

/OpenAction, /Comment, /Root and /PageLayout were among the most commonly manipulated keywords. They indeed are found mainly in legitimate PDFs, but can be easily added to malicious PDFs by the versioning mechanism. The attacker can simply insert comments inside the malicious PDF file to augment its /Comment count. Similarly, he/she can embed *legitimate* OpenAction code to add /OpenAction keywords or he/she can add new pages to insert /PageLayout keywords.

In summary, our analysis shows that even detection systems that accurately classify non-malicious data can be significantly degraded with only a few malicious modifications. This aspect highlights the importance of developing detection systems that are accurate, but also *designed to be robust* against adversarial manipulation of attack instances.

4.4.4 Discussion

In this section, we proposed a simple algorithm that allows for evasion of SVMs with differentiable kernels, and, more generally, of any classifier with a differentiable discriminant function. We investigated the attack effectiveness in the case of perfect knowledge of the attacked system. Further, we empirically showed that SVMs can still be evaded with high probability even if the adversary can only learn a classifier's copy on surrogate data (limited knowledge). We believe that the proposed attack formulation can easily be extended to classifiers with non-differentiable discriminant functions as well, such as decision trees and k-nearest neighbors.

Our analysis also suggests some ideas for improving classifier security. In particular, when the classifier tightly *encloses* the legitimate samples, the adversary must increasingly mimic the legitimate class to evade (see Fig. 4.4), and this may not always be possible; e.g., malicious network packets or PDF files still need to embed a valid exploit, and some features may be immutable. Accordingly, a guideline for designing secure classifiers is that learning should encourage a tight enclosure of the legitimate class; e.g., by using a regularizer that penalizes classifying "blind spots"—regions with low $p(\mathbf{x})$—as legitimate. Generative classifiers can be modified, by explicitly modeling the attack distribution, as in [11], and discriminative classifiers can be modified similarly by adding generated attack samples to the training set. However, these security improvements may incur higher FP rates.

In the above applications, the feature representations were *invertible*; i.e., there is a direct mapping from the feature vectors $\mathbf{x}$ to a corresponding real-world sample (e.g., a spam email, or PDF file). However, some feature mappings cannot be trivially inverted; e.g., n-gram analysis [31]. In these cases, one may modify the real-world object corresponding to the initial attack point at each step of the gradient descent to obtain a sample in the feature space that as close as possible to the sample that would be obtained at the next attack iteration. A similar technique has already been exploited in to address the pre-image problem of kernel methods [14].

Other interesting extensions include (1) considering more effective strategies such as those proposed by [46,54] to build a small but representative set of surrogate data to learn the surrogate classifier and (2) improving the classifier estimate $\hat{g}(\mathbf{x})$; e.g. using an ensemble technique like bagging to average several classifiers [16].

4.5 Poisoning Attacks Against SVMs

In the previous section, we devised a simple algorithm that allows for evasion of classifiers at test time and showed experimentally how it can be exploited to evade detection by SVMs and kernel-based classification techniques. Here we present another kind of attack, based on our work in [14]. Its goal is to force the attacked SVM to misclassify as many samples as possible at test time through *poisoning* of the training data, that is, by injecting well-crafted attack samples into the training set. Note that, in this case, the test data is assumed not to be manipulated by the attacker.

Poisoning attacks are staged during classifier training, and they can thus target *adaptive* or *online* classifiers, as well as classifiers that are being re-trained on data collected during test time, especially if in an unsupervised or semi-supervised manner. Examples of these attacks, besides our work [14], can be found in [7,9,13,39,40,53,59]. They include specific application examples in different areas, such as intrusion detection in computer networks [7,39,40,59], spam filtering [7,53], and, most recently, even biometric authentication [9, 13].

In this section, we follow the same structure of Sect. 4.4. In Sect. 4.5.1, we define the adversary model according to our framework; then, in Sects. 4.5.1.4 and 4.5.2 we, respectively, derive the optimal poisoning attack and the corresponding algorithm; and, finally, in Sects. 4.5.3 and 4.5.4 we report our experimental findings and discuss the results.

4.5.1 *Modeling the Adversary*

Here, we apply our framework to evaluate security against poisoning attacks. As with the evasion attacks in Sect. 4.4.1, we model the attack scenario and derive the corresponding optimal attack strategy for poisoning.

Notation. In the following, we assume that an SVM has been trained on a dataset $\mathscr{D}_{\text{tr}} = \{\mathbf{x}_i, y_i\}_{i=1}^n$ with $\mathbf{x}_i \in \mathbb{R}^d$ and $y_i \in \{-1,+1\}$. The matrix of kernel values between two sets of points is denoted with $\mathbf{K}$, while $\mathbf{Q} = \mathbf{K} \circ yy^\top$ denotes its label-annotated version, and α denotes the SVM's dual variables corresponding to each training point. Depending on the value of α_i, the training points are referred to as margin support vectors ($0 < \alpha_i < C$, set $\mathscr{S}$), error support vectors ($\alpha_i = C$, set $\mathscr{E}$), or reserve vectors ($\alpha_i = 0$, set $\mathscr{R}$). In the sequel, the lowercase letters s, e, r are used to index the corresponding parts of vectors or matrices; e.g., $\mathbf{Q}_{ss}$ denotes the sub-matrix of $\mathbf{Q}$ corresponding to the margin support vectors.

4.5.1.1 Adversary's Goal

For a poisoning attack, the attacker's goal is to find a set of points whose addition to $\mathscr{D}_{\mathrm{tr}}$ maximally decreases the SVM's classification accuracy. For simplicity, we start considering the addition of a single attack point $(\mathbf{x}^*, y^*)$. The choice of its label y^* is arbitrary but fixed. We refer to the class of this chosen label as *attacking* class and the other as the *attacked* class.

4.5.1.2 Adversary's Knowledge

According to Sect. 4.3.1.2, we assume that the adversary knows the training samples (*k.i*), the feature representation (*k.ii*), that an SVM learning algorithm is used (*k.iii*) and the learned SVM's parameters (*k.iv*), since they can be inferred by the adversary by solving the SVM learning problem on the *known* training set. Finally, we assume that no feedback is exploited by the adversary (*k.v*).

These assumptions amount to considering a worst-case analysis that allows us to compute the maximum error rate that the adversary can inflict through poisoning. This is indeed useful to check whether and under what circumstances poisoning may be a relevant threat for SVMs.

Although having perfect knowledge of the training data is very difficult in practice for an adversary, collecting a surrogate dataset sampled from the same distribution may not be that complicated; for instance, in network intrusion detection an attacker may easily sniff network packets to build a surrogate learning model, which can then be poisoned under the perfect knowledge setting. The analysis of this limited knowledge poisoning scenario is, however, left to future work.

4.5.1.3 Adversary's Capability

According to Sect. 4.3.1.3, we assume that the attacker can manipulate only training data (*c.i*), can manipulate the class prior and the class-conditional distribution of the attack point's class y^* by essentially adding a number of attack points of that class into the training data, one at a time (*c.ii–iii*), and can alter the feature values of the attack sample within some lower and upper bounds (*c.iv*). In particular, we will constrain the attack point to lie within a box, that is $\mathbf{x}_{\mathrm{lb}} \leq \mathbf{x} \leq \mathbf{x}_{\mathrm{ub}}$.

4.5.1.4 Attack Strategy

Under the above assumptions, the optimal attack strategy amounts to solving the following optimization problem:

Algorithm 2 Poisoning attack against an SVM

Input: $\mathscr{D}_{\text{tr}}$, the training data; $\mathscr{D}_{\text{val}}$, the validation data; y^*, the class label of the attack point; $\mathbf{x}^0$, the initial attack point; t, the step size.
Output: $\mathbf{x}^*$, the final attack point.

$\{\alpha_i, b\} \leftarrow$ learn an SVM on $\mathscr{D}_{\text{tr}}$.
$p \leftarrow 0$.
repeat
  Re-compute the SVM solution on $\mathscr{D}_{\text{tr}} \cup \{\mathbf{x}^p, y^*\}$ using the incremental SVM [20]. This step requires $\{\alpha_i, b\}$. If computational complexity is manageable, a full SVM can be learned at each step, instead.
  Compute ∇P on $\mathscr{D}_{\text{val}}$ according to Equation (4.8).
  Normalize ∇P to have unit norm.
  $p \leftarrow p+1$ and $\mathbf{x}^p \leftarrow \mathbf{x}^{p-1} + t\nabla P$
  if $\mathbf{x}_{\text{lb}} > \mathbf{x}^p$ or $\mathbf{x}^p > \mathbf{x}_{\text{ub}}$ **then**
    Project $\mathbf{x}^p$ onto the boundary of the feasible region (enforce application-specific constraints, if any).
  end if
until $P(\mathbf{x}^p) - P\left(\mathbf{x}^{p-1}\right) < \varepsilon$
return: $\mathbf{x}^* = \mathbf{x}^p$

$$\mathbf{x}^* = \text{argmax}_{\mathbf{x}} \;\; P(\mathbf{x}) = \textstyle\sum_{k=1}^{m}(1 - y_k f_{\mathbf{x}}(\mathbf{x}_k))_+ = \displaystyle\sum_{k=1}^{m}(-g_k)_+ \tag{4.2}$$

$$\text{s.t.} \;\; \mathbf{x}_{\text{lb}} \leq \mathbf{x} \leq \mathbf{x}_{\text{ub}} \;\; , \tag{4.3}$$

where the hinge loss has to be maximized on a separate validation set $\mathscr{D}_{\text{val}} = \{\mathbf{x}_k, y_k\}_{k=1}^m$ to avoid considering a further regularization term in the objective function. The reason is that the attacker aims to maximize the SVM *generalization error* and not only its empirical estimate on the training data.

4.5.2 *Poisoning Attack Algorithm*

In this section, we assume the role of the attacker and develop a method for optimizing $\mathbf{x}^*$ according to Eq. (4.2). Since the objective function is nonlinear, we use a gradient-ascent algorithm, where the attack vector is initialized by cloning an arbitrary point from the attacked class and flipping its label. This initialized attack point (at iteration 0) is denoted by $\mathbf{x}^0$. In principle, $\mathbf{x}^0$ can be any point *sufficiently deep* within the attacking class's margin. However, if this point is too close to the boundary of the attacking class, the iteratively adjusted attack point may become a reserve point, which halts further progress.

The computation of the gradient of the validation error crucially depends on the assumption that the structure of the sets $\mathscr{S}$, $\mathscr{E}$, and $\mathscr{R}$ does not change during the update. In general, it is difficult to determine the largest step t along the gradient direction ∇P, which preserves this structure. Hence, the step t is fixed to a small constant value in our algorithm. After each update of the attack point $\mathbf{x}^p$, the optimal solution can be efficiently recomputed from the solution on $\mathscr{D}_{\text{tr}}$, using the incremental SVM machinery [20]. The algorithm terminates when the change in the validation error is smaller than a predefined threshold.

4.5.2.1 Gradient Computation

We now discuss how to compute the gradient ∇P of our objective function. For notational convenience, we now refer to the attack point as $\mathbf{x}_c$ instead of $\mathbf{x}$.

First, we explicitly account for all terms in the margin conditions g_k that are affected by the attack point $\mathbf{x}_c$:

$$\begin{aligned} g_k &= \sum_j \mathbf{Q}_{kj}\alpha_j + y_k b - 1 \\ &= \sum_{j\neq c} \mathbf{Q}_{kj}\alpha_j(\mathbf{x}_c) + \mathbf{Q}_{kc}(\mathbf{x}_c)\alpha_c(\mathbf{x}_c) + y_k b(\mathbf{x}_c) - 1 \ . \end{aligned} \tag{4.4}$$

As already mentioned, $P(\mathbf{x}_c)$ is a non-convex objective function, and we thus exploit a gradient-ascent technique to iteratively optimize it. We denote the initial location of the attack point as $\mathbf{x}_c^0$. Our goal is to update the attack point as $\mathbf{x}_c^p = \mathbf{x}_c^{(p-1)} + t\nabla P$ where p is the current iteration, ∇P is a unit vector representing the attack direction (i.e., the normalized objective gradient), and t is the step size. To maximize our objective, the attack direction ∇P is computed at each iteration.

Although the hinge loss is not everywhere differentiable, this can be overcome by only considering point indices k with nonzero contributions to P; i.e., those for which $-g_k > 0$. Contributions of such points to ∇P can be computed by differentiating Eq. (4.4) with respect to $\mathbf{x}_c$ using the product rule:

$$\frac{\partial g_k}{\partial \mathbf{x}_c} = \mathbf{Q}_{ks}\frac{\partial \alpha}{\partial \mathbf{x}_c} + \frac{\partial \mathbf{Q}_{kc}}{\partial \mathbf{x}_c}\alpha_c + y_k\frac{\partial b}{\partial \mathbf{x}_c}, \tag{4.5}$$

where, by denoting the lth feature of $\mathbf{x}_c$ as x_{cl}, we use the notation

$$\frac{\partial \alpha}{\partial \mathbf{x}_c} = \begin{bmatrix} \frac{\partial \alpha_1}{\partial x_{c1}} & \cdots & \frac{\partial \alpha_1}{\partial x_{cd}} \\ \vdots & \ddots & \vdots \\ \frac{\partial \alpha_s}{\partial x_{c1}} & \cdots & \frac{\partial \alpha_s}{\partial x_{cd}} \end{bmatrix}, \text{ simil. } \frac{\partial \mathbf{Q}_{kc}}{\partial \mathbf{x}_c}, \frac{\partial b}{\partial \mathbf{x}_c} \ .$$

The expressions for the gradient can be further refined using the fact that the gradient step must preserve the optimal SVM solution. This can expressed as an adiabatic update condition using the technique introduced in [20]. In particular, for the ith *training* point, the KKT conditions of the optimal SVM solution are:

$$g_i = \sum_{j \in \mathscr{D}_{\text{tr}}} \mathbf{Q}_{ij}\alpha_j + y_i b - 1 \begin{cases} > 0;\ i \in \mathscr{R} \\ = 0;\ i \in \mathscr{S} \\ < 0;\ i \in \mathscr{E} \end{cases} \tag{4.6}$$

$$h = \sum_{j \in \mathscr{D}_{\text{tr}}} (y_j \alpha_j) = 0 \ . \tag{4.7}$$

The form of these conditions implies that an infinitesimal change in the attack point x_c causes a smooth change in the optimal solution of the SVM, under the restriction that the composition of the sets $\mathscr{S}$, $\mathscr{E}$, and $\mathscr{R}$ remains intact. This equilibrium allows us to predict the *response* of the SVM solution to the variation of x_c, as shown below.

By differentiation of the x_c-dependent terms in Eqs. (4.6) and (4.7) with respect to each feature x_{cl} $(1 \leq l \leq d)$, we obtain, for any $i \in \mathscr{S}$,

$$\frac{\partial g}{\partial x_{cl}} = \mathbf{Q}_{ss} \frac{\partial \alpha}{\partial x_{cl}} + \frac{\partial \mathbf{Q}_{sc}}{\partial x_{cl}} \alpha_c + y_s \frac{\partial b}{\partial x_{cl}} = 0$$

$$\frac{\partial h}{\partial x_{cl}} = y_s^\top \frac{\partial \alpha}{\partial x_{cl}} = 0 \ .$$

Solving these equations and computing an inverse matrix via the Sherman–Morrison–Woodbury formula [48] yields the following gradients:

$$\frac{\partial \alpha}{\partial \mathbf{x}_c} = -\frac{1}{\zeta} \alpha_c (\zeta \mathbf{Q}_{ss}^{-1} - \upsilon \upsilon^\top) \cdot \frac{\partial \mathbf{Q}_{sc}}{\partial \mathbf{x}_c}$$

$$\frac{\partial b}{\partial \mathbf{x}_c} = -\frac{1}{\zeta} \alpha_c \upsilon^\top \cdot \frac{\partial \mathbf{Q}_{sc}}{\partial \mathbf{x}_c} \ ,$$

where $\upsilon = \mathbf{Q}_{ss}^{-1} y_s$ and $\zeta = y_s^\top \mathbf{Q}_{ss}^{-1} y_s$. We thus obtain the following gradient of the objective used for optimizing our attack, which only depends on $\mathbf{x}_c$ through gradients of the kernel matrix, $\frac{\partial \mathbf{Q}_{kc}}{\partial \mathbf{x}_c}$:

$$\nabla P = \sum_{k=1}^{m} \left\{ M_k \frac{\partial \mathbf{Q}_{sc}}{\partial \mathbf{x}_c} + \frac{\partial \mathbf{Q}_{kc}}{\partial \mathbf{x}_c} \right\} \alpha_c \ , \tag{4.8}$$

where $M_k = -\frac{1}{\zeta}(\mathbf{Q}_{ks}(\zeta \mathbf{Q}_{ss}^{-1} - \upsilon \upsilon^T) + y_k \upsilon^T)$.

4.5.2.2 Kernelization

From Eq. (4.8), we see that the gradient of the objective function at iteration p may depend on the attack point $\mathbf{x}_c^p = \mathbf{x}_c^{p-1} + t\nabla P$ only through the gradients of the matrix $\mathbf{Q}$. In particular, this depends on the chosen kernel. We report below the expressions of these gradients for three common kernels (see Sect. 4.4.2.1):

- Linear kernel: $\frac{\partial K_{ic}}{\partial \mathbf{x}_c} = \frac{\partial(\mathbf{x}_i \cdot \mathbf{x}_c)}{\partial \mathbf{x}_c} = \mathbf{x}_i$
- Polynomial kernel: $\frac{\partial K_{ic}}{\partial \mathbf{x}_c} = \frac{\partial(\mathbf{x}_i \cdot \mathbf{x}_c + R)^d}{\partial \mathbf{x}_c} = d(\mathbf{x}_i \cdot \mathbf{x}_c + R)^{d-1}\mathbf{x}_i$
- RBF kernel: $\frac{\partial K_{ic}}{\partial \mathbf{x}_c} = \frac{\partial e^{-\gamma\|\mathbf{x}_i - \mathbf{x}_c\|^2}}{\partial \mathbf{x}_c} = 2\gamma K(\mathbf{x}_i, \mathbf{x}_c)(\mathbf{x}_i - \mathbf{x}_c)$

The dependence of these gradients on the current attack point, $\mathbf{x}_c$, can be avoided by using the previous attack point, provided that t is sufficiently small. This approximation enables a straightforward extension of our method to arbitrary differentiable kernels.

4.5.3 *Experiments*

The experimental evaluation presented in the following sections demonstrates the behavior of our proposed method on an artificial two-dimensional dataset and evaluates its effectiveness on the classical MNIST handwritten digit recognition dataset [32, 45].

4.5.3.1 Two-Dimensional Toy Example

Here we consider a two-dimensional example in which each class follows a Gaussian with mean and covariance matrices given by $\mu_- = [-1.5, 0]$, $\mu_+ = [1.5, 0]$, $\Sigma_- = \Sigma_+ = 0.6I$. The points from the *negative* distribution have label -1 (shown as red in subsequent figures) and otherwise $+1$ (shown as blue). The training and the validation sets, $\mathscr{D}_{\rm tr}$ and $\mathscr{D}_{\rm val}$, consist of 25 and 500 points per class, respectively.

In the experiment presented below, the red class is the *attacking* class. That is, a random point of the blue class is selected and its label is flipped to serve as the starting point for our method. Our gradient-ascent method is then used to refine this attack until its termination condition is satisfied. The attack's trajectory is traced as the black line in Fig. 4.7 for both the linear kernel (upper two plots) and the RBF kernel (lower two plots). The background of each plot depicts an error surface: hinge loss computed on a validation set (leftmost plots) and the classification error (rightmost plots). For the linear kernel, the range of attack points is limited to the box $\mathbf{x} \in [-4, 4]^2$ shown as a dashed line. This implements the constraint of Eq. (4.3).

For both kernels, these plots show that our gradient-ascent algorithm finds a reasonably good local maximum of the non-convex error surface. For the linear kernel, it terminates at the corner of the bounded region, since the error surface is unbounded. For the RBF kernel, it also finds a good local maximum of the hinge loss which, incidentally, is the maximum classification error within this area of interest.

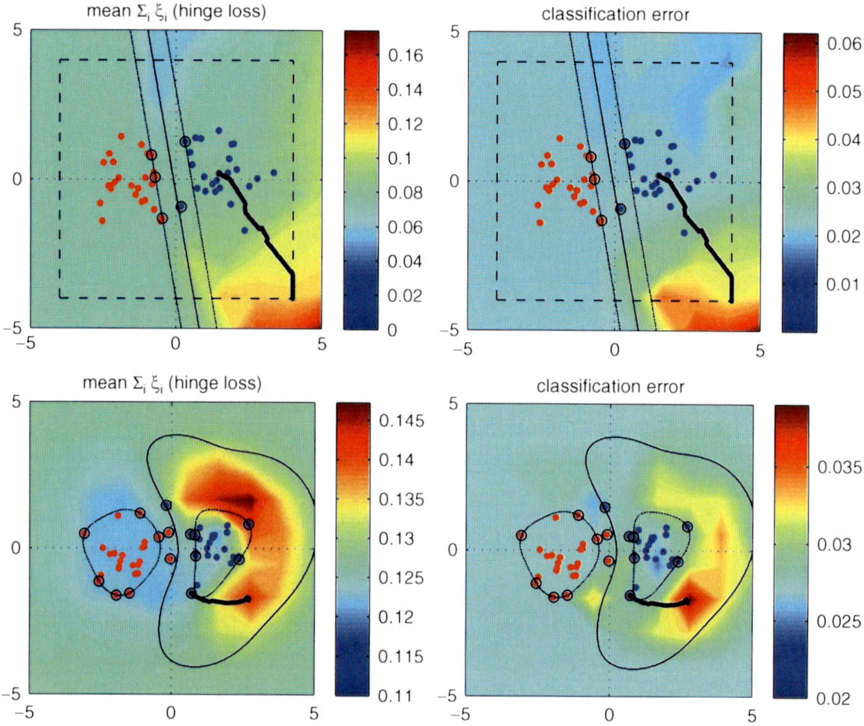

Fig. 4.7 Behavior of the gradient-based attack strategy on the Gaussian datasets, for the linear (*top row*) and the RBF kernel (*bottom row*) with $\gamma = 0.5$. The regularization parameter C was set to 1 in both cases. The *solid black line* represents the gradual shift of the attack point $\mathbf{x}_c^p$ toward a local maximum. The hinge loss and the classification error are shown in colors, to appreciate that the hinge loss provides a good approximation of the classification error. The value of such functions for each point $\mathbf{x} \in [-5,5]^2$ is computed by learning an SVM on $\mathscr{D}_{tr} \cup \{\mathbf{x}, y = -1\}$ and evaluating its performance on $\mathscr{D}_{val}$. The SVM solution on the clean data $\mathscr{D}_{tr}$ and the training data itself, are reported for completeness, highlighting the support vectors (with *black circles*), the decision hyperplane and the margin bounds (with *black lines*)

4.5.3.2 Handwritten Digits

We now quantitatively validate the effectiveness of the proposed attack strategy on the MNIST handwritten digit classification task [32, 45], as with the evasion attacks in Sect. 4.4.3. In particular, we focus here on the following two-class sub-problems: 7 vs. 1; 9 vs. 8; 4 vs. 0. Each digit is normalized as described in Sect. 4.4.3.1. We consider again a linear SVM with $C = 1$. We randomly sample a training and a validation data of 100 and 500 samples, respectively, and retain the complete testing data given by MNIST for $\mathscr{D}_{ts}$. Although it varies for each digit, the size of the testing data is about 2,000 samples per class (digit).

Figure 4.8 shows the effect of *single* attack points being optimized by our descent method. The leftmost plots of each row show the example of the attacked class used

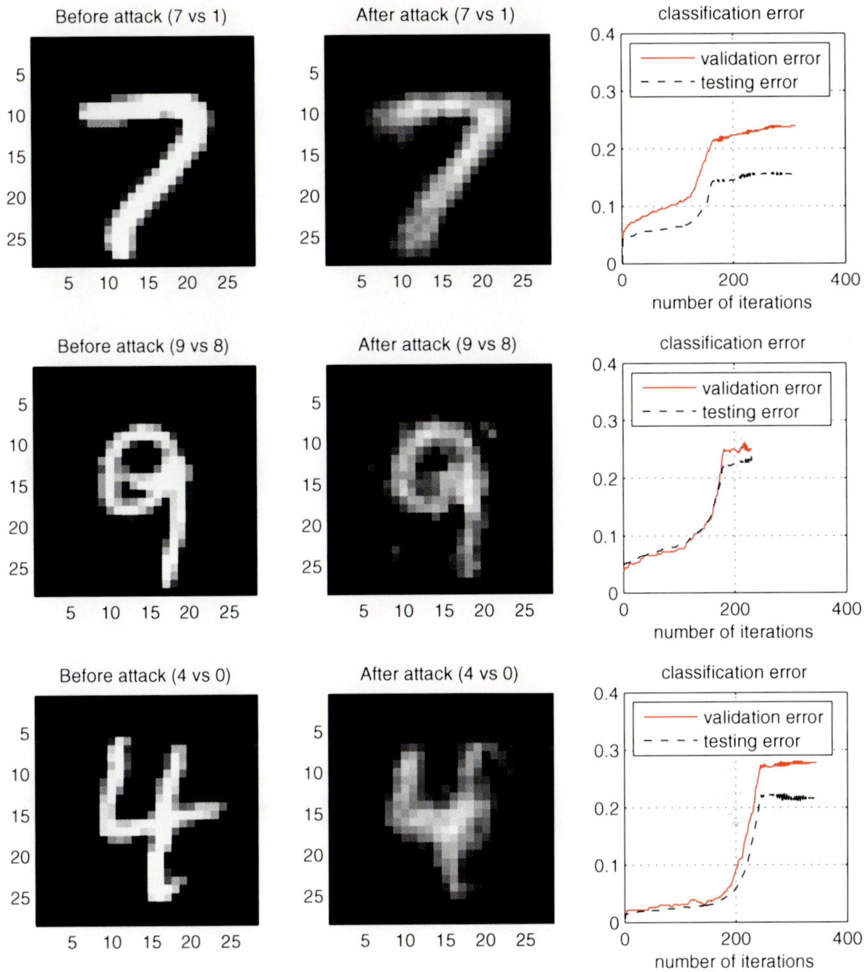

Fig. 4.8 Modifications to the initial (mislabeled) attack point performed by the proposed attack strategy, for the three considered two-class problems from the MNIST dataset. The increase in validation and testing errors across different iterations is also reported

as starting points in our algorithm. The middle plots show the final attack point. The rightmost plots depict the increase in the validation and testing errors as the attack progresses. For this experiment we run the attack algorithm five times by re-initializing the gradient-ascent procedure, and we retain the best result.

Visualizing the attack points reveals that these attacks succeed by blurring the initial prototype to appear more like examples of the attacking class. In comparing the initial and final attack points, we see that the bottom segment of the 7 straightens to resemble a 1, the lower segment of the 9 is rounded to mimicking an 8, and *ovular*

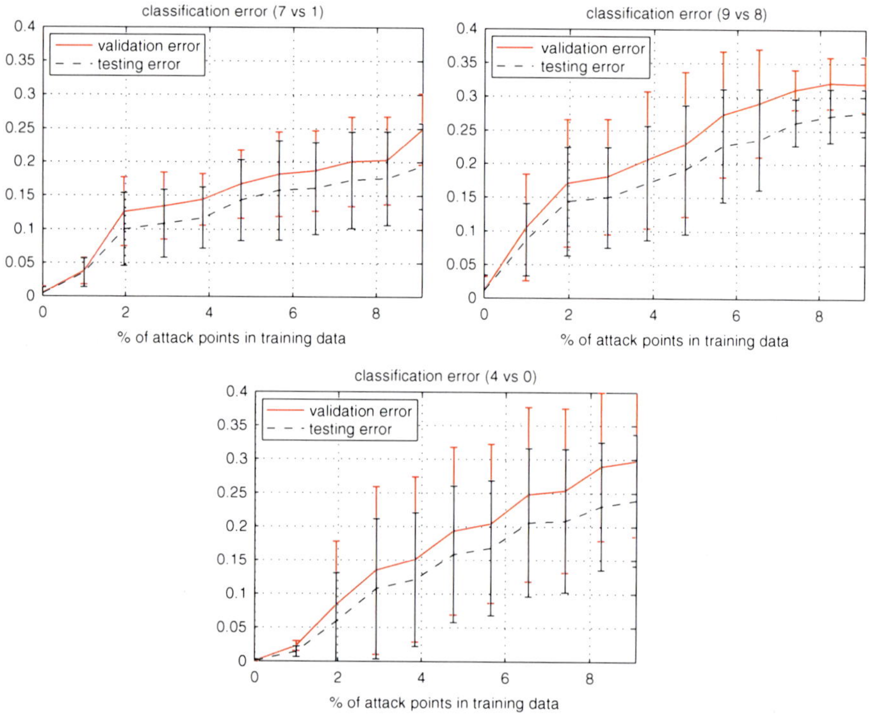

Fig. 4.9 Results of the multi-point, multi-run experiments on the MNIST dataset. In each plot, we show the classification errors due to poisoning as a function of the percentage of training contamination for both the validation (*red solid line*) and testing sets (*black dashed line*). The *top-left plot* is for the 7 vs.1 task, the *top-right plot* is for the 9 vs. 8 task, and the *bottom-middle plot* is for the 4 vs. 0 task

noise is added to the outer boundary of the 4 to make it similar to a 0. These blurred images are thus consistent with one's natural notion of visually confusing digits.

The rightmost plots further demonstrate a striking increase in error over the course of the attack. In general, the validation error overestimates the classification error due to a smaller sample size. Nonetheless, in the exemplary runs reported in this experiment, a *single* attack data point caused the classification error to rise from initial error rates of 2–5 % to 15–20 %. Since our initial attack points are obtained by flipping the label of a point in the attacked class, the errors in the first iteration of the rightmost plots of Fig. 4.8 are caused by single random label flips. This confirms that our attack can achieve significantly higher error rates than random label flips and underscores the vulnerability of the SVM to poisoning attacks.

The latter point is further illustrated in a multiple point, multiple run experiment presented in Fig. 4.9. For this experiment, the attack was extended by repeatedly injecting attack points into the same class and averaging results over multiple runs on randomly chosen training and validation sets of the same size (100 and 500 samples, respectively). These results exhibit a steady rise in classification error as the percentage of attack points in the training set increases. The variance of the error is quite high, which can be attributed to the relatively small sizes of the training and validation sets. Also note that, in this experiment, to reach an error rate of 15–20 %, the adversary needs to control at least 4–6 % of the training data, unlike in the single point attacks of Fig. 4.8. This is because Fig. 4.8 displays the best single-point attack from five restarts whereas here initial points are selected without restarts.

4.5.4 Discussion

The poisoning attack presented in this section, summarized from our previous work in [14], is a first step toward the security analysis of SVM against training data attacks. Although our gradient-ascent method is not optimal, it attains a surprisingly large impact on the SVM's classification accuracy.

Several potential improvements to the presented method remain to be explored in future work. For instance, one may investigate the effectiveness of such an attack with surrogate data, that is, when the training data is not known to the adversary, who may, however, collect samples drawn from the same distribution to learn a classifier's copy (similarly to the limited knowledge case considered in the evasion attacks of Sect. 4.4). Another improvement may be to consider the simultaneous optimization of multi-point attacks, although we have already demonstrated that greedy, sequential single-point attacks may be rather successful.

An interesting analysis of the SVM's vulnerability to poisoning suggested from this work is to consider the attack's impact under loss functions other than the hinge loss. It would be especially interesting to analyze *bounded* loss functions, like the ramp loss, since such losses are designed to limit the impact of any single (attack) point on the outcome. On the other hand, while these losses may lead to improved security to poisoning, they also make the SVM's optimization problem non-convex, and, thus, more computationally demanding. This may be viewed as another trade-off between computational complexity of the learning algorithm and security.

An important practical limitation of the proposed method is the assumption that the attacker controls the labels of injected points. Such assumptions may not hold if the labels are assigned by trusted sources such as humans, e.g., anti-spam filters use their users' labeling of messages as ground truth. Thus, although an attacker can send arbitrary messages, he/she cannot guarantee that they will have the labels necessary for his/her attack. This imposes an additional requirement that the attack data must satisfy certain side constraints to fool the labeling oracle. Further work is needed to understand and incorporate these potential side constraints into attacks.

4.6 Privacy Attacks Against SVMs

We now consider a third scenario in which the attacker's goal is to affect a breach of the training data's confidentiality. We review our recent work [60] deriving mechanisms for releasing SVM classifiers trained on privacy-sensitive data while maintaining the data's privacy. Unlike previous sections, our focus here lies primary on the study of countermeasures, while we only briefly consider attacks in the context of lower bounds. We adopt the formal framework of Dwork et al. [30], in which a randomized mechanism is said to preserve *β-differential privacy*, if the likelihood of the mechanism's output changes by at most β when a training datum is changed arbitrarily (or even removed). The power of this framework, which gained near-universal favor after its introduction, is that it quantifies privacy in a rigorous way and provides strong guarantees even against powerful adversaries with knowledge of almost all of the training data, knowledge of the mechanism (barring its source of randomness), arbitrary access to the classifier output by the mechanism, and the ability to manipulate almost all training data prior to learning.

This section is organized as follows. In Sect. 4.6.1 we outline our model of the adversary, which makes only weak assumptions. Section 4.6.2 provides background on differential privacy, presents a mechanism for training and releasing privacy-preserving SVMs—essentially a countermeasure to many privacy attacks—and provides guarantees on differential privacy and also utility (e.g., controlling the classifier's accuracy). We then briefly touch on existing approaches for evaluation via lower bounds and discuss other work and open problems in Sect. 4.6.3.

4.6.1 Modeling the Adversary

We first apply our framework to define the threat model for defending against privacy attacks within the broader context of differential privacy. We then focus on specific countermeasures in the form of modifications to SVM learning that provide differential privacy.

4.6.1.1 Adversary's Goal

The ultimate goal of the attacker in this section is to determine features and/or the label of an individual training datum. The overall approach of the adversary towards this goal is to inspect (arbitrary numbers of) test-time classifications made by a released classifier trained on the data, or by inspecting the classifier directly. The definition of differential privacy, and the particular mechanisms derived here, can be modified for related goals of determining properties of several training data; we focus on the above conventional case without loss of generality.

4.6.1.2 Adversary's Knowledge

As alluded to above, we endow our adversary with significant knowledge of the learning system, so as to derive countermeasures that can withstand very strong attacks. Indeed the notion of differential privacy, as opposed to more syntactic notions of privacy such as *k-anonymity* [65], was inspired by decades-old work in cryptography that introduced mathematical formalism to an age-old problem, yielding significant practical success. Specifically, we consider a scenario in which the adversary has complete knowledge of the raw input feature representation, the learning algorithm (the entire mechanism including the form of randomization it introduces, although not the source of randomness) and the form of its decision function (in this case, a thresholded SVM), the learned classifier's parameters (the kernel/feature mapping, primal weight vector, and bias term), and arbitrary/unlimited feedback from the deployed classifier (*k.ii–v*). We grant the attacker near complete knowledge of the training set (*k.i*): the attacker may have complete knowledge of all but one training datum, for which he/she has no knowledge of input feature values or its training label, and it is these attributes he/she wishes to reveal. For simplicity of exposition, but without loss of generality, we assume this to be the last datum in the training sample.

4.6.1.3 Adversary's Capability

Like our assumptions on the attacker's knowledge, we impose weak limitations on the adversary's capability. We assume an adversary that can manipulate both training and test data (*c.i*), although the latter is subsumed by the attacker's complete knowledge of the decision function and learned parameters—e.g., he/she may implement his/her own classifier and execute it arbitrarily, or he/she may submit or manipulate test points presented to a deployed classifier.

Our attack model makes no assumptions about the origins of the training or test data. The data need not be sampled independently or even according to a distribution—the definition of differential privacy provided below makes worst-case assumptions about the training data, and again the test data could be arbitrary. Thus the adversary may have arbitrary capability to modify class priors, training data features and labels (*c.ii–iv*) except that the adversary attacking the system may not directly modify the targeted training datum because he/she does not have knowledge of it. That said, however, differential privacy makes worst-case (no distributional) assumptions about the datum and thus one could consider even this data point as being adversarially manipulated by nature (i.e., nature does not collude with the attacker to share information about the target training datum, but that may collude to facilitate a privacy breach by selecting a "convenient" target datum).

4.6.1.4 Attack Strategy

While no practical privacy attacks on SVMs have been explored in the past—an open problem discussed in Sect. 4.6.3—a general approach would be to approximate the inversion of the learning map on the released SVM parametrization (either primal weight vector, or dual variables) around the known portion of the training data. In practice this could be achieved by taking a similar approach as done in Sect. 4.5 whereby an initial guess of a missing training point is iterated on by taking gradient steps of the differential in the SVM parameter vector with respect to the missing training datum. An interpretation of this approach is one of the simulations: to guess a missing training datum, given access to the remainder of the training set and the SVM solution on all the data, simulate the SVM on guesses for the missing datum, updating the guesses in directions that appropriately shift the intermediate solutions. As we discuss briefly in the sequel, theoretical lower bounds on achievable privacy relate to attacks in pathological cases.

4.6.2 Countermeasures with Provable Guarantees

Given an adversary with such strong knowledge and capabilities as described above, it may seem difficult to provide effective countermeasures particularly considering the complication of abundant access to side information that is often used in publicized privacy attacks [52, 65]. However, the crux that makes privacy-preservation under these conditions possible lies in the fact that the learned quantity being released is an aggregate statistic of the sensitive data; intuitively the more data being aggregated, the less sensitive a statistic should be to changes or removal of any single datum. We now present results from our recent work that quantifies this effect [60], within the framework of *differential privacy*.

4.6.2.1 Background on Differential Privacy

We begin by recalling the key definition due to Dwork et al. [30]. First, for any training set $\mathscr{D} = \{(\mathbf{x}_i, y_i)\}_{i=1}^n$ denote set $\mathscr{D}'$ to be a *neighbor* of $\mathscr{D}$ (or $\mathscr{D}' \sim \mathscr{D}$) if $\mathscr{D}' = \{(\mathbf{x}_i, y_i)\}_{i=1}^{n-1} \cup \{(\mathbf{x}'_n, y'_n)\}$ where $(\mathbf{x}_n, y_n) \neq (\mathbf{x}'_n, y'_n)$. In the present context, differential privacy is a desirable property of *learning maps*, which maps a training set $\{(\mathbf{x}_i, y_i)\}_{i=1}^n$ to a continuous discriminant function of the form $g : \mathscr{X} \to \mathbb{R}$—here a learned SVM—in some space of functions, $\mathscr{H}$. We say that a *randomized*[6] learning map $\mathscr{L}$ preserves β-differential privacy if for all datasets

[6]That is, the learning map's output is not a deterministic function of the training data. The probability in the definition of differential privacy is due to this randomness. Our treatment here is only as complex as necessary, but to be completely general, the events in the definition should be on measurable sets $G \subset \mathscr{H}$ rather than individual $g \in \mathscr{H}$.

$\mathscr{D}$, all neighboring sets $\mathscr{D}'$ of $\mathscr{D}$, and all possible functions $g \in \mathscr{H}$, the following relation holds

$$\Pr(\mathscr{L}(\mathscr{D}) = g) \leq \exp(\beta) \Pr(\mathscr{L}(\mathscr{D}') = g) \ .$$

Intuitively, if we initially fix a training set and neighboring training set, differential privacy simply says that the two resulting distributions induced on the learned functions are point-wise close—and closer for smaller β. For a patient deciding whether to submit his/her datum to a training set for a cancer detector, differential privacy means that the learned classifier will reveal little information about that datum. Even an adversary with access to the inner-workings of the learner, with access to all other patients' data, and with the ability to guess-and-simulate the learning process repeatedly with various possible values of his/her datum, cannot reverse engineer his/her datum from the classifier released by the hospital because the adversary cannot distinguish the classifier distribution on one training set, from that on neighboring sets. Moreover, variations of this definition (which do not significantly affect the presented results) allow for neighboring databases to be defined as those missing a datum; or having several varying data, not just a single one.

For simplicity of exposition, we drop the explicit bias term b from our SVM learning process and instead assume that the data feature vectors are augmented with a unit constant, and that the resulting additional normal weight component corresponds to the bias. This is an equivalent SVM formulation that allows us to focus only on the normal's weight vector.

A classic route to establish differential privacy is to define a randomized map $\mathscr{L}$ that returns the value of a deterministic, nonrandom $\hat{\mathscr{L}}$ plus a noise term. Typically, we use an exponential family in a term that matches an available Lipschitz condition satisfied by $\hat{\mathscr{L}}$: in our case, for learning maps that return weight vectors in $\mathbb{R}^d$, we aim to measure *global sensitivity* of $\hat{\mathscr{L}}$ via the L_1 norm as

$$\Delta(\hat{\mathscr{L}}) = \max_{\mathscr{D}, \mathscr{D}' \sim \mathscr{D}} \left\| \hat{\mathscr{L}}(\mathscr{D}) - \hat{\mathscr{L}}(\mathscr{D}') \right\|_1 \ .$$

With respect to this sensitivity, we can easily prove that the randomized mechanism

$$\mathscr{L}(\mathscr{D}) = \hat{\mathscr{L}}(\mathscr{D}) + Laplace(0, \Delta(\hat{\mathscr{L}})/\beta) \ ,$$

is β-differential private.[7] The well-established proof technique [30] follows from the definition of the Laplace distribution involving the same norm as used in our measure of global sensitivity, and the triangle inequality: for any training set $\mathscr{D}$, $\mathscr{D}' \sim \mathscr{D}$, response $g \in \mathscr{H}$, and privacy parameter β

[7]Recall that the zero-mean multi-variate Laplace distribution with scale parameter s has density proportional to $\exp(-\|\mathbf{x}\|_1/s)$.

$$
\begin{aligned}
\frac{\Pr(\mathscr{L}(\mathscr{D})=g)}{\Pr(\mathscr{L}(\mathscr{D}')=g)} &= \frac{\exp\left(\left\|\hat{\mathscr{L}}(\mathscr{D}')-g\right\|_1 \beta/\Delta(\hat{\mathscr{L}})\right)}{\exp\left(\left\|\hat{\mathscr{L}}(\mathscr{D})-g\right\|_1 \beta/\Delta(\hat{\mathscr{L}})\right)} \\
&\leq \exp\left(\left\|\hat{\mathscr{L}}(\mathscr{D}')-\hat{\mathscr{L}}(\mathscr{D})\right\|_1 \beta/\Delta(\hat{\mathscr{L}})\right) \\
&\leq \exp(\beta).
\end{aligned}
$$

We take the above route to develop a differentially private SVM. As such, the onus is on calculating the SVM's global sensitivity, $\Delta(\hat{\mathscr{L}})$.

4.6.2.2 Global Sensitivity of Linear SVM

Unlike much prior work applying the "Laplace mechanism" to achieving differential privacy, in which studied estimators are often decomposed as linear functions of data [15], measuring the sensitivity of the SVM appears to be nontrivial owing to the nonlinear influence an individual training datum may have on the learned SVM. However, perturbations of the training data were studied by the learning-theory community in the context of *algorithmic stability*: there the goal is to establish bounds on classifier risk, from stability of the learning map, as opposed to leveraging combinatorial properties of the hypothesis class (e.g., the VC dimension, which is not always possible to control, and for the RBF kernel SVM is infinite) [62]. In recent work [60], we showed how these existing stability measurements for the SVM can be adapted to provide the following L_1-global sensitivity bound.

Lemma 1. *Consider SVM learning with a kernel corresponding to linear SVM in a feature space with finite-dimension F and L_2-norm bounded*[8] *by κ, with hinge loss (as used throughout this chapter), and chosen parameter $C > 0$. Then the L_1 global sensitivity of the resulting normal weight vector is upper-bounded by $4C\kappa\sqrt{F}$.*

We omit the proof, which is available in the original paper [60] and which follows closely the previous measurements for algorithmic stability. We note that the result extends to any convex Lipschitz loss.

4.6.2.3 Differentially Private SVMs

So far we have established that Algorithm 3, which learns an SVM and returns the resulting weight vector with added Laplace noise, preserves β-differential privacy. More noise is added to the weight vector when either (1) a higher degree of privacy is desired (smaller β), (2) the SVM fits closer to the data (higher C) or (3) the data

---

[8] That is $\forall \mathbf{x}$, $k(\mathbf{x},\mathbf{x}) \leq \kappa^2$; e.g. for the RBF the norm is uniformly unity $\kappa = 1$; more generally, we can make the standard assumption that the data lies within some κ L_2-ball.

Algorithm 3 Privacy-preserving SVM

Input: $\mathscr{D}$ the training data; $C > 0$ soft-margin parameter; kernel k inducing a feature space with finite dimension F and κ-bounded L_2-norm; privacy parameter $\beta > 0$.
Output: learned weight vector $\mathbf{w}$.

$\hat{\mathbf{w}} \leftarrow$ learn an SVM with parameter C and kernel k on data $\mathscr{D}$.
$\mu \leftarrow$ draw i.i.d. sample of F scalars from *Laplace* $\left(0, \frac{4C\kappa\sqrt{F}}{\beta}\right)$.
return: $\mathbf{w} = \hat{\mathbf{w}} + \mu$

is more distinguishable (higher κ or F—the curse of dimensionality). Hidden in the above is the dependence on n: typically we take C to scale like $1/n$ to achieve consistency in which case we see that noise decreases with larger training data—akin to less individual influence—as expected [60].

Problematic in the above approach is the destruction to utility due to preserving differential privacy. One approach to quantifying this effect involves bounding the following notion of utility [60]. We say a privacy-preserving learning map $\mathscr{L}$ has (ε, δ)-utility with respect to non-private map $\hat{\mathscr{L}}$ if for all training sets $\mathscr{D}$,

$$\Pr\left(\left\|\mathscr{L}(\mathscr{D}) - \hat{\mathscr{L}}(\mathscr{D})\right\|_\infty \leq \varepsilon\right) \geq 1 - \delta .$$

The norm here is in the function space of continuous discriminators, learned by the learning maps, and is the point-wise L_∞ norm which corresponds to $\|g\|_\infty = \sup_{\mathbf{x}} |g(\mathbf{x})|$—although for technical reasons we will restrict the supremum to be over a set $\mathscr{M}$ to be specified later. Intuitively, this indicates that the continuous predictions of the learned private classifier are close to those predictions of the learned non-private classifier, for all test points in $\mathscr{M}$, with high probability (again, in the randomness due to the private mechanism). This definition draws parallels with PAC learnability, and in certain scenarios is strictly stronger than requiring that the private learner achieves good risk (i.e., PAC learns) [60]. Using the Chernoff tail inequality and known moment-generating functions, we establish the following bound on the utility of this private SVM [60].

Theorem 1. *The β-differentially private SVM of Algorithm 3 achieves (ε, δ)-utility with respect to the non-private SVM run with the same C parameter and kernel, for $0 < \delta < 1$ and*

$$\varepsilon \geq 8C\kappa\Phi\sqrt{F}\left(F + \log\frac{1}{\delta}\right)/\beta ,$$

where the set $\mathscr{M}$ supporting the supremum in the definition of utility is taken to be the pre-image of the feature mapping on the L_∞ ball of radius $\Phi > 0$.[9]

[9] Above we previously bounded the L_2 norms of points in features space by κ, the additional bound on the L_∞ norm here is for convenience and is standard practice in learning-theoretic results.

As expected, the more confidence δ or privacy β required, the less accuracy is attainable. Similarly, when the training data is fitted more tightly via higher C, or when the data is less tightly packed for higher κ, Φ, F, less accuracy is possible. Note that like the privacy result, this result can hold for quite general loss functions.

4.6.3 Discussion

In this section, we have provided a summary of our recent results on strong countermeasures to privacy attacks on the SVM. We have shown how, through controlled addition of noise, SVM learning in finite-dimensional feature spaces can provide both privacy and utility guarantees. These results can be extended to certain translation-invariant kernels including the infinite-dimensional RBF [60]. This extension borrows a technique from large-scale learning where finding a dual solution of the SVM for large training data size n is infeasible. Instead, a primal SVM problem is solved using a random kernel that uniformly approximates the desired kernel. Since the approximating kernel induces a feature mapping of relatively small, finite dimensions, the primal solution becomes feasible. For privacy preservation, we use the same primal approach but with this new kernel. Fortunately, the distribution of the approximating kernel is independent of the training data, and thus we can reveal the approximating kernel without sacrificing privacy. Likewise the uniform approximation of the kernel composes with the utility result here to yield an analogous utility guarantee for translation-invariant kernels.

While we demonstrated here a mechanism for private SVM learning with upper bounds on privacy and utility, we have previously also studied lower bounds that expose limits on the achievable utility of *any* learner that provides a given level of differential privacy. Further work is needed to sharpen these results. In a sense, these lower bounds are witnessed by pathological training sets and perturbation points and, as such, serve as attacks in pathological (unrealistic) cases. Developing practical attacks on the privacy of an SVM's training data remains unexplored.

Finally, it is important to note that alternate approaches to differentially private SVMs have been explored by others. Most notable is the work (parallel to our own) of Chaudhuri et al. [21]. Their approach to finite-dimensional feature mappings is, instead of adding noise to the primal solution, to add noise to the primal objective in the form of a dot product of the weight with a random vector. Initial experiments show their approach to be very promising empirically, although it does not allow for non-differentiable losses like the hinge loss.

4.7 Concluding Remarks

In security applications like malware detection, intrusion detection, and spam filtering, SVMs may be attacked through patterns that can either evade detection (evasion), mislead the learning algorithm (poisoning) or gain information about their internal parameters or training data (privacy violation). In this chapter, we demonstrated that these attacks are feasible and constitute a relevant threat to the security of SVMs, and to machine-learning systems, in general.

Evasion. We proposed an *evasion* algorithm against SVMs with differentiable kernels, and, more generally, against classifiers with differentiable discriminant functions. We investigated the attack's effectiveness in perfect and limited knowledge settings. In both cases, our attack simulation showed that SVMs (both linear and RBF) can be evaded with high probability after a few modifications to the attack patterns. Our analysis also provides some general hints for tuning the classifier's parameters (e.g., the value of γ in SVMs with the RBF kernel) and for improving classifier security. For instance, if a classifier tightly *encloses* the legitimate samples, the adversary's samples must closely mimic legitimate samples to evade it, in which case, if such exact mimicry is still possible, it suggests an inherent flaw in the feature representation.

Poisoning. We presented an algorithm that allows the adversary to find an attack pattern whose addition to the training set maximally decreases the SVM's classification accuracy. We found that the increase in error over the course of attack is especially striking. A single attack data point may cause the classification error to rise from the initial error rates of 2–5 % to 15–20 %. This confirms that our attack can achieve significantly higher error rates than random label flips and underscores the vulnerability of the SVM to poisoning attacks. As a future investigation, it may be of interest to analyze the effectiveness of poisoning attacks against non-convex SVMs with bounded loss functions, both empirically and theoretically, since such losses are designed to limit the impact of any single (attack) point on the resulting learned function. This has also been studied from a more theoretical perspective in [23], exploiting the framework of Robust Statistics [35, 51]. A similar effect is obtained by using bounded kernels (e.g., the RBF kernel) or bounded feature values.

Privacy. We developed an SVM learning algorithm that preserves *differential privacy*, a formal framework for quantifying the threat of a potential training set privacy violation incurred by releasing learned classifiers. Our mechanism involves adding Laplace-distributed noise to the SVM weight vector with a scale that depends on the algorithmic stability of the SVM and the desired level of privacy. In addition to presenting a formal guarantee that our mechanism preserves privacy, we also provided bounds on the utility of the new mechanism, which state that the privacy-preserving classifier makes predictions that are point-wise close to those of the non-private SVM, with high probability. Finally we discussed potential approaches for attacking SVMs' training data privacy, and known approaches to differentially private SVMs with (possibly infinite-dimensional feature space) translation-invariant kernels, and lower bounds on the fundamental limits on utility for private approximations of the SVM.

Acknowledgments This work has been partly supported by the project CRP-18293 funded by Regione Autonoma della Sardegna, L.R. 7/2007, Bando 2009, and by the project "Advanced and secure sharing of multimedia data over social networks in the future Internet" (CUP F71J1100069 0002) funded by the same institution. Davide Maiorca gratefully acknowledges Regione Autonoma della Sardegna for the financial support of his PhD scholarship (P.O.R. Sardegna F.S.E. Operational Programme of the Autonomous Region of Sardinia, European Social Fund 2007–2013—Axis IV Human Resources, Objective l.3, Line of Activity l.3.1.). Blaine Nelson thanks the Alexander von Humboldt Foundation for providing additional financial support. The opinions expressed in this chapter are solely those of the authors and do not necessarily reflect the opinions of any sponsor.

References

1. Adobe Systems Incorporated: PDF Reference, sixth edition, Adobe Portable Document Format, Version 1.7, November 2006.
2. Balfanz, D., Staddon, J. (eds.): Proceedings of the 1st ACM Workshop on AISec (AISec). ACM, New York (2008)
3. Balfanz, D., Staddon, J. (eds.): Proceedings of the 2nd ACM Workshop on Security and Artificial Intelligence (AISec). ACM, New York (2009)
4. Barreno, M., Nelson, B., Joseph, A., Tygar, J.: The security of machine learning. Mach. Learn. **81**, 121–148 (2010). doi:10.1007/s10994-010-5188-5
5. Barreno, M., Nelson, B., Sears, R., Joseph, A.D., Tygar, J.D.: Can machine learning be secure? In: ASIACCS '06: Proceedings of the 2006 ACM Symposium on Information. Computer and Communications Security, pp. 16–25. ACM, New York (2006). doi: http://doi.acm.org/10.1145/1128817.1128824
6. Barth, A., Rubinstein, B.I.P., Sundararajan, M., Mitchell, J.C., Song, D., Bartlett, P.L.: A learning-based approach to reactive security. IEEE Trans. Depend. Secure Comput. **9**(4), 482–493 (2012)
7. Biggio, B., Corona, I., Fumera, G., Giacinto, G., Roli, F.: Bagging classifiers for fighting poisoning attacks in adversarial environments. In: The 10th International Workshop on Multiple Classifier Systems (MCS). Lecture Notes in Computer Science, vol. 6713, pp. 350–359. Springer, Berlin (2011)
8. Biggio, B., Corona, I., Maiorca, D., Nelson, B., Šrndić, N., Laskov, P., Giacinto, G., Roli, F.: Evasion attacks against machine learning at test time. In: Blockeel, H., et al. (eds.) European Conference on Machine Learning and Principles and Practice of Knowledge Discovery in Databases (ECML PKDD), Part III. Lecture Notes in Artificial Intelligence, vol. 8190, pp. 387–402. Springer, Berlin Heidelberg (2013)
9. Biggio, B., Didaci, L., Fumera, G., Roli, F.: Poisoning attacks to compromise face templates. In: The 6th IAPR International Conference on Biometrics (ICB) (2013)
10. Biggio, B., Fumera, G., Pillai, I., Roli, F.: A survey and experimental evaluation of image spam filtering techniques. Pattern Recogn. Lett. **32**(10), 1436–1446 (2011)
11. Biggio, B., Fumera, G., Roli, F.: Design of robust classifiers for adversarial environments. IEEE Int. Conf. Syst. Man Cybern. (SMC), 977–982 (2011)
12. Biggio, B., Fumera, G., Roli, F.: Security evaluation of pattern classifiers under attack. IEEE Trans. Knowl. Data Eng. **99**, 1 (2013)
13. Biggio, B., Fumera, G., Roli, F., Didaci, L.: Poisoning adaptive biometric systems. In: Structural, Syntactic, and Statistical Pattern Recognition. Lecture Notes in Computer Science, vol. 7626, pp. 417–425 (2012)
14. Biggio, B., Nelson, B., Laskov, P.: Poisoning attacks against support vector machines. In: Proceedings of the 29th International Conference on Machine Learning (2012)
15. Blum, A., Dwork, C., McSherry, F., Nissim, K.: Practical privacy: the SuLQ framework. In: Proceedings of the 24th Symposium on Principles of Database Systems, pp. 128–138 (2005)
16. Breiman, L.: Bagging predictors. Mach. Learn. **24**(2), 123–140 (1996)
17. Brückner, M., Kanzow, C., Scheffer, T.: Static prediction games for adversarial learning problems. J. Mach. Learn. Res. **13**, 2617–2654 (2012)

18. Cárdenas, A.A., Baras, J.S.: Evaluation of classifiers: practical considerations for security applications. In: AAAI Workshop on Evaluation Methods for Machine Learning (2006)
19. Cárdenas, A.A., Nelson, B., Rubinstein, B.I. (eds.): The 5th ACM Workshop on Artificial Intelligence and Security (AISec). ACM, New York (2012)
20. Cauwenberghs, G., Poggio, T.: Incremental and decremental support vector machine learning. In: Leen, T.K., Dietterich, T.G., Tresp, V. (eds.) NIPS, pp. 409–415. MIT Press, Cambridge (2000)
21. Chaudhuri, K., Monteleoni, C., Sarwate, A.D.: Differentially private empirical risk minimization. J. Mach. Learn. Res. **12**, 1069–1109 (2011)
22. Chen, Y., Cárdenas, A.A., Greenstadt, R., Rubinstein, B. (eds.): Proceedings of the 4th ACM Workshop on Security and Artificial Intelligence (AISec). ACM, New York (2011)
23. Christmann, A., Steinwart, I.: On robust properties of convex risk minimization methods for pattern recognition. J. Mach. Learn. Res. **5**, 1007–1034 (2004)
24. Corona, I., Biggio, B., Maiorca, D.: Adversarialib: a general-purpose library for the automatic evaluation of machine learning-based classifiers under adversarial attacks. http://sourceforge.net/projects/adversarialib/ (2013)
25. Cortes, C., Vapnik, V.: Support-vector networks. Mach. Learn. **20**, 273–297 (1995)
26. Dalvi, N., Domingos, P., Mausam, Sanghai, S., Verma, D.: Adversarial classification. In: Proceedings of the 10th ACM SIGKDD International Conference on Knowledge Discovery and Data Mining (KDD), pp. 99–108 (2004)
27. Dimitrakakis, C., Gkoulalas-Divanis, A., Mitrokotsa, A., Verykios, V.S., Saygin, Y. (eds.): International ECML/PKDD Workshop on Privacy and Security Issues in Data Mining and Machine Learning (2010)
28. Drucker, H., Wu, D., Vapnik, V.N.: Support vector machines for spam categorization. IEEE Trans. Neural Netw. **10**(5), 1048–1054 (1999)
29. Duda, R.O., Hart, P.E., Stork, D.G.: Pattern Classification. Wiley-Interscience, Chichester (2000)
30. Dwork, C., McSherry, F., Nissim, K., Smith, A.: Calibrating noise to sensitivity in private data analysis. In: Proceedings of the 3rd Theory of Cryptography Conference (TCC 2006), pp. 265–284 (2006)
31. Fogla, P., Sharif, M., Perdisci, R., Kolesnikov, O., Lee, W.: Polymorphic blending attacks. In: Proceedings of the 15th Conference on USENIX Security Symposium (2006)
32. Globerson, A., Roweis, S.T.: Nightmare at test time: robust learning by feature deletion. In: Proceedings of the 23rd International Conference on Machine Learning, pp. 353–360 (2006)
33. Golland, P.: Discriminative direction for kernel classifiers. In: Neural Information Processing Systems (NIPS), pp. 745–752. MIT Press, Cambridge (2002)
34. Greenstadt, R. (ed.): Proceedings of the 3rd ACM Workshop on Artificial Intelligence and Security (AISec). ACM, New York (2010)
35. Hampel, F.R., Ronchetti, E.M., Rousseeuw, P.J., Stahel, W.A.: Robust Statistics: The Approach Based on Influence Functions. Probability and Mathematical Statistics. Wiley, New York (1986). http://www.worldcat.org/isbn/0471735779
36. Huang, L., Joseph, A.D., Nelson, B., Rubinstein, B., Tygar, J.D.: Adversarial machine learning. In: Proceedings of the 4th ACM Workshop on Artificial Intelligence and Security (AISec), pp. 43–57 (2011)
37. Joseph, A.D., Laskov, P., Roli, F., Tygar, D. (eds.): Dagstuhl Perspectives Workshop on Machine Learning Methods for Computer Security. Workshop 12371 (2012)
38. Kloft, M., Laskov, P.: Online anomaly detection under adversarial impact. In: Proceedings of the 13th International Conference on Artificial Intelligence and Statistics (AISTATS), pp. 405–412 (2010)
39. Kloft, M., Laskov, P.: A 'poisoning' attack against online anomaly detection. In: Joseph, A.D., Laskov, P., Roli, F., Tygar, D. (eds.) Dagstuhl Perspectives Workshop on Machine Learning Methods for Computer Security. Workshop 12371 (2012)
40. Kloft, M., Laskov, P.: Security analysis of online centroid anomaly detection. J. Mach. Learn. Res. **13**, 3647–3690 (2012)

41. Kolcz, A., Teo, C.H.: Feature weighting for improved classifier robustness. In: Proceedings of the 6th Conference on Email and Anti-Spam (CEAS) (2009)
42. Laskov, P., Kloft, M.: A framework for quantitative security analysis of machine learning. In: Proceedings of the 2nd ACM Workshop on Security and Artificial Intelligence, pp. 1–4 (2009)
43. Laskov, P., Lippmann, R. (eds.): NIPS Workshop on Machine Learning in Adversarial Environments for Computer Security (2007)
44. Laskov, P., Lippmann, R.: Machine learning in adversarial environments. Mach. Learn. **81**, 115–119 (2010)
45. LeCun, Y., Jackel, L., Bottou, L., Brunot, A., Cortes, C., Denker, J., Drucker, H., Guyon, I., Müller, U., Säckinger, E., Simard, P., Vapnik, V.: Comparison of learning algorithms for handwritten digit recognition. In: International Conference on Artificial Neural Networks, pp. 53–60 (1995)
46. Lowd, D., Meek, C.: Adversarial learning. In: Proceedings of the 11th ACM SIGKDD International Conference on Knowledge Discovery and Data Mining (KDD), pp. 641–647 (2005)
47. Lowd, D., Meek, C.: Good word attacks on statistical spam filters. In: Proceedings of the 2nd Conference on Email and Anti-Spam (CEAS) (2005)
48. Lütkepohl, H.: Handbook of Matrices. Wiley, New York (1996)
49. Maiorca, D., Corona, I., Giacinto, G.: Looking at the bag is not enough to find the bomb: an evasion of structural methods for malicious PDF files detection. In: Proceedings of the 8th ACM SIGSAC Symposium on Information, Computer and Communications security (ASIA CCS '13), pp. 119–130. ACM, New York (2013)
50. Maiorca, D., Giacinto, G., Corona, I.: A pattern recognition system for malicious PDF files detection. In: MLDM, pp. 510–524. Springer, Berlin (2012)
51. Maronna, R.A., Martin, R.D., Yohai, V.J.: Robust Statistics: Theory and Methods. Probability and Mathematical Statistics. Wiley, New York (2006). http://www.worldcat.org/isbn/0471735779
52. Narayanan, A., Shmatikov, V.: Robust de-anonymization of large sparse datasets. In: Proceedings of the 29th IEEE Symposium on Security and Privacy, pp. 111–125 (2008)
53. Nelson, B., Barreno, M., Chi, F.J., Joseph, A.D., Rubinstein, B.I.P., Saini, U., Sutton, C., Tygar, J.D., Xia, K.: Exploiting machine learning to subvert your spam filter. In: Proceedings of the 1st USENIX Workshop on Large-Scale Exploits and Emergent Threats, pp. 1–9 (2008)
54. Nelson, B., Rubinstein, B.I., Huang, L., Joseph, A.D., Lee, S.J., Rao, S., Tygar, J.D.: Query strategies for evading convex-inducing classifiers. J. Mach. Learn. Res. **13**, 1293–1332 (2012)
55. Perdisci, R., Gu, G., Lee, W.: Using an ensemble of one-class SVM classifiers to harden payload-based anomaly detection systems. In: Proceedings of the International Conference on Data Mining (ICDM), pp. 488–498 (2006)
56. Platt, J.: Probabilistic outputs for support vector machines and comparison to regularized likelihood methods. In: Advances in Large Margin Classifiers, pp. 61–74. MIT Press, Cambridge (2000)
57. Rizzi, S.: What-if analysis. In: Encyclopedia of Database Systems, pp. 3525–3529. Springer, New York (2009)
58. Rodrigues, R.N., Ling, L.L., Govindaraju, V.: Robustness of multimodal biometric fusion methods against spoof attacks. J. Vis. Lang. Comput. **20**(3), 169–179 (2009)
59. Rubinstein, B.I., Nelson, B., Huang, L., Joseph, A.D., Lau, S.h., Rao, S., Taft, N., Tygar, J.D.: Antidote: understanding and defending against poisoning of anomaly detectors. In: Proceedings of the 9th Conference on Internet Measurement Conference (IMC), pp. 1–14 (2009)
60. Rubinstein, B.I.P., Bartlett, P.L., Huang, L., Taft, N.: Learning in a large function space: privacy-preserving mechanisms for SVM learning. J. Privacy Confidentiality **4**(1), 65–100 (2012)
61. Schölkopf, B., Herbrich, R., Smola, A.J.: A generalized representer theorem. In: Computational Learning Theory. Lecture Notes in Computer Science, vol. 2111, pp. 416–426. Springer, Berlin (2001)

62. Schölkopf, B., Smola, A.J.: Learning with Kernels: Support Vector Machines, Regularization, Optimization, and Beyond. MIT Press, Cambridge (2001)
63. Smutz, C., Stavrou, A.: Malicious PDF detection using metadata and structural features. In: Proceedings of the 28th Annual Computer Security Applications Conference, pp. 239–248 (2012)
64. Šrndić, N., Laskov, P.: Detection of malicious PDF files based on hierarchical document structure. In: Proceedings of the 20th Annual Network & Distributed System Security Symposium (NDSS) (2013)
65. Sweeney, L.: k-anonymity: a model for protecting privacy. Int. J. Uncertain. Fuzz. Knowl. Based Syst. **10**(5), 557–570 (2002)
66. Vapnik, V.N.: The Nature of Statistical Learning Theory. Springer, New York (1995)
67. Wittel, G.L., Wu, S.F.: On attacking statistical spam filters. In: Proceedings of the 1st Conference on Email and Anti-Spam (CEAS) (2004)
68. Young, R.: 2010 IBM x-force mid-year trend & risk report. Technical Report, IBM (2010)

Chapter 5
Application of SVMs to the Bag-of-Features Model: A Kernel Perspective

Lei Wang, Lingqiao Liu, Luping Zhou, and Kap Luk Chan

Abstract The Bag-of-features model has recently achieved great success in image categorisation and become the state of the art. Support vector machines (SVMs) have played an important role in this process. This chapter first introduces the fundamentals of the Bag-of-features model in image categorisation. Following that, it is focused on how the SVM classifiers are applied to this model. In particular, we show the novel kernels developed to compare images based on a variety of representations incurred by this model. Also, how the kernels are implicitly implemented or effectively approximated to work with linear SVMs is discussed. Through this chapter, we will see that the application of SVMs not only demonstrates its elegance and efficiency but also raises new research issues to stimulate the development of SVMs.

5.1 The Bag-of-Features Model

Image categorisation is one of the fundamental tasks in the field of computer vision. It aims to classify an image to a predefined set of classes according to its visual content. By appropriately defining the classes, image categorisation can be used as an effective tool to determine the presence of objects in an image (object recognition), infer the location of an object in an image (object localisation) or in a

L. Wang (✉) • L. Zhou
University of Wollongong, Wollongong, NSW 2522, Australia
e-mail: leiw@uow.edu.au; lupingz@uow.edu.au

L. Liu
Australian National University, Acton, ACT 0200, Australia
e-mail: lingqiao.liu@cecs.anu.edu.au

K.L. Chan
Nanyang Technological University 639798, Singapore
e-mail: eklchan@ntu.edu.sg

Y. Ma and G. Guo (eds.), *Support Vector Machines Applications*,
DOI 10.1007/978-3-319-02300-7_5,
© Springer International Publishing Switzerland 2014

video sequence (object tracking), classify the scene in an image (scene recognition), determine the type of human pose in an image (pose recognition) or the action in a video clip (action recognition), detect the irregular visual patterns (unusual event detection) or search for similar images or videos from a large database (image/video retrieval), to name just a few.

Image categorisation has been researched for a long time in the fields of computer vision and pattern recognition. From the perspective of visual features, most of the research, particularly for those conducted 10 years ago or earlier, has been focused on the use of global features, for example, the features based on colour, texture or shape in a whole image. Although many significant research progresses have been made along this line, the performance of generic image categorisation is still far from being satisfactory. A powerful image categorisation model that can be generally applied is still lacking.

In the past several years, the Bag-of-features model [8,47] has attracted intensive attention in the field of visual recognition and achieved great success in a wide range of applications. It has become the state-of-the-art image categorisation model that can be generally applied. The Bag-of-features model can be viewed as a wonderful integration of the Bag-of-words model in the field of text analysis [19] and the local invariant features in the field of computer vision [32, 33]. During the last decade, a number of excellent local invariant features have been developed, for example, the well-known Scale-Invariant Feature Transform (SIFT) feature [28]. With the local invariant features, effective, reliable and robust description of visual content within a small-sized image patch can be obtained. This provides a solid basis for the transplantation of the Bag-of-words model from text analysis, giving birth to the Bag-of-features model. The work on texture classification in [25] is among the earliest work that uses the Bag-of-features model in image categorisation. Generally, the work in [8,47] is often regarded as the beginning of the Bag-of-features model in generic image categorisation.

A basic Bag-of-features model can be described as follows. First, a set of local image patches is sampled from all training images and characterised by using a local feature descriptor. After that, common visual patterns shared by these local feature descriptors are identified. They mimic the "words" in the Bag-of-words model and are called "visual words." A collection of visual words forms a "visual codebook." Following the Bag-of-words model, each image is represented by a histogram indicating the frequency of occurrences of each visual word in this image. In this way, image categorisation can be performed based on the histogram-based representation, for example, by training a classifier and performing classification. This basic Bag-of-features model has been significantly extended since its introduction and more powerful variants are being used. Figure 5.1 shows an image categorisation system based on the basic Bag-of-features model. The last step "Classification" corresponds to the application of SVMs. However, the previous steps form the basis for this application and more importantly, the development of these steps significantly reshapes the application of SVMs in the image categorisation system. To obtain a clear understanding of this system and the application of SVMs, this chapter will give a brief introduction of the four key components of this system in the following parts.

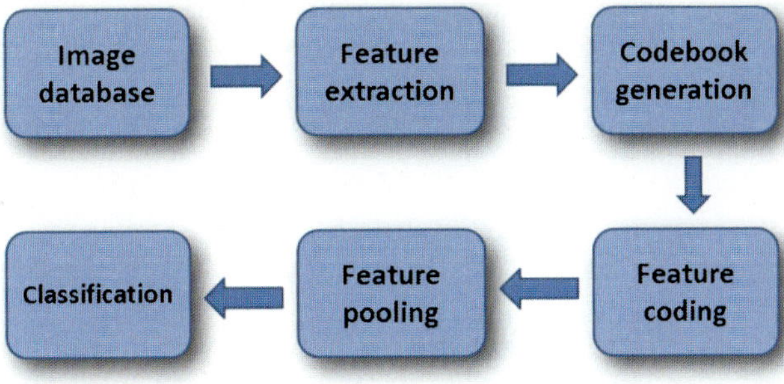

Fig. 5.1 An image categorisation system based on the basic Bag-of-features model

5.1.1 Feature Extraction

As mentioned above, the advances of local invariant features lay the foundation of the birth of the Bag-of-features model. The component "Feature extraction" in Fig. 5.1 is to extract local patches from images and then characterise the visual content therein. This component consists of two steps, feature detection and feature description. Feature detection is to identify the locations in an image where important visual information could exist, which is usually called interest point detection. A number of excellent interest point detection algorithms have been developed. They can reliably and consistently identify interest points in an image even if the image experiences the change of viewpoint, scale or illumination to some extent. The repeatability of identifying the same interest points in the varying conditions is essential to generic image categorisation. A comparative study of the commonly used interest point detectors can be found in [32]. With the recent progress of the Bag-of-features model, it is found that densely sampling local image patches can often lead to better classification performance than performing interest point detection [20]. Dense sampling could extract much more local image patches than interest point detection and thus has the advantage of avoiding missing important visual information that helps classification at the later stage. Dense sampling has now become a common way to extract local patches from images. To deal with the scaling issue, dense sampling with different-sized local patches is often used.

Once local image patches are extracted, a variety of local feature descriptors can be employed. These descriptors are designed to achieve reliable and robust description of the local patches with respect to varying conditions. The best known and the most popular descriptor may be the SIFT descriptor [28]. SIFT actually consists of both feature detector and feature descriptor, but its descriptor is often individually used by researchers to characterise densely sampled local image patches. A systematic evaluation of the performance of existing local descriptors is conducted in [33]. Depending on the size of an image, the total number of local patches extracted from an image can be in the order of thousands or even tens of thousands. Each of them will be represented by a descriptor, which is a multi-dimensional vector. Note that in this way, each image becomes a bag of orderless feature vectors or "a set of points" in a multi-dimensional vector space.

5.1.2 Visual Codebook

The Bag-of-features model originates from the Bag-of-words model. However, images do not contain words by nature. In order to use the Bag-of-words model, the concept of "visual word" is developed. Visual words can be regarded as a set of common visual patterns shared by the local patches extracted from images. For example, the patterns could be corners, T-junctions, L-junctions or any other frequently observed patterns on the changes of pixel intensities. The k-means clustering may be the most commonly used method to generate visual words. Let $\{\mathbf{x}_1, \mathbf{x}_1, \cdots, \mathbf{x}_n\}$ denote a set of local feature descriptors obtained from the step of feature extraction, where $\mathbf{x} \in \mathbb{R}^d$ is a d-dimensional vector. For example, d is 128 when the SIFT descriptor is used. The k-means clustering aims to find an optimal k-cluster-partitioning $\{C_1, C_2, \cdots, C_k\}$ of these feature descriptors by minimising the sum of the within-cluster variances. Let μ_i denote the mean of the cluster C_i. This partitioning can be shown as an optimisation problem

$$\{C_1, C_2, \cdots, C_k\} = \arg\min \sum_{i=1}^{k} \sum_{\mathbf{x}_j \in C_i} \|\mathbf{x}_j - \mu_i\|_2^2. \tag{5.1}$$

By solving this problem, the optimal means $\{\mu_1, \mu_2, \cdots, \mu_k\}$ are interpreted as visual words. A collection of the k visual words forms a visual codebook. Throughout this chapter, $\mathbf{V}$ is used to denote a visual codebook and $|\mathbf{V}|$, whose value is k here, denotes the number of visual words in the codebook.

In addition to the k-means clustering algorithm, a variety of more advanced visual codebook generation algorithms have been developed in the past few years. They generally deal with one or more of the following issues related to k-means clustering: (1) How to set the optimal k? To address this issue, methods based on multiple codebook combination, codeword selection and codeword merging have been proposed [53,55]; (2) How to conduct efficient clustering when d or k is large? To speed up clustering in this case, special data structures have been used and this leads to the work of vocabulary tree [36], randomised clustering forests [34] and fast k-means [41]; (3) Can a partitioning better than that given by the k-means clustering be obtained? In this line, fixed-radius partition and mean-shift techniques have been utilised [20]. Also, instead of using Euclidean distance, clustering with other distances has been developed to better handle the histogram structure of SIFT descriptors [56]; (4) Can supervised information be incorporated to obtain better codebooks? When the class label of each image is available, it can be used to design compact and discriminative visual codebooks. To achieve this, information-theoretic method [23] and supervised compact codebook generation [27] have recently been developed.

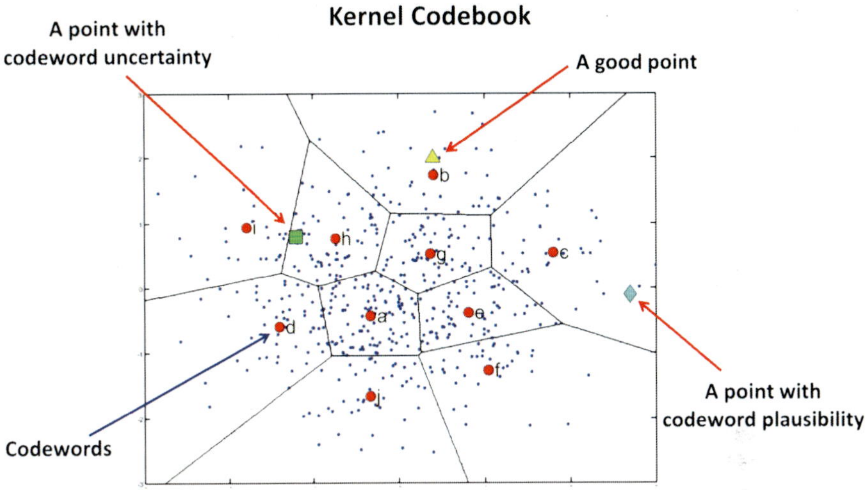

Fig. 5.2 The issues of codeword uncertainty and codeword plausibility. Image courtesy of [49]

5.1.3 Feature Coding

Once a visual codebook is created, it will be used as a basis to represent the visual content of an image. To associate the visual content with the codebook, the most common way is to assign each of the local feature descriptors extracted from an image to one of the visual words. For example, this can be done by evaluating the Euclidean distance between a local descriptor to each visual word and assigning it to the closest word. This is often known as "hard assignment," in which a local descriptor is assigned to one and only one visual word. Although this assignment is conceptually simple and computationally efficient, more advanced assignment methods have been developed in the last few years. Generally, these new methods deal with two issues: (1) how to reduce the quantisation error in the hard assignment in further? (2) how to consider the underlying manifold structure of the local feature descriptors? To resolve the first issue, a number of methods have been developed in the literature, among which kernel codebook and sparse coding are two representative methods. Kernel codebook is a "soft assignment" method, which assigns a local feature descriptor to more than one visual words to reduce quantisation error [49]. In particular, kernel codebook systematically discusses two main drawbacks, codeword uncertainty and codeword plausibility, in hard-assignment methods, as illustrated in Fig. 5.2.

Codeword uncertainty means that hard-assignment methods rigidly assign a local descriptor to one and only one visual word even if it is relevant (close) to multiple different words. For example, for a local descriptor riding on the boundary of two clusters, assigning it to either one of the two corresponding visual words could cause significant quantisation error. Codeword plausibility means that hard-assignment methods rigidly assign a local descriptor to a visual word even if this descriptor is far from all of the words. Again, this could lead to large quantisation error. To handle the two issues, the kernel codebook proposes to model the degree of relevance between a local descriptor and each visual word. Its assignment scheme can be expressed as

$$\omega_i = \frac{\kappa_\sigma(d(\mathbf{x}, \mathbf{v}_i))}{\sum_{i=1}^{|\mathbf{V}|} \kappa_\sigma(d(\mathbf{x}, \mathbf{v}_i))}, \tag{5.2}$$

where κ denotes a kernel used in kernel density estimation with the width of σ, $\mathbf{x}$ a local descriptor, $\mathbf{v}_i$ the ith visual word, $d(\mathbf{x}, \mathbf{v}_i)$ the distance between $\mathbf{x}$ and $\mathbf{v}_i$, and ω_i the coefficient assigned to $\mathbf{x}$ with respect to $\mathbf{v}_i$. As shown in [49], this soft-assignment method can well resolve the above two issues caused by hard assignment.

Sparse coding is the assignment method that represents the state of the art [57]. It is also a soft-assignment method and has an elegant theoretical framework. The sparsity enforces that only a small number of visual words can be chosen to represent a local descriptor. In addition, sparse coding not only learns the coding coefficient but can also jointly learn the visual codebook. Let $\mathbf{V}_{d\times k} = (\mathbf{v}_1, \mathbf{v}_2, \cdots, \mathbf{v}_k)$ denote a visual codebook consisting of k words. Let $\mathbf{U}_{k\times n} = (\mathbf{u}_1, \mathbf{u}_2, \cdots, \mathbf{u}_n)$ denote the coding coefficient matrix, in which $\mathbf{u}_i$ is the coding coefficient for the ith local descriptor $\mathbf{x}_i$. Sparse coding can be expressed as an optimisation problem that aims to minimise the reconstruction error, subject to a sparsity constraint on the coding coefficient.

$$\begin{aligned} \{\mathbf{V}^\star, \mathbf{U}^\star\} &= \arg\min_{\mathbf{V},\mathbf{U}} \sum_{i=1}^{n} \left(\|\mathbf{x}_i - \mathbf{V}\mathbf{u}_i\|_2^2 + \lambda \|\mathbf{u}_i\|_1\right) \\ &s.t. \ \ \|\mathbf{v}_j\| \leq 1, \ \ j = 1, 2, \cdots, k, \end{aligned} \tag{5.3}$$

where $\|\cdot\|_1$ is the ℓ_1 norm used to impose the sparsity constraint and λ is the regularisation parameter. This optimisation can be efficiently solved in an alternate manner. By fixing $\mathbf{V}$, the coding coefficient $\mathbf{u}_i$ for each local descriptor can be updated one by one by solving an ℓ_1-regularised least-squares problem. By fixing $\mathbf{U}$, the visual codebook $\mathbf{V}$ can be updated by solving a least-squares problem with quadratic constraints.

Locality-constrained Linear Coding (LLC) is another important sparse coding method, which takes the underlying manifold structure of local descriptors into account [52]. It argues that locality is more essential than sparsity and that locality induces sparsity. To enforce locality, LLC encodes a descriptor by the visual words nearby. This not only induces the sparsity but also makes similar descriptors tend to share similar coefficients. LLC solves the following optimisation problem,

$$\{\mathbf{V}^\star, \mathbf{U}^\star\} = \arg\min_{\mathbf{V},\mathbf{U}} \sum_{i=1}^{n} \left(\|\mathbf{x}_i - \mathbf{V}\mathbf{u}_i\|_2^2 + \lambda \|\mathbf{d}_i \otimes \mathbf{u}_i\|_2^2\right)$$

$$s.t.\ \mathbf{u}_i^\top \mathbf{1} = 1,\ \ \|\mathbf{v}_j\| \leq 1,\ \ j = 1,2,\cdots,k, \tag{5.4}$$

where $\mathbf{d}_i$ is a column vector indicating the distance of the descriptor $\mathbf{x}_i$ from each visual word and $\otimes$ denotes a component-wise multiplication. This optimisation problem can also be solved in an alternate manner.

Sharing the spirit of locality in LLC, a computationally more efficient coding method called Localised Soft-assignment Coding (LSC) has recently be developed [26]. LSC proposes to integrate the concept of locality into kernel codebook, by arguing that shorter Euclidean distances are more reliable in the presence of data manifold. LSC extends the kernel codebook to the following form,

$$\omega_i = \frac{\exp(-\beta d(\mathbf{x},\mathbf{v}_i))}{\sum_{i=1}^{|\mathbf{V}|} \exp(-\beta d(\mathbf{x},\mathbf{v}_i))},\ \text{ where } d(\mathbf{x},\mathbf{v}_i) = \begin{cases} \|\mathbf{x}-\mathbf{v}_i\|_2^2; & \text{if } \mathbf{v}_i \in \mathcal{N}(\mathbf{x}) \\ +\infty; & \text{otherwise.} \end{cases} \tag{5.5}$$

where $\mathcal{N}(\mathbf{x})$ denotes the local neighborhood of $\mathbf{x}$. LSC can achieve comparable coding performance as LLC but incurs much less computational cost.

The advent of sparse and localised coding schemes makes a significant change of the face of the application of SVMs to image categorisation. Together with the feature pooling scheme to be introduced, these coding schemes considerably improve the classification performance of linear SVM classifiers in image categorisation.

5.1.4 *Feature Pooling*

In order to obtain an image-level representation, the coding coefficients of the local feature descriptors extracted from an image need to be summarised. This process is often called "Pooling." Two ways are usually used, including sum-pooling and max-pooling. Sum-pooling simply adds all the coefficients for each visual word up. Let $\mathbf{z} \in \mathbb{R}^k$ denote the image-level representation for an image I. Sum-pooling can be expressed as

$$\mathbf{z} = \sum_{\mathbf{u}_i \in I} \mathbf{u}_i. \tag{5.6}$$

If $\mathbf{z}$ is set as the mean of the $\mathbf{u}_i$'s, it will be called average pooling. Max-pooling takes the maximum coefficient with respect to each visual word

$$\mathbf{z} = \max_{\mathbf{u}_i \in I} \mathbf{u}_i, \tag{5.7}$$

where the max operation is performed in a dimension-wise manner. The max-pooling can magically make linear SVM classifiers work as well as the nonlinear ones, leading to efficient image categorisation. Formal theoretical analysis has been attempted to understand why the max-pooling is superior to the sum-pooling [6,26]. More explanation on this magic in this regard will be given in the later part of this chapter from a kernel perspective.

In sum, this section introduces a basic image categorisation system using the Bag-of-features model. In particular, the four key components of this system have been discussed. This paves the way for us to gain a better understanding of the application of SVMs to image categorisation.

5.2 Application of SVMs with Histogram-Based Nonlinear Kernels

When an image-level representation is obtained for each image, an SVM classifier can be trained and used to categorise new images. Since image classes in a categorisation task are usually not linear separable, kernel-based SVM classifiers are generally used in order to obtain good classification performance, especially at the early stage of image categorisation with the Bag-of-features model. In this case the kernel function plays a pivotal role, and therefore identifying and designing appropriate kernel functions has attracted much attention at that stage. Since the histogram-based representation is widely used in that period, the kernel functions that can effectively evaluate the similarity of histograms have been researched and employed.

In the Bag-of-features model, a histogram indicates the frequency of the occurrences of each visual word in an image. It is an efficient approximation to the distribution of different visual patterns in an image. In the literature, a number of measures have been proposed to evaluate the similarity or dissimilarity between histograms. In the work of colour indexing [48], histogram intersection is proposed for object recognition. The work in [43] discusses the histogram dissimilarity measures and applies them to image retrieval. It groups the measures into bin-to-bin measures and cross-bin measures. The former includes Minkowski-form distance, Histogram intersection, Kullback-Leibler divergence, χ^2-statistics, while the latter includes Quadratic-form distance, cumulative histogram distance, and the distance based on distribution parameters. In [7], the SVMs are applied to classify generic images based on colour histograms. That work provides insightful discussion on what kind of kernel functions shall be used to evaluate the similarity of colour histograms. It shows that Non-Gaussian Radial Basis Function (RBF) kernels can achieve better classification performance than the commonly used Gaussian RBF kernels.

Although the SVM classifiers using the above histogram-based kernel functions can produce promising classification performance, training and testing nonlinear SVM classifiers incur more computational load, and this becomes a significant issue for the applications that require real-time classification. In recent years, approaches with the linear kernel have been proposed to achieve the advantage brought by histogram-based nonlinear kernels. The research in this regard generally follows two lines. The first line is to approximately identify the explicit feature mapping induced by the nonlinear kernels and then a linear SVM can be straightforwardly applied. The other line is to derive a new image representation such that a linear SVM classifier with this new representation can work as well as a nonlinear one. In doing so, these approaches not only well maintain the original classification performance but also considerably decrease the computational load in both training and test stages. In the following parts, the chapter will discuss typical histogram-based kernel functions and the approaches to approximating them via the linear kernel.

5.2.1 *Histogram-Based Nonlinear Kernels*

This part is focused on three kernel functions that are commonly used to evaluate the similarity of histograms, including the Histogram Intersection Kernel (HIK), the non-Gaussian RBF kernel and the χ^2-RBF kernel.

Recall that $\mathbf{z}$ denotes the representation of an image. Let $\phi(\cdot)$ be the mapping function implicitly induced by a kernel function $\kappa(\cdot,\cdot)$. Let $\mathbf{w}$ and b be the normal and bias of the SVM separating hyperplane. A nonlinear SVM classifier can be expressed as

$$f(\mathbf{z}) = \mathbf{w}^\top \phi(\mathbf{z}) + b = \sum_i \alpha_i y_i \kappa(\mathbf{z}, \mathbf{z}_i) + b, \tag{5.8}$$

where α_i and y_i are the coefficient and the class label for the ith training sample $\mathbf{z}_i$.

5.2.1.1 Histogram Intersection Kernel

Originally proposed in [48] to compare a given image histogram to a pre-defined model histogram for object recognition, histogram intersection has seen an important application to image categorisation with the Bag-of-features model. Let $\mathbf{z}_i$ and $\mathbf{z}_j$ denote the histograms corresponding to images i and j. Intersection of the two histograms is defined as $H(\mathbf{z}_i, \mathbf{z}_j) = \sum_{l=1}^{|\mathbf{V}|} \min(\mathbf{z}_{il}, \mathbf{z}_{jl})$, where $\mathbf{z}_{il}$ is the l-th bin of $\mathbf{z}_i$. A normalised intersection with respect to $\mathbf{z}_j$ can be defined as $H(\mathbf{z}_i, \mathbf{z}_j) = \sum_{l=1}^{|\mathbf{V}|} \min(\mathbf{z}_{il}, \mathbf{z}_{jl}) / \sum_{l=1}^{|\mathbf{V}|} \mathbf{z}_{jl}$. Usually, it is assumed that the sum of the bins in the two histograms is same. In this case, the HIK can be expressed as

$$\kappa(\mathbf{z}_i, \mathbf{z}_j) = \sum_{l=1}^{|\mathbf{V}|} \min(\mathbf{z}_{il}, \mathbf{z}_{jl}). \tag{5.9}$$

The HIK function defined in this way can be written as an inner product in a feature space and therefore it is a Mercer kernel [1]. This makes it suitable for SVM classifiers which usually need this property to achieve global optimum. Also, as indicated in [48], for two histograms $\mathbf{z}_i$ and $\mathbf{z}_j$ with $\sum_{l=1}^{|\mathbf{V}|} \mathbf{z}_{il} = \sum_{l=1}^{|\mathbf{V}|} \mathbf{z}_{jl} = T$, the HIK has an essential connection with the ℓ_1-norm distance between them. That is,

$$\kappa(\mathbf{z}_i, \mathbf{z}_j) = 1 - \frac{1}{2T} \|\mathbf{z}_i - \mathbf{z}_j\|_1. \tag{5.10}$$

This connection can help understanding the effectiveness of HIK in image categorisation with the Bag-of-features model. For a class of images containing the same object, the size of the area occupied by the object could change, leading to variation on the value of the bins for the visual words associated with the object. Also, the size of background could be different across these images and this will also cause variation on the corresponding bins. However, these variations are not essential and they do not change the object class to which the images belong to. In this case, an ℓ_1-norm distance becomes a better choice because it changes linearly with the variation while a commonly used ℓ_2-norm distance changes quadratically. It is known that the ℓ_2-norm distance corresponds to a linear kernel in the input space. This may partially explain why directly applying a linear SVM classifier to histogram representation often shows inferior classification performance. In fact, the values of the bins of a histogram are not important. Instead, whether a bin is empty or not (indicating whether the corresponding visual word appears in an image or not) matters. This case has been observed in image classification with colour histograms in [7] and the Bag-of-features model in [6].

5.2.1.2 Non-Gaussian RBF Kernel

To control the sensitivity of an RBF kernel function to the difference between two histograms, the work in [7] proposes a set of non-Gaussain RBF kernels in the following form

$$\kappa(\mathbf{z}_i, \mathbf{z}_j) = \exp\left(-\beta \sum_{l=1}^{|\mathbf{V}|} |\mathbf{z}_{il}^a - \mathbf{z}_{jl}^a|^b\right). \tag{5.11}$$

It is easy to see that when $a = 1$ and $b = 2$, the above kernel reduces to a Gaussian RBF kernel. When $a = 1$ and $b = 1$, it gives rise to a Laplacian RBF kernel

$$\kappa(\mathbf{z}_i, \mathbf{z}_j) = \exp\left(-\beta \sum_{l=1}^{|\mathbf{V}|} |\mathbf{z}_{il} - \mathbf{z}_{jl}|\right). \tag{5.12}$$

Assuming that $\mathbf{z}_j$ is obtained by perturbing one empty bin of $\mathbf{z}_i$ by Δh, the non-Gaussian kernel value between them can be written as

$$\kappa(\mathbf{z}_i, \mathbf{z}_j) = \exp\left(-\beta (\Delta h)^{ab}\right). \tag{5.13}$$

Adjusting ab can effectively change the decaying rate of the kernel value with respect to the perturbation. As can be seen, the commonly used Gaussian RBF kernel incurs a quadratic exponential decaying rate while the Laplacian RBF kernel has a linear exponential decaying rate. Experimental study is conducted in [7] to compare different settings of a and b for colour histogram-based image classification. It is found that decreasing the value of a and b can effectively improve the classification performance of SVMs on colour histograms. The best performance is obtained on the Corel image data set when $a = 0.25$ and $b = 1$, and it is significantly better than the performance obtained by a Gaussian RBF kernel. Note that the change of a does not affect the non-Gaussian RBF kernel to be a valid Mercer kernel because we can simply view $\mathbf{z}_{il}^a$ and $\mathbf{z}_{jl}^a$ as the input data. At the same time, b has to be confined between 0 and 2 to make the kernel meet the Mercer's condition. The application of SVMs with the Laplacian RBF kernel has also achieved promising classification performance in image categorisation with the Bag-of-features model.

5.2.1.3 χ^2-Radial Basis Function Kernel

The χ^2-RBF kernel can be traced back to the χ^2-test in mathematical statistics used to compare two distributions. In [44], χ^2 is used to evaluate the dissimilarity between two histograms as

$$\chi^2(\mathbf{z}_i, \mathbf{z}_j) = \sum_{l=1}^{|\mathbf{V}|} \frac{(\mathbf{z}_{il} - \mathbf{z}_{jl})^2}{\mathbf{z}_{il} + \mathbf{z}_{jl}}. \tag{5.14}$$

Based on this measure, the work in [7] defines the χ^2-RBF kernel as

$$\kappa(\mathbf{z}_i, \mathbf{z}_j) = \exp\left(-\beta \sum_{l=1}^{|\mathbf{V}|} \frac{(\mathbf{z}_{il} - \mathbf{z}_{jl})^2}{\mathbf{z}_{il} + \mathbf{z}_{jl}}\right). \tag{5.15}$$

It is not difficult to see that when $\mathbf{z}_j$ is obtained by perturbing one empty bin of $\mathbf{z}_j$ by Δh, the kernel value is $\kappa(\mathbf{z}_i, \mathbf{z}_j) = \exp(-\beta \Delta h)$, which also has a linear exponential decaying rate with respect to Δh. The χ^2-RBF kernel has been proved to be a Mercer kernel in [13]. This kernel has been experimentally compared with other kernels in [58] for image categorisation with an SVM classifier. It is found that the χ^2-RBF kernel can achieve higher classification performance than linear kernel, quadratic kernel and the Gaussian RBF kernel.

In addition to the above kernels, there is a set of special kernels used by the SVM classifiers for image categorisation, which is called "additive kernel" [50]. An additive kernel can be written as a sum of the kernels computed based on each individual dimension of data. This feature makes it be able to work directly with linear SVM classifiers after appropriate manipulation. The additive kernels will be introduced in the later parts of this chapter.

5.2.2 Approximation to Histogram-Based Nonlinear Kernels

One drawback of directly applying the nonlinear kernels introduced in the previous section is the poor scalability for large-scale data sets. However, the application of SVMs calls for highly efficient image classification algorithms in order to handle large-scale tasks.

Nowadays, it is quite often to encounter an image categorisation problem with over tens of thousands of training samples. For example, the commonly used image classification benchmarks, PASCAL [11] and Caltech 101/256 [12, 15], contain around 10,000 images and the more recently developed large-scale benchmarks such as ILSVRC [10] even have millions of images. However, nonlinear SVMs become computationally expensive when training sample size is large. In fact, merely computing the kernel matrix in a nonlinear SVM classifier will take $\mathscr{O}(n^2d)$ calculations, where n is the number of training samples and d is the dimensionality of image representation. For the state-of-the-art image representation, e.g. Bag-of-features model with a large-sized codebook and spatial pyramid [24], the value of d can be as large as hundreds of thousands. When n is also large, nonlinear SVMs will easily become computationally intractable. In addition, the computational cost of nonlinear SVMs is high in the test stage. The cost of evaluating the decision score is $\mathscr{O}(dn_{sv})$, where n_{sv} is the number of support vectors, which can be very large, for example, a few thousands in practice.

Comparing with the nonlinear SVMs, linear SVMs enjoy much higher computational efficiency. First of all, there exist very efficient training algorithms for linear SVM classifiers [16, 45]. Second, in the test stage, the cost of evaluating the decision score is just $\mathscr{O}(d)$, which can be thousands of times less than that incurred by the nonlinear SVMs. However, for the histogram-based representation in the Bag-of-features model, linear SVMs usually yield poorer performance than its nonlinear counterpart. To achieve both high computational efficiency and excellent classification performance, the recent literature has leveraged the kernel-induced feature mapping to transform a nonlinear SVM classifier to a linear one.

Recall that a kernel function κ can be written as the inner product of the feature mappings $\phi(\cdot)$, that is:

$$\kappa(\mathbf{z}_i, \mathbf{z}_j) = \langle \phi(\mathbf{z}_i), \phi(\mathbf{z}_j) \rangle. \tag{5.16}$$

If $\phi(\cdot)$ can be explicitly obtained, we can simply transform the input data by this mapping and apply a linear SVM on the mapped data. In doing so, a nonlinear SVM can be attained by solving a linear SVM. Unfortunately, the mapping function $\phi(\cdot)$ is generally implicit and may even have infinite dimensions. However, is it possible to develop a sufficiently good approximation to the feature mapping $\phi(\cdot)$? The answer is affirmative. Before systematically introducing the state-of-the-art approximation methods, this section first presents a method that is initially proposed to approximate the commonly used HIK.

5.2.2.1 An Approximation to Histogram Intersection Kernel

Recall that HIK is defined as $\kappa(\mathbf{z}_i, \mathbf{z}_j) = \sum_{l=1}^{d} \min(\mathbf{z}_{il}, \mathbf{z}_{jl})$, where d denotes the dimensionality of a histogram. This kernel is the sum of the values obtained by the nonlinear function $\min(\mathbf{z}_{il}, \mathbf{z}_{jl})$ in each dimension. Note that the function $\min(\cdot,\cdot)$ only takes two scalars as the input.

The work in [31] proposes to approximate the HIK as follows. It first develops a quantised version of the original data. More specifically, that work uniformly quantises the value of each bin into s scales. $q(\cdot)$ denotes the quantisation function and it returns the scale into which the input scalar is quantised. $\mathbf{u}$ is a function that maps the quantised scale $q(z)$ ($q(z) \in \{1, 2, \cdots, s\}$) into an s-dimensional vector. The definition of $\mathbf{u}$ is as follows

$$\mathbf{u}(z) = (u_1,\ u_2,\ \cdots\ u_k,\ \cdots\ u_s)^\top, \text{ where}$$
$$u_k = \begin{cases} 1; \text{ if } k \leq q(z) \\ 0; \text{ otherwise.} \end{cases} \tag{5.17}$$

where u_k denotes the kth dimension of $\mathbf{u}(z)$. Then the function $\min(\cdot,\cdot)$ can be approximated by

$$\min(\mathbf{z}_{il}, \mathbf{z}_{jl}) \approx \alpha \langle \mathbf{u}(\mathbf{z}_{il}), \mathbf{u}(\mathbf{z}_{jl}) \rangle, \tag{5.18}$$

where α is a constant scalar. From the definition of $\mathbf{u}$, we can see that if z is quantised into the $q(z)$th level, the first $q(z)$ dimensions of $\mathbf{u}(z)$ will be "1" and the remaining dimensions will be "0". Thus the inner product of $\mathbf{u}(\mathbf{z}_{il})$ and $\mathbf{u}(\mathbf{z}_{jl})$ will equal $\min(q(\mathbf{z}_{il}), q(\mathbf{z}_{jl}))$.

Note that the function $\mathbf{u}(\cdot)$ essentially develops an explicit feature mapping for the nonlinear function $\min(\cdot,\cdot)$. Since HIK is calculated by simply summing all $\min(\mathbf{z}_{il}, \mathbf{z}_{jl})$ $(l = 1, \cdots, d)$ up, we can define the approximate explicit feature mapping of HIK by concatenating the mapping $\mathbf{u}(\cdot)$ at each dimension of $\mathbf{z}_i$, that is

$$\phi(\mathbf{z}_i) = (\mathbf{u}^\top(\mathbf{z}_{i1})\ \cdots\ \mathbf{u}^\top(\mathbf{z}_{ik})\ \cdots\ \mathbf{u}^\top(\mathbf{z}_{id}))^\top. \tag{5.19}$$

From this method, we can see that if a nonlinear kernel can be decomposed into the sum of a set of dimension-wise nonlinear functions with only two scalars as the input, it will be convenient to find an approximate feature mapping for such a nonlinear function. In fact, this kind of kernel is called additive kernel and it has been shown in the recent literature that for this family of kernels, efficient approximate feature mappings can be developed with very good approximation accuracy.

5.2.2.2 Additive Kernel and Its Approximation

An additive kernel can be written as the sum of the kernels computed on each individual dimension of data, that is, $\kappa(\mathbf{z}_i, \mathbf{z}_j) = \sum_{l=1}^{d} k(\mathbf{z}_{il}, \mathbf{z}_{jl})$. We call the function

Table 5.1 Examples of additive kernels and their dimension-wise kernel functions

Linear kernel	$k(\mathbf{z}_{il},\mathbf{z}_{jl}) = \mathbf{z}_{il}\mathbf{z}_{jl}$
Hellinger's kernel	$k(\mathbf{z}_{il},\mathbf{z}_{jl}) = \sqrt{\mathbf{z}_{il}\mathbf{z}_{jl}}$
Histogram intersection kernel	$k(\mathbf{z}_{il},\mathbf{z}_{jl}) = \min(\mathbf{z}_{il},\mathbf{z}_{jl})$
χ^2 kernel	$k(\mathbf{z}_{il},\mathbf{z}_{jl}) = 2(\mathbf{z}_{il}\mathbf{z}_{jl})/(\mathbf{z}_{il}+\mathbf{z}_{jl})$

$k(\cdot,\cdot)$ the dimension-wise kernel function (DKF in short) and an additive kernel can be fully determined by its DKF. Note that many commonly used kernels in image classification are additive kernels. Examples of additive kernels and their DKFs are listed in Table 5.1.

The special structure of an additive kernel suggests that its approximate feature mapping can be derived by first finding the approximate mapping of its DKF and concatenating the mappings from all the dimensions. Formally, the lth DKF can be obtained as: $k_l(\mathbf{z}_{il},\mathbf{z}_{jl}) \approx \langle \phi_l(\mathbf{z}_{il}), \phi_l(\mathbf{z}_{jl})\rangle$, where $\phi_l(\cdot)$ is the approximate feature mapping for the lth dimension of $\mathbf{z}$. Note that although DKF usually takes the same form in each dimension, its approximate feature mapping $\phi_l(\cdot)$ may be different from dimension to dimension. This is because the feature distribution in each dimension can be different and to capture these differences we may need different approximate feature mappings. Once the mapping $\phi_l(\cdot)$ is obtained for each dimension, an additive kernel can be approximated by $\kappa(\mathbf{z}_i,\mathbf{z}_j) \approx \langle \phi(\mathbf{z}_i), \phi(\mathbf{z}_j)\rangle$, where $\phi(\mathbf{z}_i) = (\phi_1^\top(\mathbf{z}_{i1})\ \phi_2^\top(\mathbf{z}_{i2})\ \cdots\ \phi_d^\top(\mathbf{z}_{id}))^\top$.

In the following parts, we elaborate two representative methods to develop the approximate explicit feature mapping for the DKF: (1) additive kernel principal component analysis (PCA) and (2) homogeneous kernel map.

5.2.2.3 Kernel PCA Approximation to Additive Kernels

The additive kernel PCA method is derived from the classic kernel PCA approximation to nonlinear kernels [54]. The classic kernel PCA approximation firstly calculates the feature mapping on a finite number of samples and then uses Nyström approximation [54] to generalise this mapping to unseen data. In the following parts, we firstly introduce the classic kernel PCA approximation and then discuss its additive kernel extension.

1. Kernel PCA Approximation to Nonlinear Kernels

The quality of an approximate explicit feature mapping can be measured by the incurred approximation error. Formally, this error can be defined as follows:

$$E(\phi) = \int \left(\kappa(\mathbf{z},\mathbf{z}') - \langle \phi(\mathbf{z}), \phi(\mathbf{z}')\rangle\right)^2 p(\mathbf{z})p(\mathbf{z}')d\mathbf{z}d\mathbf{z}' \tag{5.20}$$

where $\phi(\mathbf{z})$ is the approximate mapping function which maps a d-dimensional input data to a $\tilde{d}$-dimensional vector. $p(\mathbf{z})$ is the probability density function and it is unknown in general. To make this error term tractable, we can estimate $p(\mathbf{z})$ via non-parametric density estimation method. In specific, $p(\mathbf{z})$ is estimated by using a finite number, m, of samples such that

$$p(\mathbf{z}) = \frac{1}{m}\sum_{i=1}^{m}\delta(\|\mathbf{z}-\mathbf{z}_i\|), \tag{5.21}$$

where the function $\delta(a) = 1$ when $a = 0$ and 0 otherwise. Substituting Eq. (5.21) into Eq. (5.20), we can turn the integral into the summation and it results in the following error term:

$$E(\phi) = \frac{1}{m^2}\sum_{i=1}^{m}\sum_{j=1}^{m}\left(\kappa(\mathbf{z}_i,\mathbf{z}_j) - \langle\phi(\mathbf{z}_i),\phi(\mathbf{z}_j)\rangle\right)^2. \tag{5.22}$$

As shown in [54], the approximate feature mapping for all m samples can be derived in a closed form. Let $\psi_l = (\phi_l(\mathbf{z}_1),\phi_l(\mathbf{z}_2),\cdots,\phi_l(\mathbf{z}_m))^\top$ denote a vector consisting of the lth component of $\phi(\mathbf{z}_i)$, where $i = 1,\cdots,m$. To avoid the redundancy in the approximation, constraints that $\langle\psi_i,\psi_j\rangle = 0$ and $\langle\psi_i,\psi_i\rangle = 1$ for $1 \le i < j \le m$ are imposed. As a result, the solution of minimising the error in Eq. (5.22) can be obtained by solving an eigenvalue problem, which is equivalent to the kernel PCA:

$$\mathbf{K}\psi_l = \lambda_l\psi_l, \tag{5.23}$$

where $\mathbf{K}$ denotes the kernel matrix computed on $\mathbf{z}_1,\cdots,\mathbf{z}_m$. To obtain a $\tilde{d}$-dimensional approximate feature mapping $\phi(\mathbf{z})$ for $\mathbf{z}_1,\cdots,\mathbf{z}_m$, the $\tilde{d}$ eigenvectors of $\mathbf{K}$ corresponding to the largest eigenvalues λ_l are used. For unseen samples, their mappings can be worked out via Nyström approximation:

$$\phi_l(\mathbf{z}) = \frac{\langle\kappa(\mathbf{z},:),\psi_l\rangle}{\lambda_l}, \tag{5.24}$$

where $l = 1,2,\cdots,\tilde{d}$ and $\kappa(\mathbf{z},:) = (\kappa(\mathbf{z},\mathbf{z}_1)\ \kappa(\mathbf{z},\mathbf{z}_2)\ \cdots\ \kappa(\mathbf{z},\mathbf{z}_m))^\top$.

One problem with this approximation is its computational cost. The complexity of calculating $\kappa(\mathbf{z},:)$ is $\mathcal{O}(md)$ and the complexity of calculating $\phi_l(\mathbf{z})$ for the whole $\tilde{d}$-dimensional mapping is $\mathcal{O}(m\tilde{d})$. In total, the complexity is $\mathcal{O}(m(d+\tilde{d}))$. Since the kernel function $\kappa(\cdot,\cdot)$ takes two high-dimensional vectors as input, the mapping from these two vectors to the kernel value can be complex. In this case, a large number of samples are usually needed to achieve reasonably good approximation, making m a number at the order of thousands. Consequently, the calculation of this approximate feature mapping can be very time-consuming in practice.

2. Kernel PCA Approximation to Additive Kernels

Fortunately, for additive kernels the above method can be modified to achieve much lower computational cost [40]. More specifically, we can build the dimension-wise approximation through Eq. (5.24) for each DKF of an additive kernel. The work in [40] adopts this idea and builds the approximate feature mapping by the following algorithm:

1. For each dimension l, compute the corresponding $m \times m$ kernel matrix $\mathbf{K}_l$. The (i,j)th entry of $\mathbf{K}_l$ is calculated by $\mathbf{K}_l(i,j) = k(\mathbf{z}_{il}, \mathbf{z}_{jl})$, where $k(\cdot,\cdot)$ is the DKF of the additive kernel.
2. For each dimension l, compute the $\tilde{d}_l$ (e.g. $\tilde{d}_l = 10$) largest eigenvalues of $\mathbf{K}_l$ $\{\lambda_{l,1}, \cdots, \lambda_{l,\tilde{d}_l}\}$ and their associated eigenvectors. In total, this step will generate $d \times \tilde{d}_l$ eigenvalues for all the d DKFs.
3. Sort the $d \times \tilde{d}_l$ eigenvalues and keep the $\tilde{d}$ largest ones. They are re-numbered as $\{\lambda_{11}, \lambda_{12}, \cdots, \lambda_{1n_1}; \lambda_{21}, \cdots, \lambda_{2n_2}; \cdots; \lambda_{d1}, \cdots, \lambda_{dn_d},\}$, where $\sum_{i=1}^{d} n_i = \tilde{d}$. Let $\{\psi_{11}, \psi_{12}, \cdots, \psi_{1n_1}; \psi_{21}, \cdots, \psi_{2n_2}; \cdots; \psi_{d1}, \cdots, \psi_{dn_d},\}$ be the associated eigenvectors. For a test sample $\mathbf{z}_t$, its feature mapping is obtained by concatenating the mapping for each DKF. For a given l, this mapping is defined as

$$\phi_{li}(\mathbf{z}_{tl}) = \frac{\langle k(\mathbf{z}_{tl}, :), \psi_{li} \rangle}{\lambda_{li}}, \tag{5.25}$$

where $i = 1, \cdots, n_l$ and $k(\mathbf{z}_{tl}, :) = (k(\mathbf{z}_{tl}, \mathbf{z}_{1l}), \cdots, k(\mathbf{z}_{tl}, \mathbf{z}_{ml}))^\top$. Note that in this method, the dimensions of the approximated feature mapping $n_1, \cdots, n_d$ can be different. As argued in [40], this scheme can be more adaptive to the distribution of each dimension of the input data.

The computational complexity of mapping a sample with the obtained feature mapping $\phi(\cdot)$ is still $\mathscr{O}(m(d+\tilde{d}))$. However, since the approximation is now for a much simpler kernel function (a DKF with two scalar inputs only), a much smaller number of samples and mapping dimensions are usually sufficient to attain a good approximation. In [40], it is reported that 128 samples and the same number of mapping dimensions have been sufficient, which is in contrast to the requirement of thousands of samples in the classic kernel PCA. Moreover, one can further leverage the quantisation trick to quantise $\mathbf{z}_{tl}$ into finite levels, denoted by $q(\mathbf{z}_{tl})$, and pre-compute the value of $\phi_{li}(q(\mathbf{z}_{tl}))$. In this way, the computational complexity can be further reduced.

5.2.2.4 Homogeneous Kernel Map Approximation to Additive Kernels

One drawback of the kernel PCA approximation to additive kernels lies in the fact that it is *data dependent*. That is, the approximate mapping function needs to be learned from a set of training samples. As a result, an extra training step is required before applying the approximation to an additive kernel. In the following part, we will introduce another approximation method called "homogeneous kernel map," which is *data independent*. Before introducing this method, we need to first review a method called "random Fourier kernel approximation" which inspires the homogeneous kernel map.

1. Random Fourier Kernel Approximation to Stationary Kernels

Among the commonly used kernels in the literature, there is another family of kernels called stationary or translational invariant kernel [42]. It is formally defined as a kernel satisfying the following relationship:

$$\kappa(\mathbf{z}_i,\mathbf{z}_j) = \kappa(\mathbf{z}_i+\mathbf{c},\mathbf{z}_j+\mathbf{c}). \tag{5.26}$$

A property of stationary kernels is that there always exists a function $\mathscr{K}$ to make $\kappa(\mathbf{z}_i,\mathbf{z}_j) = \mathscr{K}(\mathbf{z}_j-\mathbf{z}_i)$ valid. To prove this property, we can simply set $\mathbf{c} = -\frac{\mathbf{z}_i+\mathbf{z}_j}{2}$ and substitute it to Eq. (5.26):

$$\begin{aligned}\kappa(\mathbf{z}_i,\mathbf{z}_j) &= \kappa\Big(-\frac{\mathbf{z}_i+\mathbf{z}_j}{2}+\mathbf{z}_i,-\frac{\mathbf{z}_i+\mathbf{z}_j}{2}+\mathbf{z}_j\Big)\\ &= \kappa\Big(\frac{\mathbf{z}_i-\mathbf{z}_j}{2},-\frac{\mathbf{z}_i-\mathbf{z}_j}{2}\Big) \equiv \mathscr{K}(\mathbf{z}_j-\mathbf{z}_i).\end{aligned} \tag{5.27}$$

It has been proved that by Bochner's theorem [42] for any Positive Definite function $\mathscr{F}(\mathbf{z})$, there exists a non-negative measure p, such that $\mathscr{F}(\mathbf{z})$ is its Fourier transform:

$$\mathscr{F}(\mathbf{z}) = \int_\omega p(\omega)e^{-j\langle\omega,\mathbf{z}\rangle}d\omega. \tag{5.28}$$

Note that $\mathscr{K}(\mathbf{z}_j-\mathbf{z}_i)$ is a Positive Definite function because the corresponding $\kappa(\mathbf{z}_i,\mathbf{z}_j)$ is assumed to be a Mercer kernel, which always produces a Positive Semi-Definite kernel matrix. This suggests that $\mathscr{K}(\mathbf{z}_j-\mathbf{z}_i)$ can be represented as the Fourier transform of non-negative function p,

$$\mathscr{K}(\mathbf{z}_j-\mathbf{z}_i) = \int_\omega p(\omega)e^{-j\langle\omega,(\mathbf{z}_j-\mathbf{z}_i)\rangle}d\omega. \tag{5.29}$$

Since both $\mathscr{K}$ and p are real, we can replace $e^{-j\langle\omega,\mathbf{z}\rangle}$ by $\cos(\langle\omega,\mathbf{z}\rangle)$ and rewrite Eq. (5.29) into

$$\begin{aligned}\mathscr{K}(\mathbf{z}_j-\mathbf{z}_i) &= \int_\omega p(\omega)\cos\left(\langle\omega,(\mathbf{z}_j-\mathbf{z}_i)\rangle\right)d\omega\\ &= \int_\omega p(\omega)\left(\langle\cos(\omega,\mathbf{z}_i\rangle)\cos(\langle\omega,\mathbf{z}_j\rangle)+\sin(\langle\omega,\mathbf{z}_j\rangle)\sin(\langle\omega,\mathbf{z}_i\rangle)\right)d\omega.\end{aligned} \tag{5.30}$$

If $\mathscr{K}$ is properly scaled, $p(\omega)$ can be viewed as a probability density function. Hence, Eq. (5.30) can be seen as the expectation of $\cos(\langle\omega,(\mathbf{z}_j-\mathbf{z}_i)\rangle)$, that is, $\mathscr{K}(\mathbf{z}_j-\mathbf{z}_i) = E_\omega(\cos(\langle\omega,(\mathbf{z}_j-\mathbf{z}_i)\rangle))$. This result motivates the work in [42] to use the empirical mean to approximate the expectation. More specifically, they randomly draw m frequency components $\{\omega_i\}$, where $i = 1,\cdots,m$, from the distribution $p(\omega)$ and use the average of $\cos(\langle\omega_i,(\mathbf{z}_j-\mathbf{z}_i)\rangle)$ to approximate $E(\cos(\langle\omega,(\mathbf{z}_j-\mathbf{z}_i)\rangle))$ as

$$E(\cos(\langle\omega,(\mathbf{z}_j-\mathbf{z}_i)\rangle)) \approx \frac{1}{m}\sum_{i=1}^{m}\cos(\langle\omega_i,(\mathbf{z}_j-\mathbf{z}_i)\rangle) \triangleq \langle\phi(\mathbf{z}_i),\phi(\mathbf{z}_j)\rangle, \tag{5.31}$$

where $\phi(\mathbf{z}) = \frac{1}{\sqrt{m}}(\cos(\langle\omega_1,\mathbf{z}\rangle),\cdots,\cos(\langle\omega_m,\mathbf{z}\rangle),\sin(\langle\omega_1,\mathbf{z}\rangle),\cdots,\sin(\langle\omega_m,\mathbf{z}\rangle))^\top$. It defines an approximate explicit feature mapping and this method is called random Fourier kernel map in [42]. Note that this approximation is data independent—the only input of this approximation method is $p(\omega)$, which can be obtained via the Fourier transform of the given kernel function.

2. Homogeneous Kernel Map Approximation to Additive Kernels

Inspired by the random Fourier kernel approximation and the kernel PCA approximation to additive kernels, the work in [50] develops a unified framework to build *data-independent* approximate feature mapping for a large family of additive kernels, known as γ-homogeneous kernels. Formally, an additive kernel $\kappa(\cdot,\cdot)$ is γ-homogeneous if its DKF $k(\cdot,\cdot)$ satisfies

$$\forall c \geq 0:\ k(c\mathbf{z}_{il},c\mathbf{z}_{jl}) = c^\gamma k(\mathbf{z}_{il},\mathbf{z}_{jl}). \tag{5.32}$$

By choosing $c = 1/\sqrt{\mathbf{z}_{il}\mathbf{z}_{jl}}$, the DKF of a γ-homogeneous kernel can be written as

$$\begin{aligned} k(\mathbf{z}_{il},\mathbf{z}_{jl}) &= c^{-\gamma}k(c\mathbf{z}_{il},c\mathbf{z}_{jl}) = (\mathbf{z}_{il}\mathbf{z}_{jl})^{\frac{\gamma}{2}}k\left(\sqrt{\frac{\mathbf{z}_{il}}{\mathbf{z}_{jl}}},\sqrt{\frac{\mathbf{z}_{jl}}{\mathbf{z}_{il}}}\right) \\ &= (\mathbf{z}_{il}\mathbf{z}_{jl})^{\frac{\gamma}{2}}\mathscr{K}(\log(\mathbf{z}_{il})-\log(\mathbf{z}_{jl})), \end{aligned} \tag{5.33}$$

where the scalar function $\mathscr{K}(\cdot)$ is called "kernel signature" and it is defined as:

$$\mathscr{K}(\lambda) = k(e^{\frac{\lambda}{2}},e^{-\frac{\lambda}{2}}). \tag{5.34}$$

Note that the role of kernel signature resembles the function $\mathscr{K}$ in the random Fourier map method previously introduced. In fact, the derivation of the algorithm in [50] bears a similarity with the one for random Fourier kernel approximation.

The work in [50] proves that a γ-homogeneous kernel $\kappa(\cdot,\cdot)$ is positive definite if, and only if, its signature $\mathscr{K}(\cdot)$ is a positive definite function. With this result, the Bochner's theorem in Eq. (5.28) can be readily applied to the kernel signature:

$$k(\mathbf{z}_{il},\mathbf{z}_{jl}) = (\mathbf{z}_{il}\mathbf{z}_{jl})^{\frac{\gamma}{2}}\mathscr{K}(\lambda) = (\mathbf{z}_{il}\mathbf{z}_{jl})^{\frac{\gamma}{2}}\int_{-\infty}^{\infty} e^{-j\omega\lambda}q(\omega)d\omega$$
$$= \int_{-\infty}^{\infty}\left(e^{-j\omega\log(\mathbf{z}_{jl})}\sqrt{\mathbf{z}_{jl}^{\gamma}q(\omega)}\right)^{\star}\left(e^{-j\omega\log(\mathbf{z}_{il})}\sqrt{\mathbf{z}_{il}^{\gamma}q(\omega)}\right)d\omega, \quad (5.35)$$

where $\lambda = \log\frac{\mathbf{z}_{il}}{\mathbf{z}_{jl}}$ and $q(\omega)$ is the Fourier transform of the kernel signature $\mathscr{K}(\lambda)$. In this case, by defining

$$\phi_\omega(\mathbf{z}_{il}) = e^{-j\omega\log(\mathbf{z}_{il})}\sqrt{\mathbf{z}_{il}^{\gamma}q(\omega)}, \quad (5.36)$$

it is easy to verify that

$$k(\mathbf{z}_{il},\mathbf{z}_{jl}) = \langle\phi_\omega(\mathbf{z}_{il}),\phi_\omega(\mathbf{z}_{jl})\rangle, \quad (5.37)$$

where $\langle\cdot,\cdot\rangle$ is the inner product in a Hilbert space.

However, the feature mapping ϕ_ω has infinite dimensions. To handle this situation, the work in [50] proposes an idea that is similar to the discrete Fourier transform. They first approximate the kernel signature function $\mathscr{K}(\cdot)$ by its periodic version with the assumption that the function $\mathscr{K}(\lambda)$ has a restrictive domain of λ, which can usually be satisfied in practice. According to the theory of Fourier transform, the transform of a periodic signal is discrete. Then they simply choose the first $\tilde{d}_l$ frequency components as the feature mapping for the lth DKF.

5.2.2.5 Experimental Comparison

We quote the experimental result in [50] to give an intuition on the effectiveness of the aforementioned approximate explicit feature mappings. In this experiment, two methods, additive kernel PCA approximation in Sect. 5.2.2.3 and homogeneous kernel map in Sect. 5.2.2.4, are compared with the baseline using the original nonlinear kernel. The evaluation is carried out on Caltech-101 data set with SIFT as the local feature descriptor. A visual codebook with 600 visual words is created and a $1\times 1+2\times 2+4\times 4$ spatial pyramid [24] is used. The result is shown in Table 5.2. As seen, the approximate explicit feature mappings achieves comparable or even better classification performance than the original nonlinear kernels, but with much less training time. The speed advantage of the homogeneous kernel maps over the additive Kernel PCA approximation is due to the fact that the latter requires an extra training step to learn the mapping function. In addition, an interesting observation is that the $\gamma = 1/2$ variant of the homogeneous kernel performs much better than the original nonlinear kernel. Note that this variant is equivalent to calculating the square root of the data first and then applying the nonlinear kernel. The square rooting operation makes the value in each bin of the histogram more stable and reduces the negative impact of the "burstiness" phenomenon, which will be discussed in Sect. 5.3.

Table 5.2 This table is quoted from [50]

Appro. mapping	Dimension	SVM solver	χ^2-kernel		JS kernel		HIK	
			Acc.	Time	Acc.	Time	Acc.	Time
Original	–	LIBSVM	64.1 (± 0.6)	1769 (± 16)	64.2 (± 0.5)	6455 (± 39)	62.3 (± 0.3)	1729 (± 19)
Homogeneous	3	LIBLINEAR	64.4 (± 0.6)	312 (± 14)	64.2 (± 0.4)	483 (± 52)	64.6 (± 0.5)	367 (± 18)
Homogeneous	5	LIBLINEAR	64.2 (± 0.5)	746 (± 58)	64.1 (± 0.4)	804 (± 82)	63.7 (± 0.6)	653 (± 54)
$\frac{1}{2}$-Homogeneous	3	LIBLINEAR	67.2 (± 0.7)	380 (± 10)	67.3 (± 0.6)	456 (± 54)	67.0 (± 0.6)	432 (± 55)
Add.-kernel-PCA	3	LIBLINEAR	64.3 (± 0.5)	682 (± 67)	64.5 (± 0.6)	1351 (± 93)	62.7 (± 0.4)	741 (± 41)
Add.-kernel-PCA	5	LIBLINEAR	64.1 (± 0.5)	1287 (± 156)	64.1 (± 0.5)	1576 (± 158)	62.6 (± 0.4)	1374 (± 257)

It compares different approximate explicit feature mappings with respect to a number of original nonlinear kernels. The comparison is based on the classification performance and the incurred training time of the linear or nonlinear SVM classifiers. The "Dimension" means the dimensions of the mapping used to approximate each DFK of an additive kernel

5.3 Application of SVMs with Max-Pooling-Based Linear Kernel

The motivation of using nonlinear kernel-based SVMs is that they tend to achieve better performance than linear SVMs for image categorisation with histogram-based image representation. However, is it possible to design an image representation with which linear SVMs can achieve the performance comparable or even better than the nonlinear counterparts? If it is, the computational issue can be readily removed by adopting linear SVMs for large-scale image classification. Note that in some sense designing a new image representation for a linear kernel can also be viewed as inventing a new nonlinear kernel at the level of original image data. To show this, we can abstract the process of forming an image representation as a function $e(\cdot)$. A kernel defined based on the original image data can then be expressed as $\hat{\kappa}(I_i, I_j) = \langle e(I_i), e(I_j) \rangle$. Thus, if we change the image representation $e(\cdot)$, we virtually change the image-level kernel $\hat{\kappa}$.

The work in [57] is among the earliest ones that take the above approach. It combines sparse coding and max-pooling to obtain image representation and use it for image categorisation. That work shows that with this representation, a simple linear SVM classifier has been able to attain the performance superior to the traditional ones in which a nonlinear kernel is employed. Later, the work in [5] further discovers that the max-pooling step is the key to the success of the system in [57]. A comprehensive experimental study is conducted in [5] to compare various combinations of coding and pooling methods. Their results are quoted in Table 5.3. From the result, three conclusions can be drawn:

1. The use of max-pooling significantly improves the classification performance of linear SVMs for all the coding methods. The improvement is even significant for the simplest hard-assignment coding. In that case, a pooled coding vector only indicates the presence or absence of a visual word in the associated image;
2. By using the max-pooling, linear SVMs can achieve comparable or even better performance than the counterpart which adopts the sum-pooling and then nonlinear kernel-based SVMs.
3. Once the max-pooling is used, the classification performance obtained by linear and nonlinear SVMs becomes similar.

Table 5.3 This table is quoted from [5]

	Performance on Scene-15		Performance on Caltech-101	
Coding method and kernel	Sum-pooling	Max-pooling	Sum-pooling	Max-pooling
Hard-assignment + linear kernel	51.4 ± 0.9 %	64.3 ± 0.9 %	73.9 ± 0.9 %	80.1 ± 0.6 %
Hard-assignment + HIK kernel	64.2 ± 1.0 %	64.3 ± 0.9 %	80.8 ± 0.4 %	80.1 ± 0.6 %
Soft-assignment + linear kernel	57.9 ± 1.5 %	69.0 ± 0.8 %	75.6 ± 0.5 %	81.4 ± 0.6 %
Soft-assignment + HIK kernel	66.1 ± 1.2 %	70.6 ± 1.0 %	81.2 ± 0.4 %	83.0 ± 0.7 %
Sparse coding + linear kernel	61.3 ± 1.3 %	71.5 ± 1.1 %	76.9 ± 0.6 %	83.1 ± 0.6 %
Sparse coding + HIK kernel	70.3 ± 1.3 %	71.8 ± 1.0 %	83.2 ± 0.4 %	84.1 ± 0.5 %

It lists the classification performance obtained by combining different pooling and coding methods

In sum, we can conclude that max-pooling is an effective way to produce better image representation and that linear SVMs show excellent performance in classifying the max-pooled coding vectors.

However, why can max-pooling obtain such a "magic" performance? To understand this, several interpretations have been proposed in the literature. In the following parts, we discuss three representative ones.

The work in [4] explains the superior performance of max-pooling by showing that it can generate more discriminative image representations. Assuming the case of binary classification, that work compares the class separability of each individual feature generated through max-pooling and sum-pooling. The class separability is defined as:

$$\psi = \frac{|E(\mathbf{z}_k|C_1) - E(\mathbf{z}_k|C_2)|}{\mathrm{var}(\mathbf{z}_k)}, \tag{5.38}$$

where $\mathbf{z}_k$ denotes the value of the kth dimension of a pooled coding vector. It can be obtained by $\mathbf{z}_k = \frac{1}{n}\sum_{i=1}^{n}\mathbf{u}_{ik}$ for sum-pooling or $\mathbf{z}_k = \max_{i=1}^{n}\mathbf{u}_{ik}$, where n is the number of coding vectors in an image and $\mathbf{u}_{ik}$ denotes the coding value at the kth dimension of the ith coding vector. C_1 and C_2 denote two classes. $E(\mathbf{z}_k|C_1)$ and $E(\mathbf{z}_k|C_2)$ are the expectation of $\mathbf{z}_k$ in each class, respectively. $\mathrm{var}(\mathbf{z}_k)$ denotes the variance of $\mathbf{z}_k$. To simplify their analysis, they assume that the $\{\mathbf{u}_{1k},\cdots,\mathbf{u}_{ik},\cdots,\mathbf{u}_{nk}\}$ are i.i.d. random variables.

For hard-assignment coding, the coding value $\mathbf{u}_{ik}$ is assumed to be drawn from a Bernoulli distribution in which $\mathbf{u}_{ik} = 1$ with probability α and $\mathbf{u}_{ik} = 0$ with probability $1-\alpha$. In the context of hard-assignment coding, this means that a coding value will be "activated" ($=1$) with probability α. Based on this assumption the class separability with respect to max-pooling and sum-pooling is derived as follows:

$$\psi_{\mathrm{sum}} = \frac{|\alpha_1 - \alpha_2|\sqrt{n}}{\sqrt{\alpha_1(1-\alpha_1)} + \sqrt{\alpha_2(1-\alpha_2)}} \tag{5.39}$$

$$\psi_{\mathrm{max}} = \frac{|(1-\alpha_1)^n - (1-\alpha_2)^n|}{\sqrt{(1-(1-\alpha_1)^n)(1-\alpha_1)^n} + \sqrt{(1-(1-\alpha_2)^n)(1-\alpha_2)^n}}, \tag{5.40}$$

where α_1 and α_2 denote $P(\mathbf{u}_{ik} = 1|C_1)$ and $P(\mathbf{u}_{ik} = 1|C_2)$, respectively, that is, the probability of a coding value being "activated" in each class. As previously defined, n is the number of coding vectors in an image and it is called "pool cardinality" in Fig. 5.3. By evaluating the class separability for max-pooling and sum-pooling with different α_1 and α_2, it is found that when the activation probability α_1 and α_2 are low, the ratio of α_1 to α_2 is large, and the value of n is small, max-pooling can achieve better class separability than sum-pooling. This explains the superior performance of max-pooling.

However, the above analysis cannot be readily used to explain the better discrimination achieved by max-pooling for the coding schemes where continuous coding coefficient is used, for example, the sparse coding. [4] To refine the analysis,

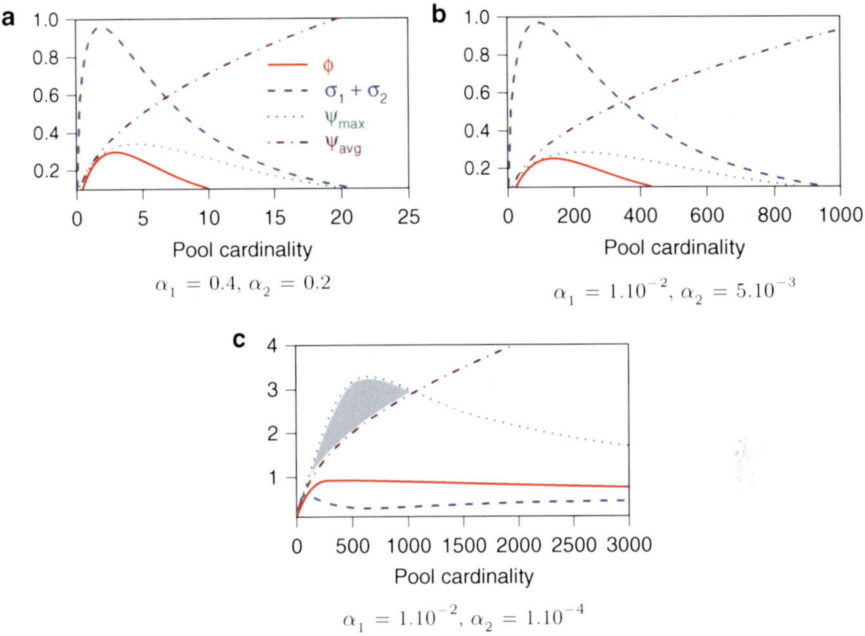

Fig. 5.3 $\phi(n) = |E(\mathbf{z}_k|C_1) - E(\mathbf{z}_k|C_2)| = |(1-\alpha_1)^n - (1-\alpha_2)^n|$ denotes the difference between the expectation of the max-pooled features in classes C_1 and C_2, and σ_1 and σ_2 are the standard deviations. ψ_{max} and ψ_{avg} are the class separability defined in Eq. (5.39). (**a**) When α_1 and α_2 are relatively large, the peak of the class separability is achieved at small pool cardinalities; (**b**) With the decreasing value of α_1 and α_2, the peak becomes wider (note the change of scale in the x axis); (**c**) When α_1 and α_2 are both small and $\alpha_1 \gg \alpha_2$, ψ_{max} becomes larger than ψ_{avg} for some cardinalities (indicated by the shaded area). Image courtesy of [4]

the work in [5] further assumes that the distribution of a coding value in an image is the mixture of two distributions: a distribution corresponding to the local descriptors from the object and a distribution corresponding to the local descriptors from the background [5]. The mixture weights of the two distributions vary from image to image. Through a more involved analysis, it is shown in [4, 5] that under certain conditions, max-pooling creates immunity to the variation in the mixture weights and thus it can lead to better classification performance.

The work in [26] points out that for soft-assignment coding, the coding coefficient can be viewed as the membership of a local feature descriptor with respect to different visual words, that is, $P(\mathbf{v}_j|\mathbf{x}_i)$, where $\mathbf{x}_i$ is the ith local descriptor in an image and $\mathbf{v}_j$ is the jth visual word. Hence, each dimension in the max-pooled coding vector can be related to the maximum membership score of the corresponding word, and this score is proved to be the lower bound of the probability of finding at least one local descriptor belonging to this word:

$$P(\mathbf{v}_j|\mathscr{A}) = 1 - \prod_{i=1}^{n}(1 - P(\mathbf{v}_j|\mathbf{x}_i)) \geq \max_{i=1,\cdots,n} P(\mathbf{v}_j|\mathbf{x}_i), \tag{5.41}$$

where $\mathscr{A}$ denotes the set of n local feature descriptors in an image. Assuming the i.i.d property for the local descriptors, the probability of finding at least one descriptor belonging to the jth visual word can be calculated by $1 - \prod_{i=1}^{n}(1 - P(\mathbf{v}_j|\mathbf{x}_i))$. As argued in [26], directly computing this probability involves the product of $(1 - P(\mathbf{v}_j|\mathbf{x}_i))$ for all $\mathbf{x}_i$ and each of them could bring in noise, making the result unreliable. In contrast, computing the lower bound via the max-pooling scheme tends to obtain a more reliable estimate since it only considers the largest term.

Based on this interpretation, the work in [26] further generalises the max-pooling to mixed-order max-pooling to model the higher order occurrence information of a visual word in an image, that is, the probability that a word occurs more than k times. As shown in that work, the classification performance can be further improved by using the mix-order max-pooling.

The analysis in [26] can also be generalised to other coding schemes with continuous coding coefficient, if the coefficient can be related to the membership score. From this viewpoint, sum-pooling will not be suitable for this coding scheme since it may accumulate low membership scores to a high one, which, however, is not a good indication of the presence of a visual word in such a case. For example, by the sum-pooling, 100 local descriptors with coding value 0.002 for a given bin will produce the pooled value of 0.2, which appears to have the same effect of observing one local descriptor with coding value 0.2. However, it is clear that the former has much weaker indication of the presence of the corresponding visual word.

Another interpretation of the good classification performance of linear SVMs with the max-pooling can be obtained from the "burstiness" phenomenon, which was initially discovered in the text classification [30] and was later observed in image retrieval with the Bag-of-features model [18]. This phenomenon indicates that "a visual pattern appears often more frequently than a statistically independent model would predict" [18]. Basically, it suggests that if one visual pattern appears once, it will be more likely to occur again in the same image. Intuitively, it can be understood in the way that the occurrence of some visual concepts will produce many repetitive local visual patterns. For example, the occurrence of a wall will produce many local patches corresponding to bricks. However, the size of an object often changes dramatically from image to image due to scale variation, which makes the occurrence frequency of its corresponding local visual patterns unstable. As a result, the sum-pooling of the coding vectors suffers from this instability because it essentially reflects the occurrence frequency of the visual words. For linear SVMs, its decision function is merely a weighted sum of the pooled coding values and thus it tends to be affected by the variation of the occurrence frequency. In contrast, nonlinear kernel SVMs can mitigate the adverse effect of burstiness phenomenon by its nonlinear operation. For example, in HIK $\kappa(\mathbf{z}_i, \mathbf{z}_j) = \sum_{l=1}^{d} \min(\mathbf{z}_{il}, \mathbf{z}_{jl})$, if a statistically less probable large value occurs at $\mathbf{z}_{il}$ due to the burstiness, its impact will be capped by $\mathbf{z}_{jl}$. However, if max-pooling is used, the linear SVMs will no longer be affected by the occurrence frequency of a visual word. In some sense, the max-pooling scheme builds a better image-level kernel and it helps the linear SVMs achieve better classification performance.

5.4 Application of SVMs with Point-Set Kernels

5.4.1 Introduction to Point-Set Kernels

Recall that in the Bag-of-features model a set of local feature descriptors is extracted from an image. Hence, an image can essentially be viewed as a point set in a multi-dimensional feature descriptor space. In addition to building a visual codebook and generating an image representation, another line of research measures the similarity between two images directly based on the associated point sets. This is desirable from the perspective of SVM application because the similarity can be used as a kernel function to perform nonlinear SVM classification. Compared with the approach of building visual codebooks, this approach can lead to a more compact and conceptually simpler classification system. Better classification result could even be achieved when appropriate point-set kernels are employed.

A variety of point-set kernels have been developed in the recent literature on machine learning and computer vision. Generally speaking, point-set kernels are constructed in two ways. One is to evaluate the similarity between the points (e.g., local feature descriptors) in two sets via a common kernel (often called a local or base kernel) and then combine the kernels to obtain a point-set kernel. The simplest one may be the sum-match kernel that sums the local kernels between every pair of points in the two sets. However, in the presence of outliers, this kernel cannot effectively reflect the set similarity because good matches between points are often buried by a large number of bad matches. In [51], a sum-max kernel is proposed, which averages the maximum local kernel value from each point in one set with respect to the points in the other set. In [29], a sum-exponent kernel is put forward. It computes the sum of the local kernels over all pairs of points in the two sets after raising each kernel value to the power p. This allows it to adjust the weight of each local kernel in the summation instead of treating them equally. The aforementioned sum-match kernel and sum-max kernel can be regarded two special cases of the sum-exponent kernel. The work in [3] suggests considering only the local kernels between the points that can be truly matched by a matching algorithm and discusses the way to make the obtained point-set kernel to satisfy the Mercer's condition. In [46], a general family of set kernels is derived based upon local kernels. The proposed kernel can combine local kernels in a linear or nonlinear way, where the nonlinear combination is achieved by mapping each point set onto a high-dimensional matrix space. In addition, the work in [37] considers the case where each point corresponds to a pixel in an image or a voxel of a video sequence. Taking into account the location information of each point, a neighborhood kernel is defined to compare each pair of points and the point-set kernel is defined as the average of all the neighborhood kernels.

The other way to construct a point-set kernel is to estimate a probability distribution of the points in each set and use the similarity between the two distributions to define a set-level kernel. The work in [21] uses the Bhattacharyya's affinity between two distributions to define a kernel for sets of vectors. To ensure the kernel to have sufficient representational power, that work maps the vectors onto a kernel-induced feature space and computes the Bhattacharyya's affinity of the probability distributions in the feature space. Similarly, in [35] a Gaussian Mixture Model (GMM) is used to obtain a probabilistic model of each set and the Kullback–Leibler divergence between the probabilistic models is used to define a point-set kernel. The work in [9] is motivated by comparing the distributions of two point sets before and after the two sets are merged. Intuitively, a strengthened distribution will be obtained if the two point sets are similar. Following this idea, that work develops point-set kernels by studying the properties of the concatenation of two point sets. In addition, the above way of constructing a point-set kernel can be related to building a kernel based on a generative model, which has been widely used as a means to combining generative models with discriminative classifiers. Fisher kernel [17] may be the most commonly used one in this regard and it has been applied to image classification recently [38, 39].

For image categorisation with the Bag-of-features model, a systematic study of point-set kernels is conducted in [51]. Among the existing point-set kernels developed in this area, the Pyramid Matching Kernel (PMK) [14], the Efficient Matching Kernel (EMK) [2] and the Fisher kernel [38] are three representative methods. This following parts will introduce the work in [51]. After that, it will be focused on the PMK, the EMK and the Fisher kernel.

5.4.2 A Kernel Recipe to Local Feature-Based Image Recognition

A general kernel-based approach is proposed in [51] for image recognition with local feature descriptors. Its motivation is to combine the representation power of local features with the excellent discriminative capability of the SVM classifiers. To achieve this goal, that work proposes a class of new kernels for point sets based on the commonly used kernels, or equally the local kernels mentioned above. Let $\mathscr{A}_i = \{\mathbf{x}_{i1}, \mathbf{x}_{i2}, \cdots, \mathbf{x}_{in_i}\}$ and $\mathscr{A}_j = \{\mathbf{x}_{j1}, \mathbf{x}_{j2}, \cdots, \mathbf{x}_{jn_j}\}$ denote the two point sets associated with images i and j, where $\mathbf{x}$ is a local feature descriptor. A measure evaluates the similarity of the set $\mathscr{A}_i$ with respect to the set $\mathscr{A}_j$ is defined as

$$m(\mathscr{A}_i|\mathscr{A}_j) = \frac{1}{n_i} \sum_{p=1}^{n_i} \max_{q=1}^{n_j} \kappa_l(\mathbf{x}_{ip}, \mathbf{x}_{jq}), \tag{5.42}$$

where $\kappa_l(\mathbf{x}, \mathbf{x}')$ plays the role of a local kernel and any existing kernel can be used for it. This measure finds the best match (in terms of the value of the local kernel) of each local feature in $\mathscr{A}_i$ in the set $\mathscr{A}_j$ and computes the average. Note that $m(\mathscr{A}_i|\mathscr{A}_j)$ does not equal $m(\mathscr{A}_j|\mathscr{A}_i)$ in general. With such a measure, the work in [51] defines a new class of kernel as

$$\kappa(\mathscr{A}_i, \mathscr{A}_j) = \frac{1}{2}\left[m(\mathscr{A}_i|\mathscr{A}_j) + m(\mathscr{A}_j|\mathscr{A}_i)\right]. \tag{5.43}$$

This kernel is a symmetrical function over the local features. Both the non-Gaussian RBF kernel and the χ^2-RBF kernel in Sect. 5.2.1 have been shown in [51] as a valid local kernel. In the experimental study, that work compares a nearest-neighbour classifier with a predefined distance metric and the SVM classifier using the corresponding point-set kernel. The results on object and face recognition data sets demonstrate the superiority of the SVM classifier with the new class of point-set kernels. However, as pointed out in [29], this class of kernels does not necessarily satisfy the Mercer's condition due to the use of the "max" operation.

5.4.3 *Pyramid Matching Kernel*

The work of PMK aims to develop a kernel function for SVM classification that can efficiently measure the similarity of two point sets [14]. It can be regarded as combining a set of local kernels to produce a point-set kernel. As indicated by its name, the PMK builds a pyramid of multi-resolution histograms to quantise the points in the two sets to be compared. At each resolution, the HIK, introduced in Sect. 5.2.1.1, is used as the local kernel to evaluate the similarity of the two sets. The similarity from different levels is then linearly combined to obtain the set-level similarity, with different weights used for the multiple resolutions.

Recall that $\mathbf{x}$ denotes a local feature descriptor in a d-dimensional space. The PMK partitions the volume occupied by the descriptors in the d-dimensional space using a set of bins with a gradually increasing size. For example, the side length of the d-dimensional bins is doubled at each resolution level. By appropriately scaling the data, the smallest bin size ensures that each individual descriptor will reside in its own bin, whereas the largest bin size ensures that all the descriptors will be contained in the same bin. Assigning the local descriptors into these bins leads to a hierarchy of multi-resolution histograms, and the bin values of this set of histograms vary with different local descriptor sets. Recall that $\mathscr{A}$ denotes a set of local descriptors extracted from an image. By concatenating the multi-resolution histograms for the set $\mathscr{A}$, a long vector can be obtained as $\phi(\mathscr{A}) = (\mathbf{h}_{-1}(\mathscr{A}), \mathbf{h}_0(\mathscr{A}), \cdots, \mathbf{h}_L(\mathscr{A}))^\top$, where $\mathbf{h}_{-1}$ indicates the histogram with the highest resolution for which no descriptor is matched (falling into the same bin) and $\mathbf{h}_L$ indicates the histogram with the lowest resolution for which all local feature descriptors stay in the same bin.

Based on the representation of $\phi(\mathscr{A})$, the PMK evaluates the similarity between the two sets $\mathscr{A}_i$ and $\mathscr{A}_j$ (or equally the similarity of images i and j) as

$$\kappa(\mathscr{A}_i, \mathscr{A}_j) = \sum_{l=0}^{L} \omega_l n_l, \tag{5.44}$$

where n_l is the number of "new" matches found at the level l and ω_l is a weight indicating the contribution of a match at level l to the final similarity. Note that a new match at level l means that a pair of local descriptors falls into the same bin of a histogram at level l but resides in different bins for any level lower than l (lower levels correspond to the histograms with higher resolutions). The weight for new matches decreases with the increase of level. This is used to emphasise that the matches at higher-resolution histograms are more important for similarity evaluation. The number of matched points at level l, denoted by $\hat{n}_l$, is computed by histogram intersection as

$$\hat{n}_l = \sum_{i=1}^{T_l} \min(\mathbf{h}_{li}(\mathscr{A}), \mathbf{h}_{li}(\mathscr{B})), \tag{5.45}$$

where T_l is the number of bins in the histogram $\mathbf{h}_l(\cdot)$. Based on $\hat{n}_l$, the number of new matches at each level l can be conveniently worked out as $n_l = \hat{n}_l - \hat{n}_{l-1}$. In addition, to remove the impact of the cardinality of a point set (e.g., a large-sized point set usually has more descriptors to be matched with the descriptors in other sets), a normalised version of the PMK is defined as

$$\kappa_{nrml}(\mathscr{A}_i, \mathscr{A}_j) = \frac{\kappa(\mathscr{A}_i, \mathscr{A}_j)}{\sqrt{\kappa(\mathscr{A}_i, \mathscr{A}_i)\kappa(\mathscr{A}_j, \mathscr{A}_j)}}. \tag{5.46}$$

The PMK is proved to be a Mercer kernel in [14] and it can be efficiently computed once the two sets of local feature descriptors are given. With the use of histogram interaction as the local kernel, the PMK works well with two sets having different cardinalities and can effectively conduct partial matching. This property is important for image categorisation because (1) the number of local descriptors from different images is often different; (2) local descriptors can disappear due to occlusion or the change of view angles, object pose, and scale; (3) irrelevant or noisy descriptors may appear due to the presence of background clutter. As experimentally demonstrated in that work, the pyramid matching used by the PMK can well approximate the result obtained by applying optimal matching between the points in two sets. The advantages of the PMK over some of the existing point-set kernels are summarised in [14] as (1) it is computationally more efficient; (2) it is proved to be positive-definite; (3) it does not need to fit a parametric model to data; (4) it can handle two sets of different cardinalities. As shown in the experimental study in [14], this kernel can demonstrate excellent classification performance in image categorisation tasks.

5.4.4 Efficient Matching Kernel

The motivation of the EMK [2] is to speed up the training and test phases in which a point-set kernel is used. Recall that a typical way to construct a point-set kernel is to combine the local kernels over all pairs of points, leading to a sum-match kernel

$$\kappa(\mathscr{A}_i,\mathscr{A}_j)=\frac{1}{|\mathscr{A}_i|}\frac{1}{|\mathscr{A}_j|}\sum_{\mathbf{x}\in\mathscr{A}_i}\sum_{\mathbf{x}'\in\mathscr{A}_j}\kappa_l(\mathbf{x},\mathbf{x}'), \tag{5.47}$$

where $\mathscr{A}_i$ denotes a set of points, $\mathbf{x}$ is a point (e.g., a local feature descriptor in a d-dimensional space) in this set and $|\mathscr{A}_i|$ is the cardinality of the set. Computing such a kernel has a computational complexity of $\mathscr{O}(|\mathscr{A}_i||\mathscr{A}_j|d)$. This complexity increases to $\mathscr{O}(|\mathscr{A}_i||\mathscr{A}_j|dn^2)$ when training an SVM classifier with a set of n training images. This makes the application of SVMs with such a kernel inefficient in handling a large-sized training set. Also, such a complexity prevents the obtained SVM classifier from efficiently classifying unseen images. Assuming that the local kernel $\kappa_l(\mathbf{x},\mathbf{x}')$ can be expressed as an inner product between $\psi(\mathbf{x})$ and $\psi(\mathbf{x}')$ in a finite-dimensional kernel-induced feature space, the above sum-match kernel can be rewritten as

$$\kappa(\mathscr{A}_i,\mathscr{A}_j)=\left\langle\frac{1}{|\mathscr{A}_i|}\sum_{\mathbf{x}\in\mathscr{A}_i}\psi(\mathbf{x}),\frac{1}{|\mathscr{A}_j|}\sum_{\mathbf{x}'\in\mathscr{A}_j}\psi(\mathbf{x}')\right\rangle\triangleq\left\langle\Psi(\mathscr{A}_i),\Psi(\mathscr{A}_j)\right\rangle, \tag{5.48}$$

where it is defined that $\Psi(\mathscr{A}_i)=\frac{1}{|\mathscr{A}_i|}\sum_{\mathbf{x}\in\mathscr{A}_i}\psi(\mathbf{x})$. As proposed in the work of EMK, if the implicit mapping $\psi(\mathbf{x})$ is approximated by an explicit mapping $\phi(\mathbf{x})$, the sum-match kernel evaluation can be avoided and a linear SVM classifier will be sufficient. This will remove the aforementioned computational issue and at the same time maintain the classification performance brought by the kernel trick.

Let $\{\mathbf{x}_1,\mathbf{x}_2,\cdots,\mathbf{x}_m\}$ be a set of predefined basis vectors in a d-dimensional space. Their images under the mapping $\psi(\cdot)$ form a matrix $\mathbf{B}=(\psi(\mathbf{x}_1),\cdots,\psi(\mathbf{x}_m))$. The work of EMK approximates $\psi(\mathbf{x})$ with its projection into the space spanned by $\mathbf{B}$. The project coefficient $\mathbf{z}$ can be obtained by minimising the reconstruction error

$$\mathbf{z}^\star=\arg\min_{\mathbf{z}\in\mathbb{R}^m}\|\psi(\mathbf{x})-\mathbf{B}\mathbf{z}\|_2^2. \tag{5.49}$$

The optimal solution $\mathbf{z}^\star$ can be analytically expressed as $\mathbf{z}^\star=\left(\mathbf{B}^\top\mathbf{B}\right)^{-1}\left(\mathbf{B}^\top\psi(\mathbf{x})\right)=\left(\mathbf{B}^\top\mathbf{B}\right)^{-1}\mathbf{k}_\mathbf{B}(\mathbf{x})$, where $\mathbf{k}_\mathbf{B}(\mathbf{x})$ is a vector consisting of the (local) kernel values between $\mathbf{x}$ and each of the m basis vectors. Let $\mathbf{K}_\mathbf{B}$ denote the $m\times m$ (local) kernel matrix computed over all the basis vectors. An approximate kernel is defined with the projection as

$$\kappa_{\text{appro}}(\mathbf{x},\mathbf{x}') = (\mathbf{B}\mathbf{z}^{\star})^{\top}(\mathbf{B}\mathbf{z}'^{\star}) = \mathbf{k}_{\mathbf{B}}(\mathbf{x})^{\top}\mathbf{K}_{\mathbf{B}}^{-1}\mathbf{k}_{\mathbf{B}}(\mathbf{x}') = \mathbf{k}_{\mathbf{B}}(\mathbf{x})^{\top}\mathbf{C}^{\top}\mathbf{C}\mathbf{k}_{\mathbf{B}}(\mathbf{x}'), \quad (5.50)$$

where $\mathbf{C}$ is a matrix satisfying $\mathbf{C}^{\top}\mathbf{C} = \mathbf{K}_{\mathbf{B}}^{-1}$. In doing so, the mapping of the approximate kernel can be explicitly obtained as $\phi(\mathbf{x}) = \mathbf{C}\mathbf{k}_{\mathbf{B}}(\mathbf{x})$. With this mapping, the approximate point-set kernel can be defined as an inner product

$$\kappa_{\text{appro}}(\mathscr{A}_i,\mathscr{A}_j) = \left\langle \frac{1}{|\mathscr{A}_i|}\sum_{\mathbf{x}\in\mathscr{A}_i}\phi(\mathbf{x}), \frac{1}{|\mathscr{A}_j|}\sum_{\mathbf{x}'\in\mathscr{A}_j}\phi(\mathbf{x}')\right\rangle \triangleq \langle\Phi(\mathscr{A}_i)\Phi(\mathscr{A}_j)\rangle, \quad (5.51)$$

where it is defined that $\Phi(\mathscr{A}_i) = \frac{1}{|\mathscr{A}_i|}\sum_{\mathbf{x}\in\mathscr{A}_i}\phi(\mathbf{x}) = \frac{1}{|\mathscr{A}_i|}\sum_{\mathbf{x}\in\mathscr{A}_i}\mathbf{C}\mathbf{k}_{\mathbf{B}}(\mathbf{x})$. Since this mapping can be explicitly obtained, it becomes unnecessary to evaluate the kernel matrix for κ_{appro} and a linear classifier can be employed. As can be seen, when the number of basis vectors, m, is not large, the computation at the training and test phases can be considerably reduced.

The last issue is to decide the matrix $\mathbf{B}$ which consists of the m basis vectors. By randomly selecting l local feature descriptors, the work of EMK jointly learns the optimal $\mathbf{B}$ and the projection coefficient $\mathbf{z}$ for each descriptor by minimising the total reconstruction error

$$\{\mathbf{B}^{\star},\mathbf{z}_1^{\star},\cdots,\mathbf{z}_m^{\star}\} = \arg\min\sum_{i=1}^{l}\|\psi(\mathbf{x}_i) - \mathbf{B}\mathbf{z}_i\|_2^2. \quad (5.52)$$

Applying the result of $\mathbf{z}^{\star} = \left(\mathbf{B}^{\top}\mathbf{B}\right)^{-1}\left(\mathbf{B}^{\top}\psi(\mathbf{x})\right)$ again, the variable $\mathbf{z}$ is removed and an optimisation problem solely for $\mathbf{B}$ is obtained as

$$\mathbf{B}^{\star} = \arg\min\left(-\sum_{i=1}^{l}\mathbf{k}_{\mathbf{B}}(\mathbf{x}_i)^{\top}\mathbf{K}_{\mathbf{B}}^{-1}\mathbf{k}_{\mathbf{B}}(\mathbf{x}_i)\right). \quad (5.53)$$

The optimal $\mathbf{B}$ is solved by a gradient descent algorithm. This completes the derivation of the EMK. As seen, the EMK method has no constraint on the type of local kernels and therefore can be widely applied.

In [2], a linear SVM classifier with the EMK is compared with a linear SVM classifier and a nonlinear SVM classifier with the Gaussian RBF kernel. For the latter two, the input is the histogram representation obtained with respect to a predefined visual codebook. In the experiment, the number of basis vectors m is 1,000 and 100,000 local feature descriptors are randomly sampled to optimise the matrix $\mathbf{B}$. The results on three benchmark data sets including Scene-15, Caltech-101 and Caltech-256 show that the proposed EMK can help the linear SVM classifier produce better classification performance than the other two SVM classifiers in comparison. Also, it leads to much higher computational efficiency in both training and test stages, especially when the number of training images is large.

5.4.5 Fisher Kernel

Fisher kernel [17] provides a way to compare samples induced by a generative model $p(\mathbf{x}|\theta)$. It maps a sample to a feature vector in the gradient space of the model parameters θ. The intuition is that similar samples induce similar log-likelihood gradients of the model parameters. Let $\mathbf{x}_i$ and $\mathbf{x}_j$ denote two samples. Fisher kernel is defined as

$$\kappa(\mathbf{x}_i, \mathbf{x}_j) = \mathbf{g}(\mathbf{x}_i)^\top \mathbf{U}^{-1}\mathbf{g}(\mathbf{x}_j), \tag{5.54}$$

where $\mathbf{g}(\mathbf{x}) = \nabla_\theta \log(p(\mathbf{x}|\theta))$ is the gradient vector describing the changing direction of θ to better fit the model. $\mathbf{U}$ is the Fisher information matrix that weights this similarity measure, for example, normalising the dynamic range of the components of the gradient vector [38].

Fisher kernel has recently been successfully applied to image categorisation with the Bag-of-features model [22, 38]. In this application, $\mathbf{x}$ ($\mathbf{x} \in \mathbb{R}^d$) denotes a local feature descriptor extracted from a set of training images. Based on these descriptors, a GMM with k components is learned

$$p(\mathbf{x}|\theta) = \sum_{i=1}^{k} w_i \mathcal{N}(\mathbf{x}|\mu_i, \Sigma_i) \tag{5.55}$$

where w_i is the mixture weight, μ_i the mean vector and Σ_i the covariance matrix of the ith component. These model parameters are compactly represented by $\theta = \{w_i, \mu_i, \Sigma_i\}_{i=1}^{k}$. All the covariance matrices Σ_i are assumed to be diagonal and expressed as $\Sigma_i = \{\sigma_{i1}^2, \sigma_{i2}^2, \cdots, \sigma_{id}^2\}$. Conceptually, each Gaussian component can be understood as a visual word.

Let $\mathscr{A} = \{\mathbf{x}_1, \mathbf{x}_2, \cdots, \mathbf{x}_n\}$ be a set of local feature descriptors extracted from an image. Let $L(\mathbf{x}|\theta) = \log(p(\mathbf{x}|\theta))$ denote the log-likelihood. The gradient vector $\mathbf{g}(\mathbf{x})$ is a concatenation of the partial derivatives of L with respect to all the parameters, that is, $\mathbf{g}(\mathbf{x}) = (\nabla_{w_i} L(\mathbf{x}|\theta);\ \nabla_{\mu_i} L(\mathbf{x}|\theta);\ \nabla_{\Sigma_i} L(\mathbf{x}|\theta))$. The partial derivatives with respect to the tth local descriptor, $\mathbf{x}_t$, are worked out as:

$$\begin{aligned}
\frac{\partial L(\mathbf{x}_t|\theta)}{\partial w_i} &= \left(\frac{\gamma_t(i)}{w_i} - \frac{\gamma_t(1)}{w_1}\right); \text{ for } i \geq 2,\\
\frac{\partial L(\mathbf{x}_t|\theta)}{\partial \mu_{ij}} &= \gamma_t(i)\left(\frac{\mathbf{x}_{tj} - \mu_{ij}}{\sigma_{ij}^2}\right),\\
\frac{\partial L(\mathbf{x}_t|\theta)}{\partial \sigma_{ij}} &= \gamma_t(i)\left(\frac{(\mathbf{x}_{tj} - \mu_{ij})^2}{\sigma_{ij}^3} - \frac{1}{\sigma_{ij}}\right),
\end{aligned} \tag{5.56}$$

where $i = 1, \cdots, k$ and $j = 1, \cdots, d$ index the Gaussian components and each component of a vector. The term $\gamma_t(i)$ is defined as $\gamma_t(i) = \frac{w_i \mathcal{N}(\mathbf{x}_t|\mu_i, \sigma_i)}{\sum_{i=1}^{k} w_i \mathcal{N}(\mathbf{x}_t|\mu_i, \sigma_i)}$. By summing the gradient vectors with respect to the n local descriptors, an image-level representation can be obtained as

$$\mathbf{z} = \mathbf{U}^{-\frac{1}{2}} \sum_{t=1}^{n} \mathbf{g}(\mathbf{x}_t), \tag{5.57}$$

where $\mathbf{U}^{-\frac{1}{2}}$ is used to normalise the dynamic range of each dimension of the gradient vector. Let $\mathbf{z}$ and $\mathbf{z}'$ denote the image representation for images I and I'. A linear kernel between two images can be defined as

$$\kappa(I, I') = \langle \mathbf{z}, \mathbf{z}' \rangle = \sum_{p=1}^{n_i} \sum_{q=1}^{n_j} \mathbf{g}(\mathbf{x}_p)^\top \mathbf{U}^{-1} \mathbf{g}(\mathbf{x}'_q), \tag{5.58}$$

where $\mathbf{x}$ and $\mathbf{x}'$ are the local descriptors from images I and I', respectively. It can be seen that this kernel can be regarded as a point-set kernel between the two images. This point-set kernel is in the form of a sum-match kernel, where the local kernel is the Fisher kernel applied to local feature descriptors.

As pointed out in [38], the gradient representation of Fisher kernel on GMM can be related to the Bag-of-features model in image categorisation. The histogram representation of the Bag-of-features model considers only the number of occurrences of each visual word, which corresponds to the zeroth-order statistics. In contrast, the Fisher kernel additionally considers the first- and second-order statistics. This produces a higher-dimensional image representation even when the number of visual words is small, which could be helpful for classification. Moreover, for this representation a linear kernel has been able to perform well, avoiding the use of costly nonlinear kernels.

The work in [38] verifies the advantage of the Fisher-kernel-induced image representation on two databases, an in-house database and PASCAL VOC2006. The GMM is trained in an unsupervised way by using the local feature descriptors extracted from all images. Linear SVMs and Sparse Logistic Regression (SLR) are investigated. Classification with the Fisher-kernel-induced image representation achieves comparable or even better performance than the results reported in the literature. In addition, the Fisher-kernel-induced image representation can bring computational advantages because it can achieve excellent classification performance with small-sized visual codebooks.

5.5 Conclusion

Kernel is the soul of the SVM classifiers and the place where the prior knowledge of an application is accommodated. The performance of SVM classifiers in an application largely depends on the appropriateness and efficiency of the employed kernels. Without exception, this is also true for the application of SVMs to image categorisation with the Bag-of-features model. This chapter takes a unique perspective to review the development of kernels in this application. Focused on two typical image representations, histogram and point set, the chapter introduces the representative kernels used by the SVM classifiers for each of them. Also, the progress of the use of kernels for each image representation has been briefly shown. For the histogram representation, we can see the trend of avoiding explicitly using kernels, with the advent of advanced coding and pooling techniques as well as powerful kernel approximation methods. This trend has also been seen in the point-set-based representation through the development of EMK. The driving force of these changes is just the applications of SVMs, which need efficient image classification methods to handle large-scale tasks, reduce system complexity and improve recognition performance. We can expect that novel kernels and the novel ways of using kernels will continue emerging with the applications of SVMs to image recognition.

References

1. Barla, A., Odone, F., Verri, A.: Histogram intersection kernel for image classification. In: ICIP, vol. 3, pp. 513–516 (2003)
2. Bo, L., Sminchisescu, C.: Efficient match kernel between sets of features for visual recognition. In: Neutral Information Proceeding Systems, pp. 135–143 (2009)
3. Boughorbel, S., Tarel, J.-P., Fleuret, F.: Non-mercer kernels for svm object recognition. In: BMVC, pp. 1–10 (2004)
4. Boureau, Y., Ponce, J., LeCun, Y.: A theoretical analysis of feature pooling in vision algorithms. In: Proceedings of International Conference on Machine learning (ICML'10), pp. 111–118 (2010)
5. Boureau, Y.-L., Bach, F., LeCun, Y., Ponce, J.: Learning mid-level features for recognition.In: Computer Vision Pattern Recognition, pp. 2559–2566 (2010)
6. Boureau, Y.-L., Bach, F., LeCun, Y., Ponce, J.: Learning mid-level features for recognition. In: Computer Vision Pattern Recognition, pp. 2559–2566 (2010)
7. Chapelle, O., Haffner, P., Vapnik, V.: Support vector machines for histogram-based image classification. IEEE Trans. Neural Netw. **10**(5), 1055–1064 (1999)
8. Csurka, G., Dance, C.R., Fan, L., Willamowski, J., Bray, C.: Visual categorization with bags of keypoints. In: ECCV, pp. 1–22 (2004)
9. Cuturi, M., Vert, J.-P.: Semigroup kernels on finite sets. In: Neutral Information Proceeding Systems, vol. 17, pp. 329–336 (2004)
10. Deng, J., Dong, W., Socher, R., Li, L.-J., Li, K., Fei-Fei, L.: ImageNet: a large-scale hierarchical image database. In: Computer Vision Pattern Recognition, pp. 248–255 (2009)
11. Everingham, M., Van Gool, L., Williams, C.K.I., Winn, J., Zisserman, A.: The PASCAL visual object classes challenge (VOC2007) results. http://www.pascal-network.org/challenges/VOC/voc2007/workshop/index.html (2007)
12. Fei-Fei, L., Fergus, R., Perona, P.: Learning generative visual models from few training examples: an incremental bayesian approach tested on 101 object categories. Comput. Vis. Image Und. **106**(1), 59–70, (2007)
13. Fowlkes, C., Belongie, S., Chung, F.R.K., Malik, J.: Spectral grouping using the nyström method. IEEE T. Pattern Anal. Mach. Intell. **26**(2), 214–225, (2004)

14. Grauman, K., Darrell, T.: The pyramid match kernel: discriminative classification with sets of image features. In: International Conference on Computer Vision, vol. 2, pp. 1458–1465 (2005)
15. Griffin, G., Holub, A., Perona, P.: Caltech-256 Object Category Dataset. Tech. Report, California Institute of Technology (2007)
16. Hsieh, C.-J., Chang, K.-W., Lin, C.-J., Keerthi, S.S., Sundararajan, S.: A dual coordinate descent method for large-scale linear svm. In: Cohen, W.W., McCallum, A., Roweis, S.T. (eds.) ICML, vol. 307 of ACM International Conference Proceeding Series, pp. 408–415. ACM (2008)
17. Jaakkola, T., Haussler, D.: Exploiting generative models in discriminative classifiers. In: Neutral Information Proceeding Systems, pp. 487–493 (1998)
18. Jégou, H., Douze, M., Schmid, C.: On the burstiness of visual elements. In: Computer Vision and Pattern Recognition 2009, pp. 1169–1176 (2009)
19. Joachims, T.: A statistical learning model of text classification for support vector machines. In: SIGIR, pp. 128–136 (2001)
20. Juriem, F., Triggs, B.: Creating efficient codebooks for visual recognition. In: International Conference on Computer Vision, vol. 1, pp. 604–610 (2005)
21. Kondor, R., Jebara, T.: A kernel between sets of vectors. In: ICML, pp. 361–368 (2003)
22. Krapac, J., Verbeek, J., Jurie, F.: Modeling spatial layout with fisher vectors for image categorization. In: International Conference on Computer Vision, pp. 1487–1494 (2011)
23. Lazebnik, S., Raginsky, M.: Supervised learning of quantizer codebooks by information loss minimization. IEEE T. Pattern Anal. Mach. Intell. **31**(7), 1294–1309 (2009)
24. Lazebnik, S., Schmid, C., Ponce, J.: Beyond bags of features: spatial pyramid matching for recognizing natural scene categories. In: Computer Vision and Pattern Recognition, vol. 2, pp. 2169–2178 (2006)
25. Leung, T.K., Malik, J.: Representing and recognizing the visual appearance of materials using three-dimensional textons. Int. J. Comput. Vision **43**(1), 29–44 (2001)
26. Liu, L., Wang, L., Liu, X.: In defense of soft-assignment coding. In: International Conference on Computer Vision, pp. 2486–2493 (2011)
27. Liu, L., Wang, L., Shen, C.: A generalized probabilistic framework for compact codebook creation. In: Computer Vision and Pattern Recognition, pp. 1537–1544 (2011)
28. Lowe, D.G.: Object recognition from local scale-invariant features. In: Proceedings of the International Conference on Computer Vision, pp. 1150–1157 (1999)
29. Lyu, S.: Mercer kernels for object recognition with local features. In: Computer Vision and Pattern Recognition, vol. 2, pp. 223–229 (2005)
30. Madsen, R.E., Kauchak, D., Elkan, C.: Modeling word burstiness using the dirichlet distribution. In: International Conference on Machine learning, pp. 545–552 (2005)
31. Maji, S., Berg, A.C., Malik, J.: Classification using intersection kernel support vector machines is efficient. In: 2008 IEEE Computer Society Conference on Computer Vision and Pattern Recognition (CVPR 2008), Anchorage, 24–26 June 2008, pp. 1–8. IEEE Computer Society (2008). http://dx.doi.org/10.1109/CVPR.2008.4587630
32. Mikolajczyk, K., Schmid, C.: Scale & affine invariant interest point detectors. Int. J. Comput. Vision **60**(1), 63–86 (2004)
33. Mikolajczyk, K., Tuytelaars, T., Schmid, C., Zisserman, A., Matas, J., Schaffalitzky, F., Kadir, T., Gool, L.V.: A comparison of affine region detectors. Int. J. Comput. Vision **65**(1–2), 43–72 (2005)
34. Moosmann, F., Triggs, B., Jurie, F.: Fast discriminative visual codebooks using randomised clustering forests. In: Neutral Information Proceeding Systems, pp. 985–992 (2006)
35. Moreno, P.J., Ho, P., Vasconcelos, N.: A kullback-leibler divergence based kernel for svm classification in multimedia applications. In: Neutral Information Proceeding Systems (2003)
36. Nistér, D., Stewénius, H.: Scalable recognition with a vocabulary tree. In: Computer Vision and Pattern Recognition, vol. 2, pp. 2161–2168 (2006)
37. Parsana, M., Bhattacharya, S., Bhattacharyya, C., Ramakrishnan, K.R.: Kernels on attributed pointsets with applications. In: Platt et al.(eds.) Advances in Neural Information Processing Systems 20, Proceedings of the Twenty-First Annual Conference on Neural Information Processing Systems, Vancouver, British Columbia, Canada, December 3–6, 2007. Curran Associates (2008)

38. Perronnin, F., Dance C.R.: Fisher kernels on visual vocabularies for image categorization. In: Computer Vision and Pattern Recognition, pp. 1–8 (2007)
39. Perronnin, F., Sánchez, J., Mensink, T.: Improving the fisher kernel for large-scale image classification. In: ECCV, vol. 4, pp. 143–156 (2010)
40. Perronnin, F., Sánchez, J., Liu, Y.: Large-scale image categorization with explicit data embedding. In: The Twenty-Third IEEE Conference on Computer Vision and Pattern Recognition (CVPR 2010), San Francisco, 13–18 June 2010, pp. 2297–2304. IEEE (2010). http://dx.doi.org/10.1109/CVPR.2010.5539914
41. Philbin, J., Chum, O., Isard, M., Sivic, J., Zisserman, A.: Object retrieval with large vocabularies and fast spatial matching. In: Computer Vision and Pattern Recognition (2007)
42. Rahimi, A., Recht, B.: Random features for large-scale kernel machines. In: Platt et al. (eds.) Advances in Neural Information Processing Systems 20, Proceedings of the Twenty-First Annual Conference on Neural Information Processing Systems, Vancouver, British Columbia, Canada, December 3–6, 2007. Curran Associates (2008)
43. Rubner, Y. Tomasi, C., Guibas, L.J.: The earth mover's distance as a metric for image retrieval. Int. J. Comput. Vision **40**(2), 99–121 (2000)
44. Schiele, B., Crowley, J.L.: Object recognition using multidimensional receptive field histograms. In: ECCV, vol. 1, pp. 610–619 (1996)
45. Shalev-Shwartz, S., Singer, Y., Srebro, N.: Pegasos: primal estimated sub-gradient solver for svm. In: ICML (2007)
46. Shashua, A., Hazan, T.: Algebraic set kernels with application to inference over local image representations. In: Neutral Information Proceeding Systems, pp. 1257–1264 (2004)
47. Sivic, J., Zisserman, A.: Video Google: a text retrieval approach to object matching in videos. In: Proceedings of the International Conference on Computer Vision, vol. 2, pp. 1470–1477 (2003)
48. Swain, M.J., Ballard, D.H.: Color indexing. Int. J. Comput. Vision **7**(1), 11–32 (1991)
49. van Gemert, J., Geusebroek, J.-M., Veenman, C.J., Smeulders, A.W.M.: Kernel codebooks for scene categorization. In: ECCV, vol. 3, pp. 696–709 (2008)
50. Vedaldi, A., Zisserman, A.: Efficient additive kernels via explicit feature maps. IEEE Trans. Pattern Anal. Mach. Intell. **34**(3), 480–492 (2012)
51. Wallraven, C., Caputo, B., Graf, A.B.A.: Recognition with local features: the kernel recipe. In: International Conference on Computer Vision, pp. 257–264 (2003)
52. Wang, J., Yang, J., Yu, K., Lv, F., Huang, T.S., Gong, Y.: Locality-constrained linear coding for image classification. In: Computer Vision and Pattern Recognition, pp. 3360–3367 (2010)
53. Wang, L.: Toward a discriminative codebook: codeword selection across multi-resolution. In: Computer Vision and Pattern Recognition, pp. 1–8 (2007)
54. Williams, C., Seeger, M.: Using the nystrm method to speed up kernel machines. In: Neutral Information Proceeding Systems (2001)
55. Winn, J.M.: Criminisi, A.: Minka, T.P.: Object categorization by learned universal visual dictionary. In: International Conference on Computer Vision, pp. 1800–1807 (2005)
56. Wu, J., Rehg, J.M.: Beyond the euclidean distance: creating effective visual codebooks using the histogram intersection kernel. In: International Conference on Computer Vision, pp. 630–637 (2009)
57. Yang, J., Yu, K., Gong, Y., Huang, T.S.: Linear spatial pyramid matching using sparse coding for image classification. In: Computer Vision and Pattern Recognition, pp. 1794–1801 (2009)
58. Zhang, J. Marszalek, M., Lazebnik, S., Schmid, C.: Local features and kernels for classification of texture and object categories: a comprehensive study. Int. J. Comput. Vision **73**(2), 213–238 (2007)

Chapter 6
Support Vector Machines for Neuroimage Analysis: Interpretation from Discrimination

Luping Zhou, Lei Wang, Lingqiao Liu, Philip Ogunbona, and Dinggang Shen

Abstract Support vector machines (SVMs) have been widely used in neuroimage analysis as an effective multivariate analysis tool for group comparison. As neuroimage analysis is often an exploratory research, it is an important issue to characterize the group difference captured by SVM with anatomically interpretable patterns, which provides insights into the unknown mechanism of the brain. In this chapter, SVM-based methods and applications are introduced for neuroimage analysis from this point of view. The discriminative patterns are decoded from SVMs through distinctive feature selection, SVM decision boundary interpretation, and discriminative learning of generative models.

6.1 Introduction

Neuroimage analysis works on the digital imaging scans to study the relationship between the brain anatomies and functions. With the development of medical imaging techniques, we are now able to "see through" the brain. Analyzing brain images can provide fundamental insights into how the brain is organized and responds to the damage such as mental diseases. Pattern recognition plays an important role in neuroimage analysis to quantify structural and functional differences between diverse groups, identify disease precursors or structural trajectories in

L. Zhou (✉) • L. Wang • P. Ogunbona
Department of Computer Science and Software Engineering, University of Wollongong, Wollongong, NSW, Australia
e-mail: lupingz@uow.edu.au; leiw@uow.edu.au; philipo@uow.edu.au

L. Liu
School of Engineering, Australian National University, Canberra, ACT, Australia
e-mail: liulq83@gmail.com

D. Shen
Department of Radiology, University of North Carolina at Chapel Hill, Chapel Hill, NC, USA
e-mail: dgshen@med.unc.edu

Y. Ma and G. Guo (eds.), *Support Vector Machines Applications*, DOI 10.1007/978-3-319-02300-7_6,
© Springer International Publishing Switzerland 2014

neurodevelopment diseases, or detect brain states associated with a stimulus, etc. Traditional research in this field is dominated by mass univariate methods. Take the prevailing voxel-based morphometry (VBM) analysis [1], for example. After aligning brain images to a common stereotaxic space, significant group differences are identified by performing univariate statistic tests at each image voxel separately. Such methods potentially fail to capture the complex group differences induced by the combination of image voxels [2]. Therefore, support vector machines (SVMs) [3] have been applied in this field to facilitate multivariate analysis and achieved many successful applications [4–8].

Different from the traditional classification tasks in machine learning, neuroimage analysis cares about not only a good classification performance but also the nature of the difference between classes. This is because neuroimage analysis is often an exploratory research upon the unknown mechanism of the brain. Therefore a scientific insight into the problem is sometimes more important than simply learning a "black-box" predictor. For instance, we want to localize where the discrimination is with regard to the anatomical or functional brain regions. In this chapter, we will introduce the applications of SVM for neuroimage analysis from this specific perspective. How to identify and decode discriminative information not only characterizes neuroimage analysis but also touches some fundamental issues of the learning theory.

In this chapter, three major issues are discussed to achieve interpretable discriminations. First, SVM-based feature selection methods are introduced to identify the most distinctive features for discrimination (Sect. 6.2). Feature selection is the most widely used method for localizing discrimination for neuroimage analysis possibly due to its conceptual simplicity. Second, the optimal separating hyper-plane obtained by SVMs is utilized for the identification of the anatomically meaningful group difference (Sect. 6.3). This issue is considered to respect the spatial and anatomical constraints (Sect. 6.3.1) and to deal with the usage of nonlinear kernels (Sect. 6.3.2). Third, instead of obtaining the interpretation from the discriminative classifiers as mentioned above, generative models are built to directly represent the neuroimaging data, by which the interpretation and further inference can be conveniently made. The parameters of the generative models are discriminatively learned via SVMs in order to incorporate the essential class differences (Sect. 6.4).

6.2 Identification of Distinctive Features

The most common approach in pattern recognition used for identifying distinctive features is arguably feature selection. It can effectively infer which features are more important for a given recognition task and separate them from the irrelevant and noisy features. This property fits the application of neuroimage analysis very well because images are often characterized by a set of predefined features but which features are indeed useful for a given task cannot be known in advance.

Feature selection has been intensively researched in the field of pattern recognition and many methods have been proposed in the literature. Usually, a feature selection method can be categorized into one of the following three groups. The first group, namely filter methods, uses general purpose criteria that are independent of a specific classifier. Examples of methods in this group include distance or separability measures, correlations, and information-theoretic measures. The statistic tests used in the traditional VBM methods for neuroimage analysis also belong to this category. The second group depends on a specified classifier and is referred to as wrapper methods. Examples of wrapper methods include recursive feature elimination (RFE), C4.5, naive Bayes. The last group is the embedded methods in which feature selection is essentially a part of the classifier. Sparse SVMs and the LASSO method are two examples in this group. Generally speaking, filter-based methods tend to give more generic selection of features that can work reasonably well with a large family of classifiers. Also, they have the least computational load among the three groups of methods. The wrapper and embedded methods can usually achieve better selection performance since they are tightly coupled with the classifier that will be used for classification. However, they are usually computationally more intensive.

Feature selection methods from all the above three groups have been applied to neuroimage analysis in the literature. This section will be focused on the wrapper and the embedded methods coupled with SVM classifiers. The selected features not only encode the class discrepancy but also enhance the performance of SVM classifiers in turn. We will introduce three feature selection methods, namely, RFE with SVMs (wrapper method), sparse SVMs (embedded method), and multiple kernel learning (MKL) with SVMs (embedded method).

6.2.1 *Recursive Feature Elimination with SVMs*

RFE is a backward sequential feature elimination method. It iteratively removes the irrelevant and noisy features to identify the distinctive ones. In the method of SVM-RFE, the procedure of RFE is wrapped around SVM classifiers to take advantage of their excellent classification performance. SVM-RFE is first proposed in [9] for gene selection and has been successfully applied to neuroimage analysis. In the following part, RFE with linear and nonlinear SVMs are introduced respectively.

Linear SVM-RFE. Recall that an SVM decision function is $f(\mathbf{x}) = \mathbf{w}^\top \mathbf{x} + b$, where $\mathbf{w}$ is the weight vector and b the bias. In each iteration, SVM-RFE removes the feature that has incurred the least magnitude change in the SVM object function $J(\mathbf{w})$, which is the squared norm of the weight vector, $\|\mathbf{w}\|^2$, in a linear case. It is shown that such a change can be evaluated by $\Delta J(i) = w_i^2$, where w_i is the ith component of $\mathbf{w}$ and also the weight corresponding to the ith feature. The algorithm of SVM-RFE works as follows:

1. Given a set of n training samples $\{\mathbf{x}, y\}_{i=1}^n$, where $\mathbf{x}$ ($\mathbf{x} \in \mathbb{R}^d$) is a vector consisting of d features and y is the class label of $\mathbf{x}$;
2. Let t denote the number of remaining features, which is initialized by d. Let d' be the total number of features to be selected;

3. Train a linear SVM classifier with all the t features and obtain the weight vector $\mathbf{w} = (w_1, w_2, \cdots, w_t)^\top$;
4. Sort the t features based on the change $\Delta J(i) = w_i^2$ $(i = 1, 2, \cdots, t)$ and remove the feature with the least value;
5. Set $t = t - 1$. If $t > d'$, go to step 3 and terminate otherwise.

Note that the order with which the features are removed also induces a ranking of the d features based on their importance.

Nonlinear SVM-RFE. The above linear SVM-RFE has been extended to the nonlinear case to fully take advantage of the power of nonlinear SVMs and effectively identify distinctive features for classes that cannot be linearly separated. In the nonlinear case, the objective function of SVMs is often expressed in a dual form as $J(\alpha) = \frac{1}{2}\alpha^\top \mathbf{K}\alpha - \mathbf{1}^\top \alpha$ subject to appropriate constraints on α. The vector α ($\alpha \in \mathbb{R}^n$) consists of the coefficient for each of the n training samples. The $\mathbf{K}$ is an $n \times n$ matrix and its (i, j) entry is $y_i y_j k(\mathbf{x}_i, \mathbf{x}_j)$, where $k(\cdot, \cdot)$ is a kernel function employed by the nonlinear SVM classifier. Removing the ith feature changes $\mathbf{K}$ and let $\mathbf{K}(-i)$ denote the resulting matrix. To avoid entirely retraining the nonlinear SVM classifier, α is assumed to be unchanged with the removal of a feature. In this way, the change of the SVM object function can be expressed as $\Delta J(i) = \frac{1}{2}\alpha^\top \mathbf{K}\alpha - \frac{1}{2}\alpha^\top \mathbf{K}(-i)\alpha$. Applying the same procedure of linear SVM-RFE as shown above, features can then be selected and ranked based on a nonlinear SVM classifier. Note that when the number of features is large, both linear and nonlinear SVM-RFE can be generalized to remove multiple features together at each iteration to improve the computational efficiency of feature selection.

SVM-RFE has been adopted by many neuroimage applications for feature selection. For example, in [10] it is used to identify the brain regions affected by the mild cognitive impairment (MCI) based on white matter connectivity extracted from the diffusion tensor imaging (DTI). In [11] SVM-RFE is used to identify the shape features related to the Alzheimer's disease (AD) from hippocampal surfaces. It is also used to identify the distinctive features for a number of mental disorders, such as AD [12], autism spectrum disorder (ASD) [13], and attention deficit hyperactivity disorder (ADHD) [14], from structural and functional magnetic resonance imaging (MRI). In all these applications, SVM-RFE appears to be an efficient tool to select a small subset of sufficiently discriminative features, with which the SVM classifier can achieve better classification performance than using all the original features. The feature ranking obtained by SVM-RFE indicates the relative importance of features to the discrimination. However, the number of features that should be selected from the ranking usually needs to be experimentally determined.

6.2.2 *Sparse Support Vector Machines*

Different from SVM-RFE, sparse SVM can automatically determine which features to be selected. In a linear SVM classifier, each feature is weighted by the corresponding component of the weight vector $\mathbf{w}$. When a component is suppressed to zero, it means that the corresponding feature will not be used for classification. This idea is used by sparse SVMs to perform feature selection, among which 1-norm SVM is a representative that has been used for neuroimage analysis [15]. The optimization problem of 1-norm SVM is expressed as

$$\begin{aligned} &\min_{\mathbf{w},b,\xi} \|\mathbf{w}\|_1 + C\textstyle\sum_{i=1}^n \xi_i \\ &s.t. \;\; y_i(\mathbf{w}^\top \mathbf{x}_i + b) \geq 1 - \xi_i, \;\; \xi_i \geq 0, \;\; \forall i. \end{aligned} \tag{6.1}$$

Note that the mere (but critical) difference of 1-norm SVMs from the conventional SVMs lies in the fact that in the objective function $\|\mathbf{w}\|_2$ is replaced by $\|\mathbf{w}\|_1$, which is the 1-norm of $\mathbf{w}$. Minimizing this norm makes $\mathbf{w}$ tend to be sparse; that is, encouraging the components of the weight vector to become zero. The above optimization problem can be conveniently converted to a linear optimization problem

$$\begin{aligned} &\min_{\mathbf{w},b,\xi,\mathbf{v}} \sum_{j=1}^{d} v_j + C\sum_{i=1}^{n} \xi_i \\ &s.t. \;\; y_i(\mathbf{w}^\top \mathbf{x}_i + b) \geq 1 - \xi_i, \;\; \xi_i \geq 0, \;\; \forall i; \\ &\qquad -v_j \leq w_j \leq v_j, \;\; j = 1, 2, \cdots, d, \end{aligned} \tag{6.2}$$

where d is the dimensionality of $\mathbf{w}$. Solving this linear optimization problem will produce a sparse $\mathbf{w}$, which can then be used to identify the distinctive features for classification.

The 1-norm SVMs have been used to conduct feature selection for classifying single photon emission computed tomography (SPECT) images from the AD patients and the healthy subjects [15]. In that work, the normalized voxel intensities in the SPECT images for each subject are used as the features for classification. However, the number of training samples is much less than the number of voxels in that work (less than 100 training samples vs. more than 1,000 voxels). This causes the notorious "small sample problem" which could hurt the generalization performance of the obtained classifier. To address this issue, the work in [15] proposes the 1-norm SVMs to select a small number of important features for classification. Moreover, that work argues that spatial information should be taken into consideration so that the results would be consistent with the clinician's intuition: rather than selecting scattered individual voxels, a region of contiguous voxels should be selected together. To incorporate the spatial information, that work defines a binary matrix $\mathbf{R}$ whose (i, j)-th entry indicates the spatial relationship between the ith and jth features. This matrix is then incorporated into the above 1-norm SVMs as a regulariser for the weight vector $\mathbf{w}$. Accordingly, the optimization problem in Eq. (6.2) becomes

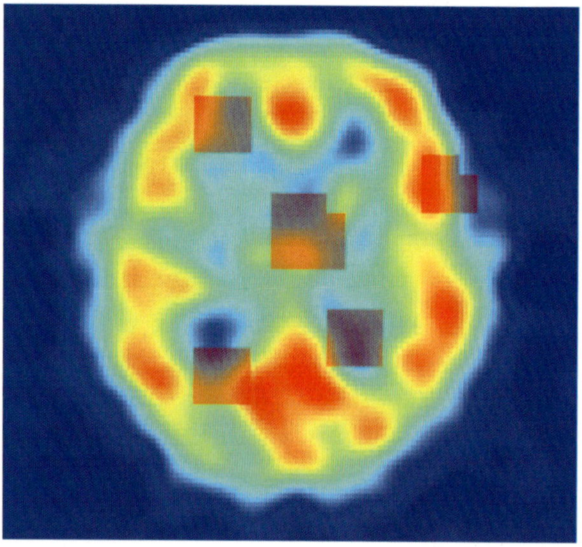

Fig. 6.1 Image courtesy of [15]: this image visualizes part of the features (regions of contiguous voxels) selected by the 1-norm SVMs with spatial information in a 2D view. The selected features are overlaid on the SPECT image of an Alzheimer's disease patient

$$\min_{\mathbf{w},b,\xi,\mathbf{v}} \sum_{j=1}^{d} v_j + C\sum_{i=1}^{n} \xi_i \tag{6.3}$$

$$s.t. \quad y_i(\mathbf{w}^\top \mathbf{x}_i + b) \geq 1 - \xi_i, \;\; \xi_i \geq 0, \;\; \forall i;$$

$$-\mathbf{Rv} \leq \mathbf{w} \leq \mathbf{Rv},$$

where the "$\leq$" in the last constraint is applied component-wisely. Imposing the last constraint makes the value of w_i depend on not only the ith feature but also the features that are spatially close. The idea of spatial regularization used by the 1-norm SVM also appears in [16] as introduced in Sect. 6.3.1, where a general framework of incorporating spatial and anatomical proximities into the traditional 2-norm SVMs is proposed. That work cannot perform feature selection as the sparse SVMs do via using 1-norm.

The effectiveness of the 1-norm SVMs with spatial information is experimentally verified by comparing with human experts, the Fisher's linear discriminant classifier and the statistical parametric mapping commonly used in neuroimage analysis. It is found that the 1-norm SVMs achieve higher classification performance than human experts and the statistical parametric mapping method. Also, it shows comparable performance as the Fisher's linear discriminant classifier on the generalization performance when the training and the test data are from different institutions. To demonstrate the effectiveness of the 1-norm SVMs on feature selection, the work in [15] visualizes a subset of the selected features, as given in Fig. 6.1. As indicated in that work, the proposed method selects those meaningfully grouped features that are more consistent with the clinician's intuition than the traditional feature selection algorithms.

6.2.3 *Multiple Kernel Learning with SVMs*

MKL has been an active research topic in the field of machine learning and pattern recognition in the last decade. It aims to automatically learn the most suitable kernel for a given learning task by linearly combining multiple base kernels. The weights of the combination can naturally reflect the importance of each base kernel when they are appropriately normalized to become comparable. When a sparsity constraint is imposed upon the combination weights, MKL can be used to select the distinctive kernels for a given learning task. MKL can be readily related to feature selection because a base kernel is often built upon one or a group of features. In this case, selecting base kernels also selects the corresponding features. Moreover, MKL brings advantages to feature selection because (1) a group of features can be conveniently and jointly considered via a base kernel and (2) the potential importance of features can be more comprehensively evaluated by MKL through adaptively selecting the base kernels that work best with features. In neuroimage analysis, MKL-SVM has been used to identify the distinctive features or the distinctive modality modeled by a set of features.

The MKL algorithm can be briefly described as follows. Let $k_1, \cdots, k_m$ be a set of base kernels. Let $\lambda = \{\lambda_1, \cdots, \lambda_m\}$ be a set of weights with the constraints of $\lambda \geq 0$ and $\sum_{i=1}^{m} \lambda_i = 1$. Let $k(\cdot,\cdot) = \sum_{i=1}^{m} \lambda_i k_i(\cdot,\cdot)$ denote the kernel obtained by a weighted combination of the m base kernels. Since the weights are all nonnegative, the kernel k will still be a Mercer kernel as long as all the base kernels are. In MKL-SVM, a nonlinear SVM using the kernel k is trained and the kernel weight λ_i $(i = 1, \cdots, m)$ is jointly learned with the SVM coefficients α. This leads to the following optimization problem:

$$\min_{\lambda} J(\lambda) \qquad (6.4)$$
$$s.t. \;\; \lambda_i \geq 0, \;\; \sum_{i=1}^{m} \lambda_i = 1,$$

where the objective function $J(\lambda)$ is

$$J(\lambda) = \max_{\alpha} \left(-\frac{1}{2} \sum_{i=1}^{n} \sum_{j=1}^{n} \alpha_i \alpha_j y_i y_j \sum_{l=1}^{m} \lambda_l k_l(\mathbf{x}_i, \mathbf{x}_j) + \sum_{i=1}^{n} \alpha_i \right) \qquad (6.5)$$
$$s.t. \;\; \sum_{i=1}^{n} \alpha_i y_i = 0, \;\; 0 \leq \alpha_i \leq C, \forall i.$$

This optimization can be efficiently solved, for example, by the reduced gradient method proposed in the work of SimpleMKL [17]. Note that the constraint $\sum_{i=1}^{m} \lambda_i = 1$ will make the optimal solution set, $\lambda_1, \cdots, \lambda_m$, sparse. A number of kernel combination weights will be set to zero, indicating that the corresponding base kernels are essentially not used.

MKL has been successfully applied to neuorimage analysis recently. For example, it is used to integrate the measurements from different modalities including MRI, positron emission tomography (PET), cerebrospinal fluid (CSF) parameters, and genotype [18–20]. Also, it has been applied to fuse different types of features from the same imaging modality [21]. MKL can efficiently handle a large number of kernels. When applied to neuroimage analysis, it assigns multiple kernels of varying types and hyperparameters to each feature group and automatically determines the optimal kernel weight combination [18,19]. Take the work in [18] for example, where MKL is used to differentiate MCI (a prodromal of AD) subjects who progressed to AD (called as MCI converter) and who remained stable. Both imaging data (MRI and fluorodeoxyglucose PET (FDG-PET)) and biological measures (CSF assays, NeuroPsychological Status Exams (NPSE), and APOE genotype) are used in the integrated classification framework provided by MKL. For the imaging data, different types of kernels are constructed based on a varying number of voxel-wise features. It is found that the combination of kernels from image modalities can achieve better classification accuracy than any individual kernel based on imaging features, showing the advantage of the combination with MKL. Although the major purpose of [18] is to improve the prediction accuracy of MCI converters with features from multimodality, it also points out that when linear kernel is used, the decision boundary of SVMs can be interpreted as a set of voxel weights with the values of $\mathbf{w}_m = \lambda_m \sum_i \alpha_i \phi_m(\mathbf{x}_i)$, which indicates the relative importance of various brain regions to the disease and can be visualized for inspection. It is found that MKL classifiers place greater weights on the brain regions highly correlated with AD, which reinforces the effectiveness of MKL for feature selection.

6.3 Anatomical Interpretation of Decision Boundary

In addition to feature selection, the normal vector of the optimal separating hyperplane in SVM has been used in literature to identify the spatial or functional patterns of discrimination when linear kernels are employed [8]. Despite the simplicity, these initial efforts in this regard have some limitations. Firstly, these methods tend to produce scattered patterns that lack spatial coherence even with linear kernel, giving rise to the difficulty for anatomical interpretation [7, 16, 22]. Secondly, these methods cannot deal with nonlinear kernels that have to be used for many complex linearly non-separable neuroimaging data. In such cases, the kernel mapping function is too complicated to be resolved explicitly, which impedes the anatomical interpretation of the discrimination captured in the high dimensional feature space [23–26]. Solutions to deal with these issues have been proposed in recent research works, which have significantly advanced the state of the art. They are introduced as follows.

6.3.1 *Spatial and Anatomical Regularization of Linear SVM*

Recall that, given N training samples $\{\mathbf{x}_i, y_i\}$, where $\mathbf{x}_i \in \mathbb{R}^d$ is the feature vector with d dimensions and y_i the label of the i-th sample, a hard margin linear SVM solves the following optimization:

$$\min_{\mathbf{w},b} \|\mathbf{w}\|^2 + \lambda \sum_{i=1}^{N} \max(0, y_i(\mathbf{w}^\top \mathbf{x}_i + b)). \tag{6.6}$$

The variables $\mathbf{w}$ and b represent the linear and the bias terms of the decision function $f(\mathbf{x}) = \mathbf{w}^\top \mathbf{x} + b$, respectively. The parameter λ is a weight of penalty to the misclassification. Geometrically, $\mathbf{w}$ is the normal vector of the optimal separating hyperplane, pointing to the direction where the maximal group difference happens. With a linear kernel, $\mathbf{w}$ has the same dimension as the original feature vector $\mathbf{x}$, and thus can be used as a pattern template encoding group discrimination.

In neuroimage applications, $\mathbf{w}$ is often visualized to reflect the class difference of interest. When a linear SVM is used, $\mathbf{w}$ lies in the same space as the input features. Take image voxels as the features, for example. Each component of $\mathbf{w}$, denoted as $\mathbf{w}_i$, corresponds to an image voxel. A large value of $\mathbf{w}_i$ indicates that the corresponding image voxel has a large contribution to the classification, and vice versa. Therefore, visualizing $\mathbf{w}$ against the brain image can provide an intuitive and concise visual summary of the complex discriminative patterns for exploration. In [16, 22], it is pointed out that in order to be appropriately employed for group comparison, the variable $\mathbf{w}$ should agree with some spatial and anatomical considerations for neuroimage analysis. For example, the components $\mathbf{w}_i$ should have similar values if the corresponding image voxels are in a neighborhood or connected anatomically or functionally. However, as shown in Eq. (6.6), the standard linear SVM only employs a norm-based constraint on $\mathbf{w}$, which cannot guarantee the satisfaction of those spatial or anatomical constraints. Consequently, the optimal $\mathbf{w}$ induced by a standard linear SVM often leads to scattered discriminative patterns that are noisy for interpretation. In the following, we introduce the work in [16, 22], which endeavors to improve this situation.

6.3.1.1 Incorporation of Prior into SVM

Based on the observations mentioned above, spatial and anatomical information are explicitly introduced into SVM models in [16, 22]. As pointed out in [16, 22], incorporating prior information into a standard SVM model is often conducted via kernel design or regularization. That work follows the line of regularization. In particular, a regularization operator P is introduced into Eq. (6.6) as follows:

$$\min_{\mathbf{w},b} \|P\mathbf{w}\|^2 + \lambda \sum_{i=1}^{N} \max(0, y_i(\mathbf{w}^\top \mathbf{x}_i + b)), \tag{6.7}$$

where P is a linear projection. Let $\tilde{\mathbf{x}} = P^{-1}\mathbf{x}$. It is straightforward to see that Eq. (6.7) can be rewritten as a standard linear SVM:

$$\min_{\mathbf{w},b} \|\mathbf{w}\|^2 + \lambda \sum_{i=1}^{N} \max(0, y_i(\mathbf{w}^\top \tilde{\mathbf{x}}_i + b)). \tag{6.8}$$

That shows, the regularized SVM can be viewed as the standard SVM working on the transformed raw data $\tilde{\mathbf{x}}$, or equally, the standard SVM working on the raw data $\mathbf{x}$ with a kernel $k(\mathbf{x}_1, \mathbf{x}_2) = \langle P^{-1}\mathbf{x}_1, P^{-1}\mathbf{x}_2 \rangle$.

In [16, 22], a framework using Laplacian regularization for P is proposed to flexibly handle various requirements of spatial and anatomical proximities. Such regularization is considered in both discrete and continuous cases. They are elaborated as follows.

In the discrete case, the regularization is exerted via graphs. For example, when image voxels are considered, the voxel connectivity (6, 18, or 26) of the image can be used as the regularization graph. Under this circumstance, the regularization operator P is defined as $P : \mathbf{w} \mapsto e^{\frac{1}{2}\beta L}\mathbf{w}$, where L is the Laplacian matrix of the graph and β controls the degree of the regularization. The graph Laplacian L is determined by the degree matrix D and the adjacency matrix A of the graph: $L = D - A$. The adjacency matrix A represents the connections of the nodes in the graph. If node i and j is connected, $A_{ij} = 1$; otherwise, $A_{ij} = 0$. The degree matrix D is a diagonal matrix $D_{ii} = \sum_j A_{ij}$. Such a regularization leads to an SVM classification with a new kernel $k_\beta(\mathbf{x}_1, \mathbf{x}_2) = \mathbf{x}_1^\top e^{-\beta L}\mathbf{x}_2$.

In the continuous case, the regularization is exerted via compact Riemannian manifolds. For example, when cortical surfaces are considered, the discretization of the surface highly depends on the construction of the surface mesh. In order to circumvent the possible problems caused by this step, a cortical surface is treated as a 2D Riemannian manifold in [16, 22]. Compared with the discrete case, the regularization operator P is now defined as $P : \mathbf{w} \mapsto e^{\frac{1}{2}\beta\Delta_g}\mathbf{w}$, where Δ_g denotes the Laplacian–Beltrami operator.

The above is a general framework to incorporate prior, which may take different forms with different requirements of constraints, such as spatial proximity, tissue types, anatomical brain regions, and brain connectivities. Take the discrete case, for example. The regularization graph has to be carefully designed to satisfy the prior constraints. If spatial proximity is to be enforced, the local voxel connectivity can be used as the regularization graph and hence a common Laplacian matrix L as mentioned above is employed. As pointed out in [22], this is equivalent to preprocessing the image with a Gaussian smoothing kernel of $\sigma = \sqrt{\beta}$. The problem becomes a bit complicated when anatomical proximity needs to be considered. For example, two voxels may be linked even when they are in distance because the anatomical brain regions they belong to are likely to be connected by white matter tracts. Therefore, the "adjacency" of two voxels is determined by two probabilities: the probability for a voxel to be located in a specific region of interest (ROI), and the probability for two ROIs to be connected. Let $\mathbf{E} \in \mathbb{R}^{d \times R}$ be a matrix, where d is the number

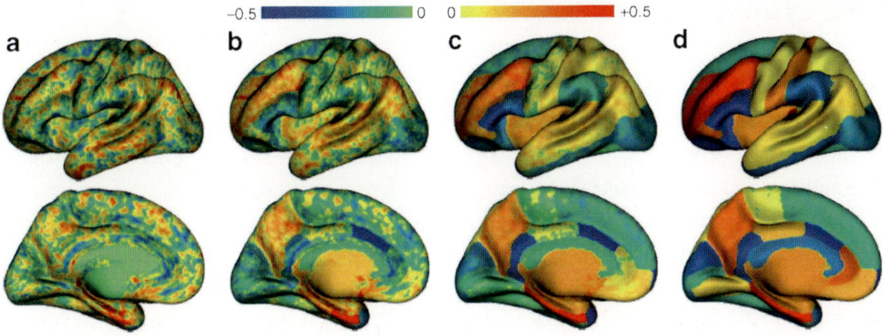

Fig. 6.2 Image courtesy of [16]: Anatomically regularized $\mathbf{w}$ for cortical thickness. Color code indicates the value of the component $\mathbf{w}_i$. From (**a**) to (**d**), $\beta = 0, 2, 4, 6$

of voxels and R is the number of ROIs. The (i, j)-th entry of $\mathbf{E}$ corresponds to the probability for the i-th voxel to be located in the j-th ROI. Let the matrix $\mathbf{C} \in \mathbb{R}^{R \times R}$ be positive semidefinite. The (i, j)-th entry of $\mathbf{C}$ corresponds to the probability for the i-th and the j-th ROIs to be connected. The adjacency matrix $\mathbf{A}$ therefore is $\mathbf{A} = \mathbf{ECE}^\top$. In [16,22], a normalized graph Laplacian $\tilde{L}$ is used: $\tilde{L} = \mathbf{I} - \tilde{\mathbf{E}}^\top \tilde{\mathbf{E}}$, where $\tilde{\mathbf{E}} = \mathbf{D}^{-\frac{1}{2}} \mathbf{E} \mathbf{C}^{\frac{1}{2}}$. In addition, the work in [16, 22] also points out that the spatial and the anatomical proximities can be combined on graphs or on statistical manifolds (via Fisher metric).

6.3.1.2 Applications

In [16], the regularized SVMs are applied for the classification of the patients with AD and the healthy controls by both the gray matter concentration and the cortical thickness extracted from MRI. Both spatial and anatomical proximities are considered in the experiment. For the anatomical regularization, atlases are used to parcellate the brain into a set of ROIs. The image voxels or cortical surface vertices within the same ROI should contribute similarly to the classification, which could be reflected by the values of the components of $\mathbf{w}$. A large absolute value of the i-th component $\mathbf{w}_i$ indicates a significant contribution to the discrimination, and vice versa. An example result quoted from [16] on cortical thickness is given in Fig. 6.2. As shown, the $\mathbf{w}$ without regularization (Fig. 6.2a, $\beta = 0$) produces noisy and scattered patterns, while the regularized $\mathbf{w}$ produces patterns consistent with the anatomical partitions. The atrophies in the regions with large absolute values of $\mathbf{w}_i$ suggest a high likelihood to be AD, which include the ROIs of hippocampus, amygdala, parahippocampal gyrus, cingulum, etc. These ROIs agree well with the neuropathology of AD, indicating that the regularized SVM is a promising tool for group comparison.

6.3.2 *Discriminative Direction for Nonlinear Kernels*

When data are linearly non-separable, kernel SVMs have to be used for accurate classifications. They work as follows. A kernel mapping function $\Phi : \mathbb{R}^d \to \mathscr{F}$ is implicitly defined by a kernel trick. It maps the feature vectors from their original input space $\mathbb{R}^d$ to a high dimensional feature space $\mathscr{F}$ where the originally linearly non-separable data become linearly separable so that a linear SVM can be performed there. When kernel SVMs are employed, the group discrimination stored in $\mathbf{w}$ (the normal vector of the separating hyperplane) is only mathematically meaningful in the higher dimensional space $\mathscr{F}$, which needs to be projected back into the original input space $\mathbb{R}^d$ to obtain anatomical interpretation. However, computing the inverse mapping of Φ is intractable because Φ is too complicated to be explicitly defined. To deal with this problem, Golland et al. put forward a concept called "*discriminative direction*" in the $\mathbb{R}^d$ space [23,24]. When a sample in one class travels along the discriminative direction, it will gradually become akin to the samples in the opposite class. During this course, only group difference appears, whereas the differences caused by individual variation within the same class are excluded. Comparing the sample before and after such a change can locate the group difference. Note that different samples in a population may have different discriminative directions. For example, let us consider two classes of points lying on two concentric annuli with different radii, respectively. The discriminative directions are different from point to point, coinciding with the radial direction at that point. In [23], Golland et al. propose a method to approximate the discriminative direction and apply it to detect the distinctive difference in hippocampal shapes between the healthy controls and the schizophrenia patients. Some other studies also use the discriminative directions to localize the group difference between the healthy controls and the patients with the AD for ROI-based or voxel-based whole brain analysis [27, 28]. However, by revisiting Golland's discriminative direction in [25, 26], Zhou et al. point out that Golland's method may have a problem of generating spurious group difference by neglecting the underlying distribution of the data. Therefore, they propose a "*regularized discriminative direction*" to overcome this problem. This method improves the fidelity of the detected group difference by confining the solution to the sub-dimensional manifold in the original $\mathbb{R}^d$ space. In the following, we unfold the details about how to infer the discriminative direction and the regularized discriminative direction, respectively.

6.3.2.1 Discriminative Direction

Let $\mathscr{D} = \{\mathbf{x}_1, \mathbf{x}_2, \cdots, \mathbf{x}_n\}$ $(\mathbf{x}_i \in \mathbb{R}^d)$ denote a set of n training samples labeled in two groups. As mentioned earlier, training a kernel SVM implicitly performs a mapping $\Phi(\cdot)$ from an input space $\mathbb{R}^d$ to a Hilbert space (known as the feature space) $\mathscr{F}$ with higher or even infinite dimensionality. Then a linear SVM is performed in $\mathscr{F}$

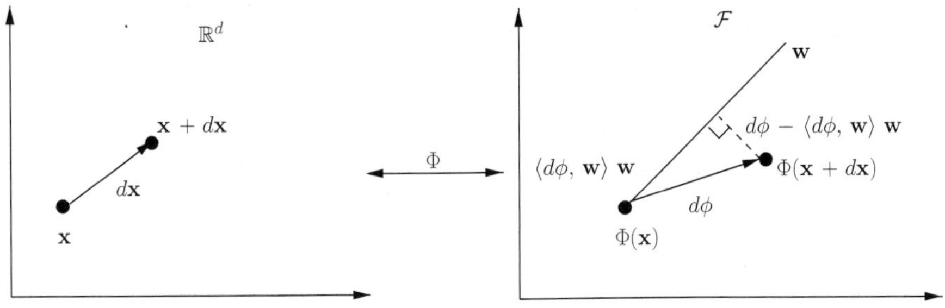

Fig. 6.3 Image courtesy of [25]: explanation of how to infer the "discriminative direction" by Golland's method [23]. A point $\mathbf{x}$ in the input space $\mathbb{R}^d$ is mapped to a point $\Phi(\mathbf{x})$ in the feature space $\mathscr{F}$. When $\mathbf{x}$ moves a small step $d\mathbf{x}$ in $\mathbb{R}^d$, a displacement $d\phi$ will be induced in $\mathscr{F}$ accordingly. The divergence of $d\phi$ from $\mathbf{w}$ is $d\phi - \langle d\phi, \mathbf{w}\rangle\ \mathbf{w}$, which is perpendicular to $\mathbf{w}$. Minimizing this divergence with respect to $d\mathbf{x}$ can make $\Phi(\mathbf{x})$ move along $\mathbf{w}$ as much as possible in $\mathscr{F}$. Such a direction $d\mathbf{x}$ is the "discriminative direction" at point $\mathbf{x}$ in $\mathbb{R}^d$

to infer the optimal separating hyperplane with the corresponding normal vector $\mathbf{w}$. The dimensionality of $\mathbf{w}$ equals that of $\mathscr{F}$, which might also be infinite.

To only reflect the group difference between two classes, ideally we want to strictly move a data point $\Phi(\mathbf{x})$ in $\mathscr{F}$ along the direction $\mathbf{w}$ that encodes the group discrimination, and find the corresponding changes in $\mathbb{R}^d$. However this is problematic, because the kernel mapping $\Phi(\cdot)$ often maps the input space $\mathbb{R}^d$ onto a lower dimensional manifold in $\mathscr{F}$, which restricts the possible movement along $\mathbf{w}$. For example, when a Gaussian radial basis function (RBF) kernel is used, the whole $\mathbb{R}^d$ will only be mapped onto a unit hypersphere in $\mathscr{F}$ instead of the whole $\mathscr{F}$. Forcing $\Phi(\mathbf{x})$ to move strictly along the normal vector $\mathbf{w}$ will deviate from this manifold, causing the resulting new $\Phi(\mathbf{x}')$ to have no preimage in $\mathbb{R}^d$ anymore. In other words, by forcing the preimage of $\Phi(\mathbf{x})$ to exist, $\Phi(\mathbf{x})$ cannot move strictly along $\mathbf{w}$.

In [23, 24], Golland et al. propose to solve this problem by searching for a direction $d\mathbf{x}$ in $\mathbb{R}^d$, whose corresponding movement in $\mathscr{F}$ causes minimal divergence from the normal vector $\mathbf{w}$. The idea is illustrated in Fig. 6.3. Specifically, when $\mathbf{x}$ moves along $d\mathbf{x}$, a displacement $d\phi = \Phi(\mathbf{x} + d\mathbf{x}) - \Phi(\mathbf{x})$ is induced in $\mathscr{F}$ accordingly. The displacement component of $d\phi$ that is perpendicular to $\mathbf{w}$ represents the deviation from $\mathbf{w}$ and can be computed as $d\phi - \langle d\phi, \mathbf{w}\rangle\ \mathbf{w}$. Minimizing the divergence from $\mathbf{w}$ makes the movement of $\Phi(\mathbf{x})$ maximally agree with $\mathbf{w}$. For this purpose, the following optimization is solved:

$$\begin{aligned} \text{Find } d\mathbf{x} = \arg\min_{d\mathbf{x}\in\mathbb{R}^d} \|d\phi - \langle d\phi, \mathbf{w}\rangle \mathbf{w}\|^2 &= \arg\min_{d\mathbf{x}\in\mathbb{R}^d} \langle d\phi, d\phi\rangle - \langle d\phi, \mathbf{w}\rangle^2 \\ \text{such that } \|d\mathbf{x}\|^2 &= \varepsilon \\ \text{and } \ d\phi &= \Phi(\mathbf{x}+d\mathbf{x}) - \Phi(\mathbf{x}), \end{aligned} \tag{6.9}$$

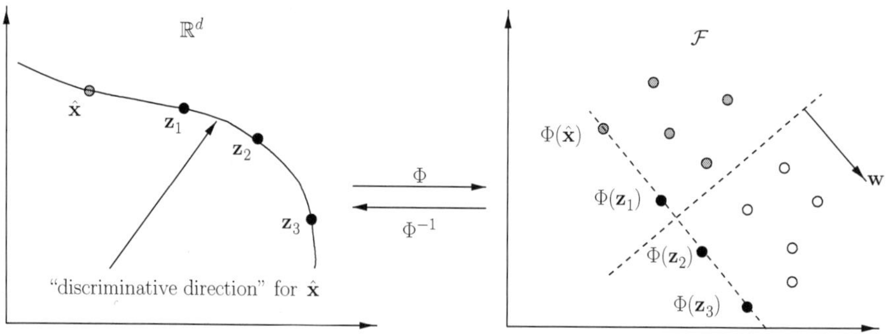

Fig. 6.4 Image courtesy of [25]: discriminative direction at a point $\hat{\mathbf{x}}$. A nonlinear mapping $\mathbf{\Phi}$ maps the shape descriptor space $\mathbb{R}^d$ onto a feature space $\mathscr{F}$, where the two classes (*gray and white dots*) become linearly separable. The vector $\mathbf{w}$ is the normal of the separating hyperplane found in $\mathscr{F}$. Move $\Phi(\hat{\mathbf{x}})$ along $\mathbf{w}$ to a new position $\Phi(\mathbf{z}_1)$ and project it back to $\mathbb{R}^d$ as $\mathbf{z}_1$. The vector $\mathbf{z}_1 - \hat{\mathbf{x}}$ is the "discriminative direction" at $\hat{\mathbf{x}}$

where ε is a preset small positive real number. It is pointed out in [25] that the constraint of $\|d\mathbf{x}\|^2 = \varepsilon$ allows $d\mathbf{x}$ to be searched identically along *all* directions in $\mathbb{R}^d$. In other words it implicitly assumes that the original feature vectors are distributed in the whole space of $\mathbb{R}^d$, which may not necessarily be true in practice.

Different from the method in [23], Zhou et al. propose a conceptually simpler way to approximate the discriminative direction [25,26]. They take advantage of the preimage techniques [29–31] to reconstruct the approximation of $\mathbf{x}$ when moving $\Phi(\mathbf{x})$ strictly along $\mathbf{w}$ in $\mathscr{F}$. They minimize a cost function that can be formulated only in terms of the inner product $\langle\Phi(\mathbf{x}_i),\Phi(\mathbf{x}_j)\rangle$, avoiding the explicit manipulation of the unknown $\Phi(\mathbf{x})$.

The idea in [26] is illustrated in Fig. 6.4. For a given feature vector $\hat{\mathbf{x}}$, in $\mathscr{F}$ its image $\Phi(\hat{\mathbf{x}})$ is moved along $\mathbf{w}$ to a new position $\Phi(\hat{\mathbf{x}}) + s\mathbf{w}$, where s is a given step and $\mathbf{w}$ has been normalized to a unit vector. Let $\mathbf{z}$ denote the best estimate of the preimage of $\Phi(\hat{\mathbf{x}}) + s\mathbf{w}$. Then $\mathbf{z} - \hat{\mathbf{x}}$ forms the discriminative direction at $\hat{\mathbf{x}}$. The residual error $\rho(\mathbf{z})$ for reconstructing the preimage $\mathbf{z}$ is minimized:

$$\mathbf{z}^\star = \arg\min_{\mathbf{z}\in\mathbb{R}^d} \rho(\mathbf{z}) = \arg\min_{\mathbf{z}\in\mathbb{R}^d} \|\Phi(\hat{\mathbf{x}}) + s\mathbf{w} - \Phi(\mathbf{z})\|^2. \tag{6.10}$$

Let us assume a Gaussian RBF kernel is used, e.g., $k(\mathbf{x}_i,\mathbf{x}_j) = \langle\Phi(\mathbf{x}_i),\Phi(\mathbf{x}_j)\rangle = \exp(-\|\mathbf{x}_i - \mathbf{x}_j\|^2/2\sigma^2)$, so that $\langle\Phi(\mathbf{z}),\Phi(\mathbf{z})\rangle$ is constant. Completing the square term in Eq. (6.10) leads to a simpler cost function

$$\mathbf{z}^\star = \arg\max_{\mathbf{z}\in\mathbb{R}^d} \langle\Phi(\hat{\mathbf{x}}) + s\mathbf{w}, \Phi(\mathbf{z})\rangle. \tag{6.11}$$

As mentioned, in SVM it holds that $\mathbf{w} = \sum_i \alpha_i \Phi(\mathbf{x}_i)$. Eq. (6.11) can be rewritten as

$$\mathbf{z}^\star = \arg\max_{\mathbf{z}\in\mathbb{R}^d} k(\hat{\mathbf{x}},\mathbf{z}) + s\sum_i \alpha_i k(\mathbf{x}_i,\mathbf{z}), \tag{6.12}$$

which is only relevant to the inner products $k(\cdot)$. Therefore, the unsolvable mapping Φ does not have to be computed.

Note that both Golland's and Zhou's methods implicitly assume the original feature vectors occupying the whole input space $\mathbb{R}^d$. Inferring discriminative direction in these ways may induce spurious group difference as observed in [25], which is suggested to possibly come from the ignorance of the underlying data distribution. An example is given in [25] to illustrate this problem. Sharing the same hypotenuse, two classes of right-angled triangles are given with their right-angle vertex located in the first and the second quadrants, respectively. All the right-angle vertices are located on a semi-circle taking the hypotenuse as its diameter. When moving along the discriminative direction, the right-angle vertex deviates from the semi-circle where it resides, causing the loss of right-angleness for the newly generated triangles. This introduces a spurious angle discrepancy when comparing the triangles before and after the change to locate the group difference. Therefore, in [25], a regularized discriminative direction is proposed to improve the fidelity of the detected group difference, and the process is explained as follows.

6.3.2.2 Regularized Discriminative Direction

Instead of looking for the preimage $\mathbf{z}$ within the whole input space, in [25] Zhou et al. propose to compute $\mathbf{z}$ within a restricted subspace Ω ($\Omega \in \mathbb{R}^d$), that is,

$$\mathbf{z}^\star = \arg\min_{\mathbf{z}\in\Omega} \rho(\mathbf{z}). \tag{6.13}$$

In [25], two methods are proposed to infer the subspace Ω from data, one from a local distribution perspective and the other from a geometric point of view.

Recall that $\hat{\mathbf{x}}$ denotes a sample in $\mathbb{R}^d$, whose image $\Phi(\hat{\mathbf{x}})$ in $\mathscr{F}$ is moved along the direction of $\mathbf{w}$; and $\mathbf{z}$ denotes the best estimate of the preimage of $\Phi(\hat{\mathbf{x}})$ during this movement. Let $N_\varepsilon(\hat{\mathbf{x}}) = \{\mathbf{x} \mid \|\mathbf{x}-\hat{\mathbf{x}}\| \leq \varepsilon\}$ be a neighborhood of $\hat{\mathbf{x}}$ determined by the small positive value ε, and $p(\mathbf{x} \mid \mathbf{x} \in N_\varepsilon(\hat{\mathbf{x}}))$ be an empirical probability density function estimated from $N_\varepsilon(\hat{\mathbf{x}})$.

The local distribution-based regularization requires that within a small movement step, the obtained preimage $\mathbf{z}$ should comply with the local distribution $p(\mathbf{x} \mid \mathbf{x} \in N_\varepsilon(\hat{\mathbf{x}}))$, which is modeled as a normal distribution with the mean $\mu = \hat{\mathbf{x}}$ and the covariance matrix $\Sigma = \frac{1}{|N_\varepsilon(\hat{\mathbf{x}})|-1} \sum_{\mathbf{x}_i \in N_\varepsilon(\hat{\mathbf{x}})} (\mathbf{x}_i - \hat{\mathbf{x}})(\mathbf{x}_i - \hat{\mathbf{x}})^\top$. Here $|N_\varepsilon(\hat{\mathbf{x}})|$ denotes the number of the training samples within $N_\varepsilon(\hat{\mathbf{x}})$. Requiring $\mathbf{z}$ to comform to this distribution is to require an adequately large value of $p(\mathbf{z}|\mu,\Sigma)$, or equally an adequately small value of $(\mathbf{z}-\mu)^\top\Sigma^{-1}(\mathbf{z}-\mu)$. Therefore, in [25] the following objective function is minimized:

$$\mathbf{z}^\star = \arg\min_{\mathbf{z}\in\mathbb{R}^d} \; \rho(\mathbf{z}) + 2\eta\,(\mathbf{z}-\mu)^\top\Sigma^{-1}(\mathbf{z}-\mu), \tag{6.14}$$

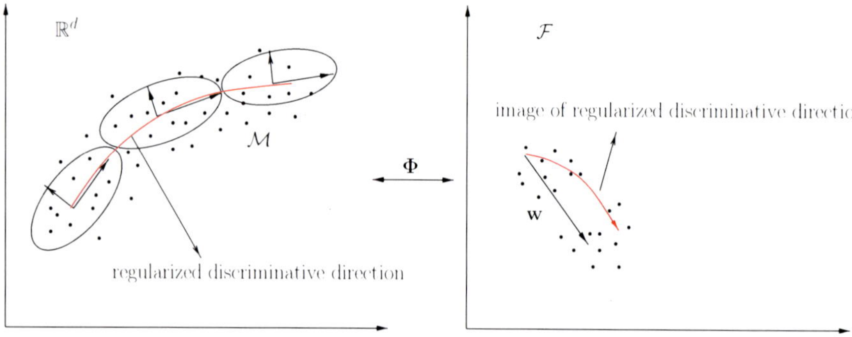

Fig. 6.5 Image courtesy of [25]: the regularized discriminative direction is confined on the sub-dimensional manifold formed by the shape descriptors (the *dark dot*) in $\mathbb{R}^d$ space. It is a compromise between following the classifier provided discriminative direction **w** and conforming to the shape distribution

where $\eta \geq 0$ is a regularization parameter. When η is 0, this problem reduces to the minimization of $\rho(\mathbf{z})$ in Eq. (6.10).

Equation (6.14) can be extended to handle rank-deficient Σ by decomposing Σ as $\Sigma = \Gamma\Lambda\Gamma^\top$, where each column of Γ is an eigenvector and Λ is $\text{diag}\{\lambda_1, \cdots, \lambda_r, 0, \cdots, 0\}$. The λ_i is the i-th positive eigenvalue and r is the rank of Σ. An optimal solution $\mathbf{z}^\star$ should satisfy:

$$\begin{aligned} \mathbf{z}^\star \in\ & \{\mathbf{z} \mid (\Gamma^\top(\mathbf{z}-\mu))_i = 0 \ \text{ for } \ i = r+1, \cdots, d\} \\ & = \{\mathbf{z} \mid \mathbf{z} = \mu + \hat{\Gamma}\hat{\Lambda}^{\frac{1}{2}}\mathbf{u}\}, \end{aligned} \tag{6.15}$$

where $\mathbf{u}$ is a vector in $\mathbb{R}^r$. The $r \times r$ matrix $\hat{\Lambda}$ is $\text{diag}\{\lambda_1, \cdots, \lambda_r\}$. The $d \times r$ matrix $\hat{\Gamma} = (\Gamma_1, \cdots, \Gamma_r)$ contains the eigenvectors corresponding to $\hat{\Lambda}$. Therefore, instead of optimizing $\mathbf{z} \in \mathbb{R}^d$, in [25] it is proposed to optimize $\mathbf{u} \in \mathbb{R}^r$, where $r \ll d$ in practice:

$$\mathbf{u}^\star = \arg\min_{\mathbf{u} \in \mathbb{R}^r} \ \rho(\mu + \hat{\Gamma}\hat{\Lambda}^{\frac{1}{2}}\mathbf{u}) + 2\eta\langle \mathbf{u}, \mathbf{u}\rangle. \tag{6.16}$$

This significantly reduces the number of parameters to be estimated. Once $\mathbf{u}$ is obtained, the preimage $\mathbf{z}$ can be computed by Eq. (6.15).

Equation (6.15) is geometrically explained in Fig. 6.5 by [25]. Let $\mathcal{M}$ be the manifold where the original feature vectors reside and $T_\mu(\mathcal{M})$ be a tangent plane of $\mathcal{M}$ at μ. The preimage $\mathbf{z}$ is restricted to the tangent plane $T_\mu(\mathcal{M})$ which is spanned by the eigenvectors in $\hat{\Gamma}$. Since a manifold can be locally approximated by its tangent plane, Eq. (6.15) requires the preimage $\mathbf{z}$ to conform to the manifold $\mathcal{M}$.

In [25], an approximation (first order Taylor expansion) of Eq. (6.16) is solved with an analytical solution. It can be theoretically proved that when the regularization parameter $\eta = 0$, this analytical solution equals to that of Golland's method in Eq. (6.9), indicating Golland's method is a special case of the local distribution regularized discriminative direction.

In addition, local geometric constraints are also proposed in [25] to require $\mathbf{z}$ lying in the convex hull of $N_\varepsilon(\hat{\mathbf{x}})$. That is, $\mathbf{z}$ is a convex combination of its neighbors $N_\varepsilon(\hat{\mathbf{x}})$: $\mathbf{z} = \sum_{i=1}^{n} \omega_i \mathbf{x}_i$, where $\omega_i \geqslant 0$, $\sum_{i=1}^{n} \omega_i = 1$, $\mathbf{x}_i \in N_\varepsilon(\hat{\mathbf{x}})$, and n is the number of neighbors. In order to find the regularized discriminative direction, the following optimization is proposed to solve:

$$\min_{\omega_i} \left\| \Phi(\hat{\mathbf{x}}) + s\mathbf{w} - \Phi\left(\sum_{i=1}^{n} \omega_i \mathbf{x}_i\right) \right\|^2$$

$$\text{such that } \omega_i \geqslant 0$$

$$\text{and } \sum_{i=1}^{n} \omega_i = 1, \tag{6.17}$$

and $\mathbf{z}$ is then recovered by $\mathbf{z} = \sum_{i=1}^{n} \omega_i \mathbf{x}_i$.

The convex combination method is closely related to the local distribution constrained method as follows. Recall that in local distribution constrained method, $\mathbf{z}$ follows a normal distribution $N(\mathbf{z}|\mu,\Sigma)$. Therefore, $E[\mathbf{z}]_{\text{dist}} = \mu$ and $Cov(\mathbf{z})_{\text{dist}} = \Sigma$. In the convex combination method, it is easy to show that $E[\mathbf{z}]_{\text{comb}} = \mu$ and $Cov(\mathbf{z})_{\text{comb}} = (\sum_i \omega_i^2)\Sigma$. Since $\sum_i \omega_i^2 \leq 1$, $Cov(\mathbf{z})_{\text{comb}}$ has the same eigenvectors as $Cov(\mathbf{z})_{\text{dist}}$ but smaller eigenvalues. This indicates that the convex combination method implicitly confines $\mathbf{z}$ to the local distribution of its neighbors but in a more strict manner.

6.3.2.3 Remarks

A common practice for group comparison is to take the group mean as the representative and reveal the group difference by comparing the group means. This would be difficult to handle the cases when the complex group difference varies across the group. The discriminative direction approach effectively deals with this problem. When SVM is used, instead of comparing the group means, both the regularized and the unconstrained discriminative directions are applied on the support vectors for analysis. This is because it is the support vectors that determine the optimal separating hyperplane in SVM, and thus encode the group difference.

Either the unconstrained or the regularized discriminative direction can be applied to different kernel classifiers including kernel SVMs. It is a method to decode group difference that has been already captured by the classifiers. Therefore the robustness of its result is influenced by the performance of the classifiers. A too low generalization performance of the classifier may cast doubt on the fidelity of the detected group difference.

6.3.2.4 Applications

Discriminative direction has been applied in neuroimage analysis for understanding the anatomical and developmental aspects of mental disorders. Example applications include: detecting the morphological changes of hippocampi for the

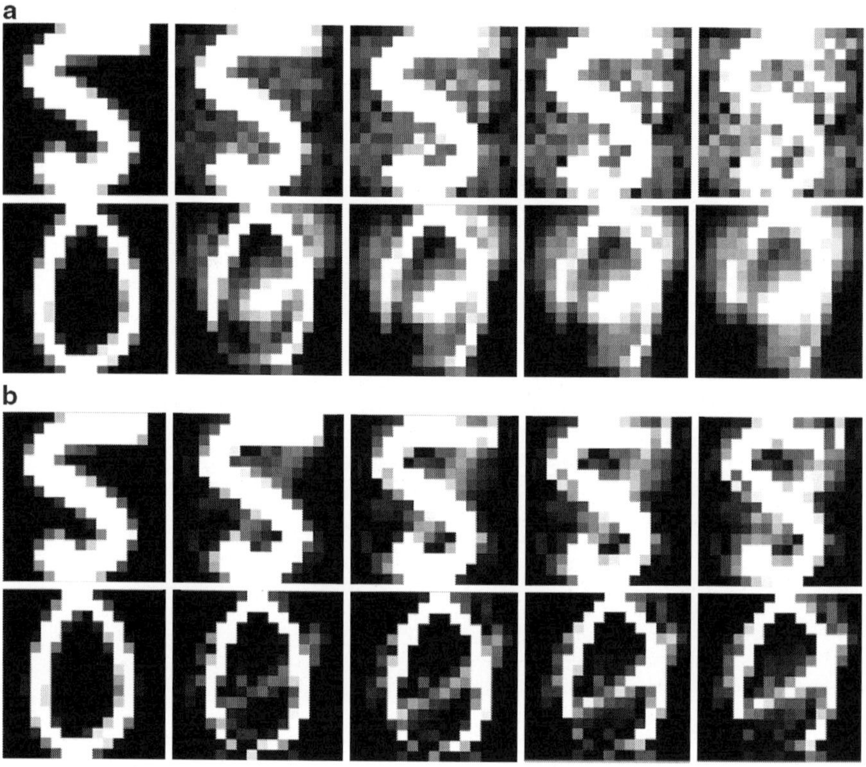

Fig. 6.6 Image courtesy of [25]: group difference captured by (**a**) the unconstrained discriminative direction in [24] and (**b**) the regularized discriminative direction in [25]. The *top row* shows the deformation from digit 5 to digit 8. The *bottom row* shows the deformation from digit 0 to digit 9

schizophrenia [23] and the AD [25], detecting the morphological changes of corpus callosum in affective disorder [23], detecting brain spatial pattern changes with ageing or between sexes [27], and detecting the morphological changes of the whole brain for the schizophrenia [28]. The movements of support vectors along the discriminative direction are visualized to reflect the typical patterns of group difference. In the following, the results of the regularized discriminative direction applied on hippocampal shape analysis are directly quoted from [25], demonstrating how this method performs for group comparison.

In [25], the normal morphology of hippocampi is compared between sexes using the regularized discriminative direction. Since there is a lack of ground truth for this open problem, the fidelity of the regularized discriminative direction is first validated on a handwritten digit image database and a facial image database, where the group differences are clear and visually perceptible.

Two classification tasks are conducted by SVMs with RBF kernel: classifying two digits from the handwritten images (Fig. 6.6); and classifying left-side view and right-side view facial images (Fig. 6.7). Selected support vectors (samples) are

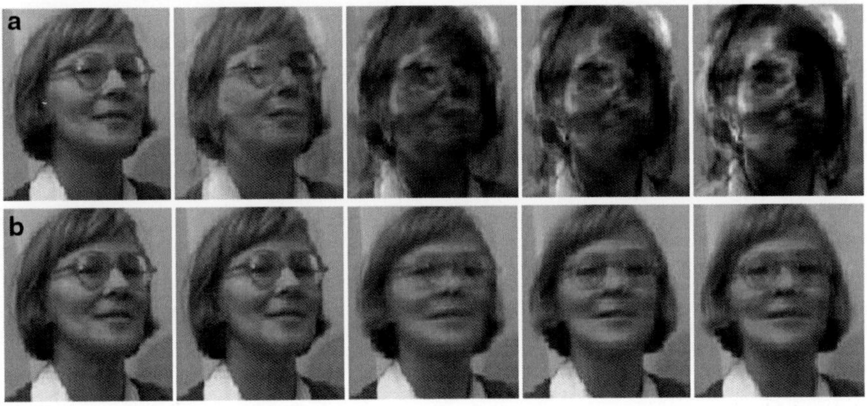

Fig. 6.7 Image courtesy of [25]: group difference captured by (**a**) the unconstrained discriminative direction in [24] and (**b**) the regularized discriminative direction in [25]. During the deformation, a right-side-view face (the *leftmost image*) turns towards left gradually (keeping adding class difference)

moved from one class towards the other along the discriminative direction to reflect the class difference. It can be seen that the unconstrained discriminative direction in [24] usually introduces more noisy or spurious differences, while the regularized discrimination direction appropriately decodes the discrimination by introducing only the minimal necessary group changes. The individual variability within each group (i.e., the different handwritten images of the same digit, or the different owners of the faces for the same view) is excluded by the regularized method from the detected class discrepancy. It is noticed that in Fig. 6.7 the regularized method generates *facial* images during the whole process, while the unconstrained method cannot guarantee that, indicating the importance of considering only the underlying data distribution.

In [25], the fidelity of the detected changes is quantitatively measured by estimating how well the newly generated image (by gradually adding class difference) belongs to the distribution of the training set. This is achieved by the one-class SVM [32] that estimates the probability density for abnormality detection. One-class SVMs infer a function f whose values decrease from the densely distributed areas to the sparsely distributed areas. Figure 6.8 shows the f-values computed during the course of gradually changing a left-side view facial image into a right-side view image. As shown, the regularized discriminative direction consistently achieves larger f values than the unconstrained discriminative direction in [24] at every step. This demonstrates that moving along the regularized discriminative direction yields new samples conforming to the distribution of the training set.

After sanity check, the regularized discriminative direction is applied on hippocampal shapes belonging to the male and the female groups of the healthy elderly. The shapes (normalized by volume) are represented by the surface landmarks reconstructed from the spherical harmonics (SPHARM) representation [33] of degree five. An SVM classifier with RBF kernel is employed for classification.

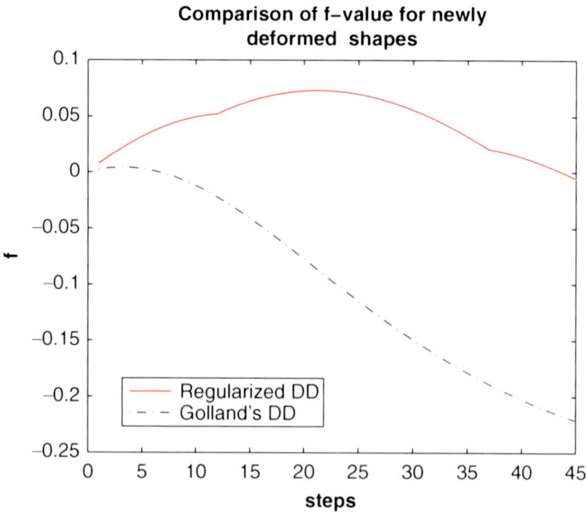

Fig. 6.8 Image courtesy of [25]: comparing the values of the f function in a one-class SVM for the deformation of a left-side view facial image towards the right-side view

Support vectors are selected to study the discriminative direction as shown in Fig. 6.9. The color code indicates the deformation that a hippocampus undergoes to become alike to the opposite class. From green to red, the amount of protrusion increases. From green to blue, the amount of shrinkage increases. Consistent with the results from the sanity check, the regularized discriminative direction generates more compact changes concentrated in fewer regions but at greater magnitude in the hippocampal head and tail, while the unconstrained discriminative direction shows more scattered patterns with usually a compression next to an expansion. The hippocampal shape difference detected by the regularized discriminative direction can be cross-validated by the documented findings [34] where the hippocampus volume loss in the lateral areas of the head and tail is found in males, but not observed in females. Figure 6.10 shows the f-values of the one-class SVM applied on the hippocampal shapes. As shown, the regularized discriminative direction achieves higher values for most of the cases, indicating that the deformed shapes better agree with the underlying distribution than those obtained by the unconstrained method.

6.4 Generative Models with Discriminative Learning

Detecting the difference between clinical groups is typically formulated as a classification problem that aims to optimally separate two groups from each other. Discriminative methods such as SVMs are often involved into this process. Despite their efficiency in distinguishing different groups, these methods usually focus on

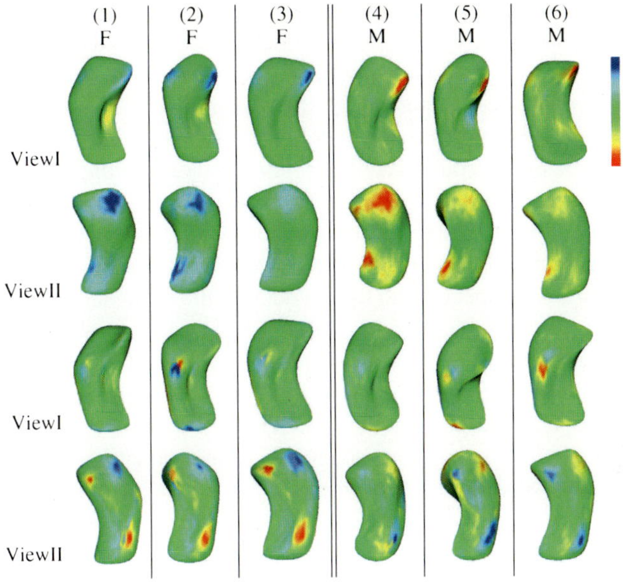

Fig. 6.9 Image courtesy of [25]: localized discrimination for sexes on hippocampi of six individuals (three females on the *left*, namely individual 1, 2, 3; three males on the *right*, namely individual 4, 5, 6) from two perspective views. The *top two rows* are generated by the regularized method in [25], and the *bottom two rows* are generated by Golland's method in [24]

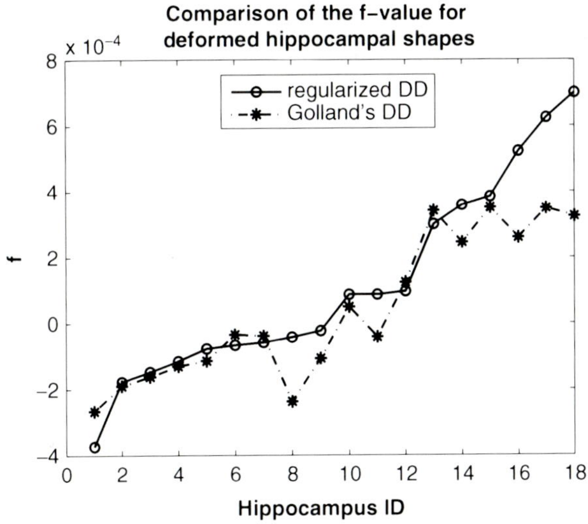

Fig. 6.10 Image courtesy of [25]: the values of the f function in a one-class SVM on 18 support vectors in the hippocampus database obtained by the regularized method in [25] and Golland's method in [24]

the boundary of the separation, which does not lend themselves to good interpretation, especially when the classification models are complicated [25]. However, in neuroimage analysis the understanding of the group difference is very important, which may reveal the mechanism behind the mental diseases. Therefore, a number of research works in this field turn to generative models that are developed to explain how the neuroimage data could have been generated. The group difference is then identified by comparing the generative models learned separately from each class. Due to their capacity of interpretation, generative models are widely used in the newly emerging field of brain network analysis that aims to understand the alteration of brain network with respect to the factors such as the brain damage. For example, group-level brain networks have been built from FDG-PET images by compressed sensing in [35] and sparse inverse covariance estimation in [36], built from MRI by pair-wise correlations of cortical thickness in ROIs in [37], and built from functional MRI images by sparse Markov Random Field in [38].

Although generative methods are amenable to interpretations as they explicitly infer a representative model for each group, it is well known that generative methods are not necessarily discriminative and usually have inferior performance in classification than discriminative methods. For example, when applied to infer brain networks, generative methods are prone to emphasizing major structures shared within each group and neglecting subtle structures such as the disease-induced changes. This may not only make subsequent investigations less meaningful but can also lead to spurious conclusions.

In this chapter, a recent method [39] that integrates the benefits of generative and discriminative (SVMs) methods for neuroimage analysis is introduced. Although that method is based on the application of brain network analysis, the underlying idea is general and can be adapted to other neuroimage applications. In the following, the background of the application is first introduced, followed by the introduction of the associated generative model and the discriminative learning of the generative model for both discrimination and representation.

6.4.1 Background: Brain Network Analysis

Studying the brain as a complex network of inter-connected brain regions is becoming a new research trend in neuroimage analysis. It has been applied to a variety of imaging modalities [36–38, 40, 41] at various spatial and temporal scales to provide new insights into the brain. The reorganization of brain networks has been found in many age-related mental diseases, such as the Alzheimer's disease, schizophrenia, and stroke, as a response of the brain to the damage.

Mathematically, a brain network can be modeled as a graph. Figure 6.11 illustrates a common process of how to build a brain network. The brain images belonging to different subjects are aligned (deformably registered) to the same stereotaxic space to establish correspondence. They are then parcellated into common regions of interest (ROI) where local image features are extracted. A brain network is modeled by a graph with each node corresponding to a brain region

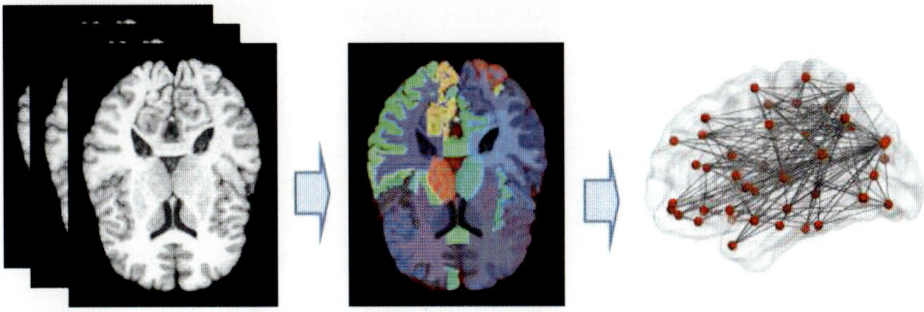

Fig. 6.11 Illustration of brain network construction

and each edge corresponding to the connectivity between regions. The connectivity could be statistical dependencies (functional connectivity) or causal relationships (effective connectivity) [42]. Regardless of imaging modality, brain networks can always be mathematically abstracted as either *undirected* (for functional connectivity) or *directed* (for effective connectivity) graphs. The *directionality* is also of interest because it may indicate the pathways of how the disease spreads.

The existing works for brain network analysis can be generally categorized as generative methods [37, 41] or discriminative methods [43–45]. In generative methods, a representative network is explicitly inferred for *each group* (e.g., patients or healthy controls) independently and compared with each other to locate the differences. In discriminative methods, an individual network is constructed for *each subject*. Based on these networks, a discriminative classifier is trained to directly classify a given subject to one of the groups. As mentioned above, the group-level brain networks obtained by the generative methods are representative and good for interpretation, but may not be sufficiently discriminative. To bridge the gap, the work in [39] proposes a method to combine the advantages of both the generative and the discriminative methods to learn brain "effective connectivity" for the Alzheimer's disease. The bridge is Fisher kernel, which induces sample-specific feature vectors from generative models for SVMs to use. These Fisher-kernel-induced feature vectors are functions of the parameters of the generative models. Therefore, learning these parameters is converted to learn Fisher kernels via minimizing a generalization error bound of SVMs in [39]. In this way, the improvements of discrimination in both the generative models and the SVM classifiers can be simultaneously achieved.

6.4.2 *Generative Model: Sparse Gaussian Bayesian Network*

In [39], brain effective connectivity is built to investigate the directional effect of one brain region over another from MRI and FDG-PET images, respectively. That work is based on the generative model of sparse Gaussian Bayesian network (SGBN) originally proposed in [41].

Traditional methods to construct directional brain network from neuroimages usually require prior models of connectivity, which prevents their use from exploratory research [46, 47]. SGBN provides a data-driven method for large-scale directional brain network construction. It employs Gaussian Bayesian network (BN) model and imposes sparsity on the network connectivity.

Specifically, let $\mathbf{x} = [x_1, x_2, \cdots, x_m]^\top$ be a sample of m features (variables). A BN $\mathcal{G}$ is a directed acyclic graph (DAG) that expresses the factorization of a joint distribution $p(\mathbf{x}) = \prod_{i=1,\cdots,m} p(x_i|\mathbf{Pa}(x_i))$, where $\mathbf{Pa}(x_i)$ denotes the parent nodes of x_i. In a linear Gaussian BN, each node x_i is regressed over its parent nodes: $x_i = \theta_i^\top \mathbf{Pa}(x_i) + \varepsilon_i$, where θ_i is the regression coefficients, and ε_i is the Gaussian noise. The matrix $\Theta = [\theta_1, \cdots, \theta_m]$ denote the parameters of a GBN. Let $\mathbf{Pa}(\mathbf{x}_i)$ be a matrix whose j-th column represents a realization of the vector $\mathbf{Pa}(x_i)$ on the j-th sample. To learn an SGBN from the data, a constrained sparse regression is solved in [41] as follows:

$$\min_{\Theta} \sum_{i=1}^{m} \|\mathbf{f}_i - \mathbf{Pa}(\mathbf{x}_i)^\top \theta_i)\|_2^2 + \lambda_1 \|\theta_i\|_1 \tag{6.18}$$

$$s.t. \;\; \Theta_{ji} \times \mathbf{P}_{ij} = 0, \forall i, j = 1, \cdots, m, \;\; i \neq j.$$

where $\mathbf{f}_i$ corresponds to a realization of the i-th random variable x_i on the n samples. Each variable x_i regresses over all the other variables $\mathbf{Pa}(x_i)$. The matrix $\mathbf{P}$ is an $m \times m$ matrix. If there is a directed *path* from x_i to x_j, $\mathbf{P}_{ij} = 1$; otherwise, $\mathbf{P}_{ij} = 0$. The constraint enforces the DAG property of the SGBN, that is, there should be no directed cycles in the graph.

A simple way to use generative SGBNs for prediction is to train, for each class, an SGBN individually and classify a new sample by assigning it to the class with a higher likelihood. However, this may ignore some subtle but critical network differences that distinguish the two classes. Therefore, the work in [39] proposes to learn the parameters of the generative model from the two classes jointly to keep the essential discrimination.

6.4.3 *Bridge: Fisher Kernel*

As pointed out in [48], there are three categories of approaches to combine generative and discriminative models: blending, iterative, and staged methods. The last category currently attracts the most attentions. In staged methods, the discriminative models are usually trained on the features provided by the generative models. Fisher kernel [49] is a representative method in this family, and adopted in [39]. Compared with most Fisher-kernel-based methods in the literature, the work in [39] goes one step further. The Fisher kernel is utilized not only to induce feature vectors to train SVM classifiers but also to optimize SGBN parameters based on the performance of SVM classifiers.

Fisher kernel, introduced in [49], provides a way to compare samples induced by a generative model $p(\mathbf{x}|\Theta)$ with model parameters Θ. It is defined as $K(\mathbf{x},\mathbf{x}') = \mathbf{g}_{\mathbf{x}}^\top \mathbf{U}^{-1}\mathbf{g}_{\mathbf{x}'}$. The Fisher vector $\mathbf{g}_{\mathbf{x}} = \nabla_\theta \log(p(\mathbf{x}|\theta))$ maps a sample $\mathbf{x}$ to a feature vector in the gradient space of the model parameters Θ. The Fisher information metric $\mathbf{U}$ weights the similarity measure, but is often set as an identity matrix in practice [49]. The intuition of Fisher kernel is that similar objects induce similar log-likelihood gradients of Θ.

Applying Fisher kernel to SGBN and jointly considering two classes, the sample-specific feature vector in [39] is defined as $\Phi_\Theta(\mathbf{x}) = [\nabla_{\Theta_1}\mathscr{L}(\mathbf{x}|\Theta_1)^\top, \nabla_{\Theta_2}\mathscr{L}(\mathbf{x}|\Theta_2)^\top]^\top$, where Θ_1 and Θ_2 are the SGBN parameters of the two classes, respectively. For a given sample $\mathbf{x}$, it holds that:

$$\begin{aligned}\frac{\partial\mathscr{L}(\mathbf{x}|\Theta)}{\partial\theta_i} &= \frac{\partial\sum_{i=1}^m \log p(x_i|\mathbf{Pa}(x_i),\theta_i)}{\partial\theta_i} \\ &= \frac{\partial\sum_{i=1}^m\left(\frac{-(x_i-\theta_i^\top\mathbf{Pa}(x_i))^2}{2\sigma_i^2} - \log(2\pi\sqrt{\sigma_i})\right)}{\partial\theta_i} \\ &\triangleq \mathbf{S}(x_i)\theta_i + \mathbf{s}_0(x_i),\end{aligned} \tag{6.19}$$

where $\mathbf{S}(x_i)$ is a matrix and $\mathbf{s}_0(x_i)$ is a vector. As shown, the SGBN-induced Fisher vector $\Phi_\Theta(\mathbf{x})$ is a linear function of the SGBN parameters Θ.

6.4.4 Discriminative Learning via SVM

In order to separate the two classes of the SGBN-induced feature vectors effectively, the work in [39] jointly learns the parameters of SGBN and the separating hyperplane of SVMs via Fisher kernel. An efficient approximation of radius-margin bound, the upper bound of the Leave-One-Out error, is minimized to keep good generalization of the SVMs. Meanwhile, to maintain reasonable capacity of representation, the work in [39] explicitly controls the fitting errors of the learned model during the optimization and ensures the DAG property of the learned SGBN.

In order to incorporate radius-margin bound, 2-SVM with soft margin has to be used:

$$\begin{aligned}&\min_{\mathbf{w},\xi}\ \frac{1}{2}\|\mathbf{w}\|_2^2 + C\xi^\top\xi \\ &s.t.\ \ y_i(\mathbf{w}^\top\Phi(\mathbf{x}_i)+b) \geq 1-\xi_i,\ \ \xi_i \geq 0,\ \ \forall i\end{aligned} \tag{6.20}$$

where ξ is the slack variables and C the regularization parameter, which are used to control mis-classification. In [39] the radius-margin bound is introduced into the SVM classifiers as follows. First, the 2-SVM is rewritten as SVM with hard margin by slightly modifying the kernel $\mathbf{K} := \mathbf{K} + \mathbf{I}/C$, where $\mathbf{I}$ is identity matrix and $\mathbf{K}(i,j) = \langle \Phi(\mathbf{x}_i), \Phi(\mathbf{x}_j) \rangle$. Then the radius of minimal enclosing ball (MEB), denoted as R^2, is multiplied by the term $\frac{1}{2}\|\mathbf{w}\|^2$ in the objective function. Since the estimate of R^2 may become noisy and unstable when the sample size is small, it is replaced by the trace-based scatter matrix $\mathrm{tr}(\mathbf{S}_T)$ [50], where $\mathbf{S}_T = \sum_{i=1}^{n}(\mathbf{x}_i - \mathbf{m})^\top(\mathbf{x}_i - \mathbf{m})$, and $\mathbf{m}$ is the mean of total n samples. Therefore, the optimization problem for the discriminative learning in [39] becomes:

$$\min_{\Theta,\mathbf{w}} \frac{1}{2}\mathrm{tr}(\mathbf{S}_T)\|\mathbf{w}\|_2^2 \tag{6.21}$$

$$\begin{aligned} s.t.\ & y_i(\mathbf{w}^\top \Phi_\Theta(\mathbf{x}_i) + b) \geq 1, \ \forall i \\ & h(\mathbf{D}_1, \Theta_1) \leq T_1, \ h(\mathbf{D}_2, \Theta_2) \leq T_2, \\ & \Theta_1 \in DAG, \ \Theta_2 \in DAG. \end{aligned}$$

Here $\Phi_\Theta(\mathbf{x}_i)$ is the Fisher-kernel-induced feature vector. The function $h(\cdot)$ measures the squared fitting errors of the corresponding SGBNs for the data $\mathbf{D}_1$ and $\mathbf{D}_2$ from the two classes. The same DAG constraint is used as in Eqn. (6.18). This optimization problem can be solved by alternately optimizing the separating hyperplane $\mathbf{w}$ and the parameter Θ.

6.4.5 *Applications*

In [39], the above method is applied to identify discriminative brain effective connectivity for the MCI subjects from MR and FDG-PET images. These images are spatially normalized and segmented into tissues of gray matter (GM), white matter (WM), and cerebrospinal fluid (CSF) and parcellated into ROIs. The network nodes are constituted by the gray matter volume inside each ROI for MRI, and the average tracer uptake inside each ROI for PET images. Two MRI data sets and one PET data set from the publicly accessible database ADNI [51] are used for the validation. Compared with the generative SGBN models learned in [41], the method in [39] simultaneously improves the discriminative power of the generative SGBN models and the discriminative SVM classifier significantly. Specifically, the discriminative learning step alone increases the test classification accuracy of the generative SGBNs by 3–5 %. The test classification accuracy of the SVM classifier using the SGBN-induced Fisher vectors has been increased about 10 % for all three data sets. These improvements of discrimination are achieved with the cost of at most 1 % increase of squared fitting errors as explicitly controlled by the optimization, which ensures the reasonable maintenance of representation for the SGBN models. With the learned SGBN models, the structure of the brain network can be visualized by binarizing the edges Θ with a predefined threshold. An example result from [39] is given in Fig. 6.12. The change of connections in

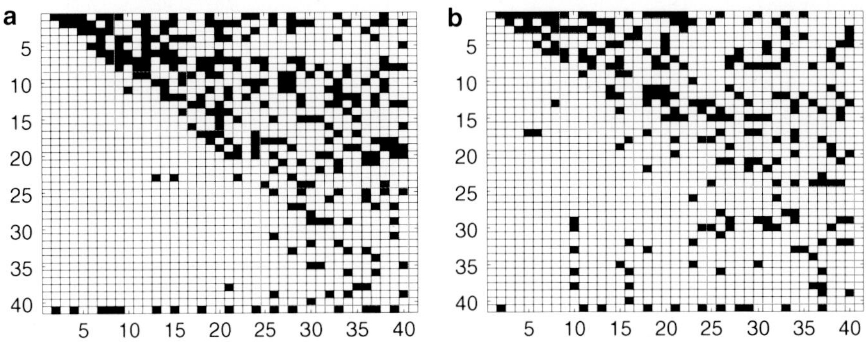

Fig. 6.12 Image courtesy of [39]: structure of connectivity for (**a**) Normal controls, (**b**) MCI patients

the MCI group from the normal control group can be identified by the comparison of the network structures from the two classes. Moreover, with Fisher kernel, new features reflecting the changing rate of brain connection strength are introduced, which enriches the investigation of brain connectivity.

6.5 Conclusion

For the classification tasks in neuroimage analysis, the interpretation of the results is as important as the discrimination of the classifier. Researchers have spent significant efforts in this regard to obtain distinctive patterns that are related to mental disorders or brain cognition. This is not an easy task especially when the underlying relationships of the data are complex. Three kinds of methods are introduced in this chapter. The SVM-based feature selection methods aim to identify a discriminative subset of features that can optimize the performance of SVMs. The selection of features gives insight into the importance of brain regions related to the research problem. Alternatively, the interpretation can be obtained from the SVM decision boundary that encodes the class discrimination. Methods in this category should respect certain anatomical constraints, such as the ROIs from brain atlases and the manifold where the data inhabit, in order to improve the fidelity of the interpretation. The above methods are built directly on the discriminative classifiers of SVMs that are not designed for data representation, especially for those having complex structures, such as graphs. In such cases, generative models may be used for representation, whose discriminative power is obtained from discriminatively learning the model parameters via SVMs. The methods introduced in this chapter are representative of the several possible ways to interpret classification results. With the advent of new neuroimage data and analysis tasks, more research work and advances can be expected in this area.

References

1. Mechelli, A., Price, C.J., Friston, K.J., Ashburner, J.: Voxel-based morphometry of the human brain: methods and applications. Curr. Med. Imaging Rev. **11**, 105–113 (2005)
2. Davatzikos, C.: Why voxel-based morphometric analysis should be used with great caution when characterizing group differences. Neuroimage **23**(1), 17–20 (2004)
3. Cristianini, N., Shawe-Taylor, J.: An Introduction to Support Vector Machines. Cambridge University Press, Cambridge (2000)
4. Magnin, B., Mesrob, L., Kinkingnhun, S., Plgrini-Issac, M., Colliot, O., Sarazin, M., Dubois, B., Lehricy, S., Benali, H.: Support vector machine-based classification of alzheimer's disease from whole-brain anatomical mri. Neuroradiology **51**(2), 73–83 (2009)
5. Lao, Z., Shen, D., Liu, D., Jawad, A., Melhem, E., Launer, L., Bryan, R., Davatzikos, C.: Computer-assisted segmentation of white matter lesions in 3d mr images using support vector machine. Acad. Radiol. **15**(3), 300–313 (2008)
6. Bauer, S., Nolte, L., Reyes, M.: Fully automatic segmentation of brain tumor images using support vector machine classification in combination with hierarchical conditional random field regularization. In: Medical Image Computation and Computer Assisted Intervention (MICCAI) 2011. pp. 354–361 (2011)
7. Cuingnet, R., Rosso, C., Lehricy, S., Dormont, D., Benali, H., Samson, Y., Colliot, O.: Spatially regularized svm for the detection of brain areas associated with stroke outcome. In: Proceedings of Medical Image Computing Computer Assisted Intervention (MICCAI). pp. 316–323 (2010)
8. Kloppel, S., Stonnington, C.M., Chu, C., Draganski, B., Scahill, R.I., Rohrer, J.D., Fox, N.C., Jack, C.R., Ashburner, J., Frackowiak, R.: Automatic classification of mr scans in alzheimer's disease. Brain **131**, 681–689 (2008)
9. Isabelle, G., André, E.: An introduction to variable and feature selection. J. Mach. Learn. Res. **3** 1157–1182 (2003)
10. Wee, C., Yap, P., Li, W., Denny, K., Browndyke, J., Potter, G., Welsh-Bohmer, K., Wang, L., Shen, D.: Enriched white matter connectivity networks for accurate identification of mci patients. Neuroimage **54**(3), 1812–1822 (2011)
11. Li, S., Shi, F., Pu, F., Li, X., Jiang, T., Xie, S., Wang, Y.: Hippocampal shape analysis of alzheimer disease based on machine learning methods. Am. J. Neuroradiol. **28**, 1339–1345 (2007)
12. Mesrob, L., Magnin, B., Colliot, O., Sarazin, M., Hahn-Barma, V., Dubois, B., Gallinari, P., Lehericy, S., Kinkingnehun, S., Benali, H.: Identification of atrophy patterns in alzheimer's disease based on svm feature selection and anatomical parcellation. Med. Imaging Augmented Reality **5128**, 124–132 (2008)
13. Calderoni, S., Retico, A., Biagi, L., Tancredi, R., Muratori, F., Tosetti, M.: Female children with autism spectrum disorder: an insight from mass-univariate and pattern classification analyses. Neuroimage **59**(2), 1013–1022 (2012)
14. Colby, J., Rudie, J., Brown, J., Douglas, P., Cohen, M., Shehzad, Z.: Insights into multimodal imaging classification of adhd. Front. Syst. Neurosci. **6**(59), 1–18 (2012)
15. Stoeckel, J., Fung, G.: Svm feature selection for classification of spect images of alzheimer's disease using spatial information. In: ICDM, pp. 410–417 (2005)
16. Cuingnet, R., Glauns, J., Chupin, M., Benali, H., Colliot, O., The Alzheimer's Disease Neuroimaging Initiative: spatial and anatomical regularization of svm: a general framework for neuroimaging data. IEEE Trans. Pattern Anal. Mach. Intell. (TPAMI) **35**(3), 682–696 (2012)
17. Rakotomamonjy, A., Bach, F., Grandvalet, Y., Canu, S.: Simplemkl. J. Mach. Learn. Res. **9**, 2491–2521 (2008)
18. Hinrichs, C., Singh, V., Xu, G., Johnson, S., ADNI: Predictive markers for ad in a multi-modality framework: an analysis of mci progression in the adni population. Neuroimage **55**(2), 574–589 (2011)

19. Hinrichs, C., Singh, V., Xu, G., Johnson, S.: Mkl for robust multi-modality ad classification. In: Proceedings of Medical Image Computation and Computer Assisted Intervention (MICCAI), pp. 786–794 (2009)
20. Zhang, D., Wang, Y., Zhou, L., Yuan, H., Shen, D.: Multimodal classification of alzheimer's disease and mild cognitive impairment. Neuroimage **55**(3), 856–867 (2011)
21. Cuingnet, R., Gerardin, E., Tessieras, J., Auzias, G., Lehericy, S., Habert, M., Chupin, M., Benali, H., Colliot, O.: Automatic classification of patients with alzheimer's disease from structural mri: a comparison of ten methods using the adni database. Neuroimage **56**(2), 766–781 (2010)
22. Cuingnet, R., Chupin, M., Benali, H., Colliot, O.: Spatial and anatomical regularization of svm for brain image analysis. In: Proceedings of the Neural Information Processing Systems conference (NIPS), pp. 460–468 (2010)
23. Golland, P., Grimson, W., Shenton, M., Kikinis, R.: Detection and analysis of statistical differences in anatomical shape. Med. Image Anal. **9**(1), 69–85 (2005)
24. Golland, P.: Discriminative direction for kernel classifiers. In: Advances in Neural Information Processing Systems (NIPS), pp. 745–752 (2001)
25. Zhou, L., Hartley, R., Wang, L., Lieby, P., Barnes, N.: Identifying anatomical shape difference by regularized discriminative direction. IEEE Trans. Med. Imaging **28**(6), 937–950 (2009)
26. Zhou, L., Hartley, R., Wang, L., Lieby, P., Barnes, N.: Regularized discriminative direction for shape difference analysis. In: Medical Image Computing and Computer-Assisted Intervention (MICCAI), pp. 628–635 (2008)
27. Lao, Z., Shen, D., Xue, Z., Karacali, B., Resnick, S.M., Davatzikos, C.: Morphological classification of brains via high-dimensional shape transformations and machine learning methods. Neuroimage **21**, 46–57 (2004)
28. Fan, Y., Shen, D., Gur, R., Gur, R., Davatzikos, C.: Compare: classification of morphological patterns using adaptive regional elements. IEEE Trans. Med. Imaging **21**, 46–57 (2007)
29. Mika, S., Schoelkopf, B., Smola, A., Mueller, K., Scholz, M., Raetsch, G.: Kernel PCA and de-noising in feature spaces. In: Proceedings of Advances in Neural Information Processing Systems, pp. 536–542 (1999)
30. Kwok, J.T., Tsang, I.W.: The pre-image problem in kernel methods. IEEE Trans. Neural Netw. **15**(6), 1517–1525 (2004)
31. Rathi, Y., Dambreville, S., Tannenbaum, A.: Statistical shape analysis using kernel PCA. In: Proceedings of SPIE Electronic Imaging 2006, pp. 425–432 (2006)
32. Scholkopf, B., Platt, J.C., Shawe-Taylor, J.C., Smola, A.J., Williamson, R.C.: Estimating the support of a high-dimensional distribution. Neural Comput. **13**, 1443–1471 (2001)
33. Shen, L., Ford, J., Makedon, F., Saykin, A.: Hippocampal shape analysis surface-based representation and classification. In: Proceedings of SPIE-Medical Imaging, pp. 253–264 (2003)
34. Styner, M., Lieberman, J., Pantazis, D., Gerig, G.: Boundary and medial shape analysis of the hippocampus in schizophrenia. Med. Image Anal. **8**(3), 197–2003 (2004)
35. Lee, H., Lee, D., Kang, H., Kim, B., Chung, M.: Sparse brain network recovery under compressed sensing. IEEE Trans. Med. Imaging **30**(5), 1154–1165 (2011)
36. Huang, S., Li, J., Sun, L., Ye, J., Fleisher, A., Wu, T., Chen, K., Reiman, E.: Learning brain connectivity of alzheimer's disease by sparse inverse covariance estimation. Neuroimage **50**(3), 935–949 (2010)
37. He, Y., Chen, Z., Evans, A.: Small-world anatomical networks in the human brain revealed by cortical thickness from mri. Cereb. Cortex **17**(10), 2407–2419 (2007)
38. Cecchi, G., Rish, I., Thyreau, B., Thirion, B., Plaze, M., Paillere-Martinot, M., Martelli, C., Martinot, J., Poline, J.: Discriminative network models of schizophrenia. In: Proceedings of Advances in Neural Information Processing Systems (NIPS) 2009, pp. 252–260 (2009)
39. Zhou, L., Wang, L., Liu, L., Ogunbona, P., Shen, D.: Discriminative brain effective connectivity analysis for alzheimers disease: a kernel learning approach upon sparse gaussian bayesian network. In: IEEE Conference on Computer Vision and Pattern Recognition (CVPR) 2013, pp. 2243–2250 (2013)

40. Li, X., Coyle, D., Maguire, L., Watson, D., McGinnity, T.: Gray matter concentration and effective connectivity changes in alzheimers disease: a longitudinal structural mri study. Neuroradiology **53**(10), 733–748 (2011)
41. Huang, S., Li, J., Ye, J., Fleisher, A., Chen, K., Wu, T., Reiman, E.: A sparse structure learning algorithm for gaussian bayesian network identification from high-dimensional data. IEEE Trans. Pattern Anal. Mach. Intell. **35**(6), 1328–1342 (2013)
42. Sporns, O.: Brain connectivity. Scholarpedia **2**(10), 4695 (2007)
43. Zhou, L., Wang, Y., Li, Y., Yap, P., Shen, D.: Hierarchical anatomical brain networks for mci prediction by partial least square analysis. In: IEEE Computer Society Conference on Computer Vision and Pattern Recognition (CVPR) (2011)
44. Li, Y., Wang, Y., Wu, G., Shi, F., Zhou, L., Lin, W., Shen, D.: Discriminant analysis of longitudinal cortical thickness changes in alzheimer's disease using dynamic and network features. Neurobiol. Aging **33**(2), e15–e30 (2012)
45. Fan, Y., Liu, Y., Wu, H., Hao, Y., Liu, H., Liu, Z., Jiang, T.: Discriminant analysis of functional connectivity patterns on grassmann manifold. Neuroimage **56**(4), 2058–2067 (2011)
46. Bullmore, E., Horwitz, B., Honey, G., et al.: How good is good enough in path analysis of fmri? Neuroimage **11**, 289–301 (2000)
47. Friston, K., Harrison, L., Penney, W.: Dynamic causal modeling. Neuroimage **19**, 1273–1302 (2003)
48. Perina, A., Cristani, M., Castellani, U., Murino, V., Jojic, N.: Free energy score spaces: using generative information in discriminative classifiers. IEEE Trans. Pattern Anal. Mach. Intell. **34**(7), 1249–1262 (2012)
49. Jaakkola, T., Haussler, D.: Exploiting generative models in discriminative classifiers. In: NIPS (1998)
50. Liu, X., Wang, L., Yin, J., Zhu, E., Zhang, J.: An efficient approach to integrating radius information into multiple kernel learning. IEEE Trans on Cybernetics. **43**(2), 557–569 (2013)
51. ADNI: http://www.adni-info.org

Chapter 7
Kernel Machines for Imbalanced Data Problem in Biomedical Applications

Peng Li, Kap Luk Chan, Sheng Fu, and Shankar M. Krishnan

Abstract Kernel machines such as the support vector machines (SVMs) have been reported to perform well in many applications. However, the performance of a binary SVM can be adversely affected by an imbalanced set of training samples, known as the imbalanced data problem. One-class SVMs, as a recognition-based approach, can be used to train and recognize the majority class and such kernel machines have already been developed. In this chapter, we review and study the effects of imbalanced datasets on the performance of both one-class SVMs and binary SVMs. We show that a hybrid kernel machine comprising one-class SVMs and binary SVMs in a multi-classifier system alleviates the imbalanced data problem. We also report the deployment of such hybrid kernel machines in two biomedical applications where the imbalanced data problem exists.

The research presented in this chapter was carried out when all authors were with the Nanyang Technological University, Singapore.

P. Li
University of Bristol, Bristol, UK
e-mail: lipeng@ieee.org

K.L. Chan (✉)
Nanyang Technological University, Singapore
e-mail: eklchan@ntu.edu.sg

S. Fu
KK Hospital, Singapore
e-mail: fu.sheng@kkh.com.sg

S.M. Krishnan
Wentworth Institute of Technology, Boston, USA
e-mail: krishnans@wit.edu

Y. Ma and G. Guo (eds.), *Support Vector Machines Applications*,
DOI 10.1007/978-3-319-02300-7_7,
© Springer International Publishing Switzerland 2014

7.1 Introduction

Kernel machines are algorithms in which kernels are employed to conceptually map data from an input space into a higher-dimensional feature space where the data can be processed using linear methods. The mapping is usually nonlinear and is implemented implicitly through the kernel trick. Many kernel methods have been developed by the machine learning community, such as support vector machines (SVMs) [51], kernel-based principal component analysis (KPCA) [36], kernel-based linear discriminant analysis (KLDA) [35], kernel-based independent component analysis (KICA) [2] and kernel-based nearest neighbour classifier [38].

SVM, a most widely used kernel machine, was originally developed for two-class classification. Based on the principle of structural risk minimization, discriminative binary SVMs, referred to as Binary Support Vector Classifier (*BSVC*) in this chapter, have been reported to perform well in many real applications [9, 12, 37]. However, SVM also suffers from some fundamental problems in statistical pattern recognition, such as the imbalanced data problem [19], in which the size of the training data from one class is significantly larger than that of the other class in a two-class classification task. Such a problem is frequently encountered in many biomedical applications where data from both positive and negative diagnosis categories are not available equally. For example, the data kept by a hospital can be mostly on positive diagnoses where data for negative diagnoses are not all kept. Another scenario can be in screening or patient monitoring where most cases are diagnosed as negative and only a small number of cases are diagnosed as positive. This means the collected data for the two categories are highly imbalanced and they will impact on the performance of binary classifiers, such as the *BSVCs*.

One possible solution to the imbalanced data problem is to use "recognition"-based approach instead of the conventional discriminative two-class classification approach [18]. "Recognition"-based approach is based on a one-class classification model in which only the data from one class (usually the class with more training samples, known as the majority class) are used to train a classifier [47] as opposed to using data from both classes in traditional two-class classifier training. This can prevent the adverse influence due to using a less representative smaller dataset of the minority class and hence avoiding the problem of imbalanced datasets. Two examples of such one-class classification kernel machines are the one-class Support Vector Classifier called the (νSVC) [43] and the Support Vector Data Description (*SVDD*) [48]. These one-class SVMs (*OSVC*) are trained using the data from the majority class only. However, the performance of one-class classifiers is reported to be seldom superior to the traditional two-class classifiers in real applications [41]. One reason might be that the data distribution of majority and minority classes is not suitable to be modeled as a one-class classification problem. Another reason may be due to the fact that only the data from one-class are used in one-class classifier training and no information about the other class is used. Hence, the one-class classifiers are to "recognize" the trained class rather than discriminating two classes.

To complement the strengths of these two types of kernel machines, this chapter shows how the hybrid kernel machines, in which the one-class SVMs and binary SVMs work in tandem as a multi-classifier system, handle the imbalanced data problem. We also report the use of such kernel machines in a couple of biomedical applications, namely, abnormal heart beat annotation from ECG waveform and tumor region detection from colonoscopic images.

In the following sections, we first introduce the one-class and binary SVMs and investigate how their training can be affected by the imbalanced data problem. We then present the hybrid kernel machines and show that the classifiers' performance can be improved. After that, we include two biomedical applications and show how the hybrid kernel machines can be used in these applications.

7.2 One-Class and Binary SVMs

In this section, the fundamentals of both discriminative two-class *SVM*s and recognition-based one-class *SVM*s are introduced. Their classification performance on a particular type of imbalanced data problem is investigated in Sect. 7.5 using an artificial dataset.

7.2.1 *Discriminative Support Vector Machines for Binary Classification*

SVM, a method based on the principles of statistical learning theory [52], can be applied to classification, regression and concept learning. The SVM is originally developed for two-class classification (or binary classification) task and it has been extended for multiple-classification [44]. For the sake of completeness, we briefly introduce the binary SVM in this subsection.

In two-class classification, an SVM classifier is trained using a training set of labeled samples in which the two classes are labelled as $+1$ and -1, respectively,

$$X = \{x_i \in R^d | i = 1,2,\cdots,N\} \tag{7.1}$$

where N is the number of samples in the training set. Each sample x_i is represented by a feature vector of d dimensions and labelled as $y_i \in \{+1,-1\}$. The classifier can be represented by a function $f(x) : x \to y$. The label y can be obtained for each pattern x by the classifier. It is assumed that the training and test data are drawn from the same distribution $P(x,y)$. The optimal function f can be found by minimizing the expected risk

$$R(f) = \int r(f(x),y)dP(x,y) \tag{7.2}$$

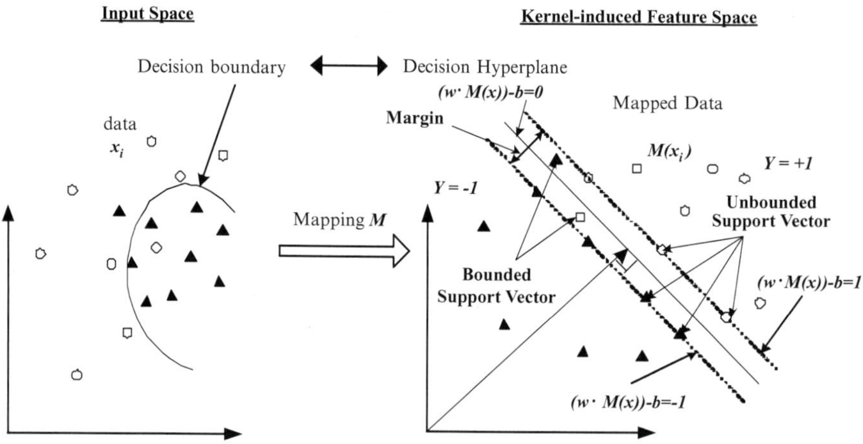

Fig. 7.1 Kernel mapping and optimal separating hyperplane of SVM

where r is a loss function. Conveniently, $r(f(x),y) = |f(x) - y|$ can be defined as 0 for correct classification and 1 for incorrect classification, known as 0/1 loss.

In practice, the underlying probability distribution $P(x,y)$ is usually unknown. Therefore, the risk R cannot be minimized directly. However, the risk can be approximated by minimizing the empirical risk

$$R_{em}(f) = \frac{1}{N}\sum_{i=1}^{N} r(f(x_i), y_i). \tag{7.3}$$

The empirical risk R_{em} converges to the expected risk R when the number of training samples tends to infinity ($N \to \infty$). However, overfitting may occur when the number of training samples is small [3]. Instead, the expected risk can be estimated while avoiding overfitting using the Vapnik Chervonenkis (VC) theory and the structural risk minimization (SRM) principle [51].

In practice, the bound on the expected risk is often difficult to compute. Fortunately, the decision functions in SVMs are restricted to hyperplanes whose VC-dimension can be bounded in terms of another quantity, called the "margin" [51].

Given that the two-class training set X with N samples are not linearly separable, the data are mapped to another feature space using a mapping M by which the mapped data $M(x)$ can be separated by an optimal separating hyperplane expressed as

$$f(x) = (w \cdot M(x)) - b \tag{7.4}$$

in which w is a weight vector, b is a bias item. (.) is an inner product. Such a mapping is illustrated in Fig. 7.1.

The "margin" is defined as the minimal distance of a sample to the decision hyperplane $f(x)$. w and b can be scaled so that the closest point to the hyperplane satisfies $|w \cdot M(x) - b| = 1$. Then the margin can be calculated using two samples from opposite classes $M(x_1)$ and $M(x_2)$ which have $w \cdot M(x_1) - b = 1$ and $w \cdot M(x_2) - b = -1$, respectively, and thus,

$$\frac{w}{||w||} \cdot (M(x_1) - M(x_2)) = \frac{2}{||w||} \tag{7.5}$$

The optimization problem then becomes:

$$\min_{w,b} \frac{1}{2} ||w||^2 \tag{7.6}$$

subject to constraints

$$y_i((w \cdot M(x_i)) - b) \geq 1, i = 1, 2, \cdots, N \tag{7.7}$$

For noisy data, some slack variables θ_i can be introduced to relax the constraints in (7.7):

$$y_i((w \cdot M(x_i)) - b) \geq 1 - \theta_i, \quad \theta_i \geq 0, \quad i = 1, 2, \cdots, N \tag{7.8}$$

The optimization problem in (7.6) can be reformulated as

$$\min_{w,b,\theta} \left\{ \frac{1}{2} ||w||^2 + C \sum_{i=1}^{N} \theta_i \right\} \tag{7.9}$$

where $C > 0$ is a regularization parameter to control the trade-off of the empirical error and the capacity terms.

The minimization problem in (7.9) is called primal in optimization theory. Its first item is related to the model complexity and the second item is the empirical risk R_{em}. Therefore, minimizing (7.9) can minimize the expected risk R. This problem can be solved by introducing Lagrange Multipliers $\beta_i \geq 0$ and $\gamma_i \geq 0$, $i = 1, 2, \cdots, N$, and with the constraints in (7.8), this leads to the dual problem:

$$\max_{\beta} \left\{ \sum_{i=1}^{N} \beta_i - \frac{1}{2} \sum_{i,j=1}^{N} y_i y_j \beta_i \beta_j (M(x_i) \cdot M(x_j)) \right\} \tag{7.10}$$

with constraints

$$\sum_{i=1}^{N} y_i \beta_i = 0 \tag{7.11}$$

and

$$C \geq \beta_i \geq 0, i = 1, 2, \cdots, N \tag{7.12}$$

This is a quadratic programming problem, which can be solved using standard algorithms, such as sequential minimization optimization [39]. Related codes can be found in [1].

In fact, only the inner product is calculated in (7.10) and (7.4). No explicit mapping M is needed. Such an inner product can be replaced using a kernel function

$$K(x_i, x_j) = M(x_i) \cdot M(x_j) \tag{7.13}$$

provided that this kernel $K(x_i, x_j)$ satisfies the Mercer's theorem. Then, equations (7.10) and (7.4) can be reformulated as follows

$$\max_{\beta} \left\{ \sum_{i=1}^{N} \beta_i - \frac{1}{2} \sum_{i,j=1}^{N} y_i y_j \beta_i \beta_j K(x_i, x_j) \right\} \tag{7.14}$$

$$f(x) = \sum_{i=1}^{N} y_i \beta_i K(x_i, x) - b \tag{7.15}$$

Among all possible kernels, the Radial Basis Function (RBF) kernel is a widely used one,

$$K(x_i, x_j) = e^{-\frac{\|x_i - x_j\|^2}{\sigma^2}} \tag{7.16}$$

7.2.2 *Recognition-Based One-Class Support Vector Machines*

As mentioned at the beginning of this chapter, recognition-based one-class SVMs can be used to handle the imbalanced data problem by learning from the majority class samples. We give a brief review of this type of SVMs in the following subsection.

7.2.2.1 One-Class Classification

One-class classification is also known as novelty detection, outlier detection and concept learning [47]. The problem formulation in one-class classification is different from conventional two-class classification. In one-class classification, it is assumed that only information of one of the classes, **the target class**, is available, and no information is available from the other class, known as **the outlier class**. The task of one-class classification is to define a boundary around the target class such that it accepts as much of the targets as possible and excludes the outliers as much as possible (Fig. 7.2b).

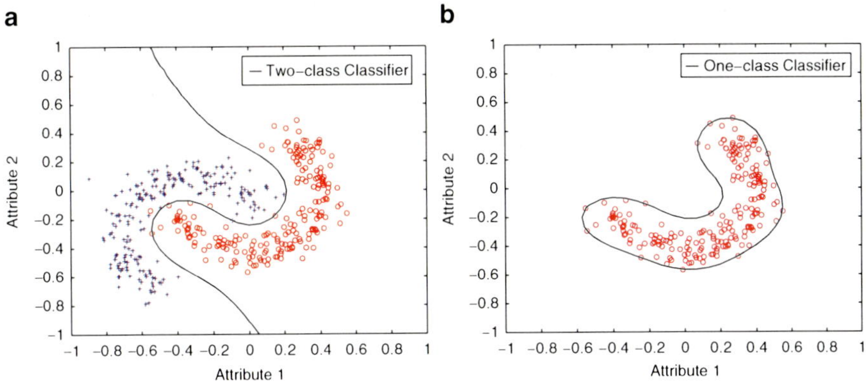

Fig. 7.2 Typical decision boundaries of (**a**) Two-class classifier and (**b**) One-class classifier in a 2-D toy problem

The philosophy behind one-class classification is in agreement with the way that human beings learn a concept. Suppose one expects to teach a child the concept of "car." One only needs to give him or her some examples of cars and it is not necessary to give the examples of non-"car," such as truck, bus or train. This is to say, people can learn a concept using only the examples of the target class. Of course, the information about non-target or outliers is helpful to improve the discrimination between the target and the non-target classes. However, using the examples from only the target class is sufficient to learn the concept of the target and recognize whether a new pattern belongs to the concept of the "target class."

To sum up, as illustrated in Fig. 7.2, the decision boundary of two-class classifier is supported by the samples of both classes and it utilizes the information from both classes while the decision boundary of one-class classifier is formed using only the data from one class. The two-class classifier is trained for "discrimination" purpose but the one-class classifier is trained to "recognize" the target samples rather than for "discrimination" purpose. Therefore, classification performance of one-class classifiers is usually worse than two-class classifiers when the data from both classes are available [41].

In one-class classifiers, a threshold is usually set so that the decision boundary of the classifier can enclose the target samples as much as possible. This is usually difficult when no information is available from the other class. One way is to reject some target to form a tighter boundary. The threshold can be determined based on the errors of classifying the target class only [47].

One-class classification has been used in many fields. Hojjatoleslami et al. employed a RBF network for density estimation in the detection of micro-calcifications in mammograms [14]. Manevitz and Yousef used One-Class Support Vector Machine for document classification [32]. Tax et al employed Support Vector Data Description in pump failure detection [48] and image retrieval [49]. A survey of the one-class classifiers can be found in [33, 34].

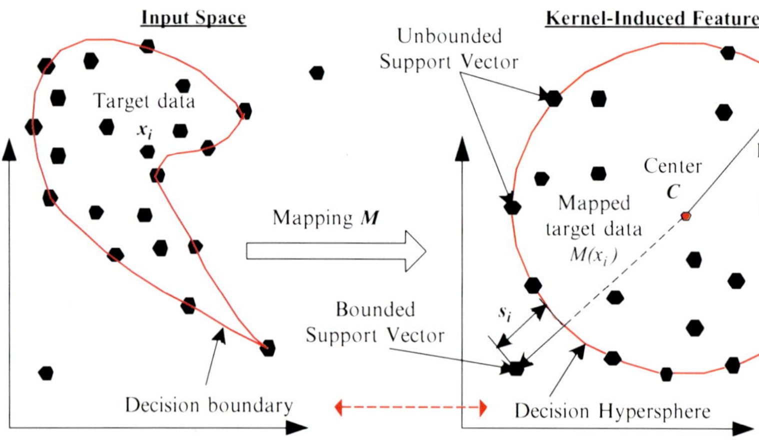

Fig. 7.3 Illustration of the kernel mapping of SVDD

7.2.2.2 Support Vector Data Description

The formulation of SVDD is as follows. Given a set of target data X with N samples, a nonlinear mapping M is sought to map X into some high dimensional kernel-induced feature space in which a hypersphere is sought to enclose the mapped target data $M(X)$ with smallest radius R centered at c. Figure 7.3 illustrates the nonlinear kernel mapping. The problem becomes

$$\min_{\{R,c,S\}} \left\{ R^2 + \frac{1}{\nu N} \sum_{i=1}^{N} S_i \right\} \tag{7.17}$$

subject to

$$\| M(x_i) - c \|^2 \leq R^2 + S_i, \quad i = 1, 2, \cdots, N \tag{7.18}$$

where S_i $(S_i \geq 0)$ are some slack variables to allow soft boundaries, i.e. some target data are allowed to lie outside of the hypersphere so as to control the trade-off between two types of errors. $\nu \in (0, 1]$ is a regularization parameter used to control the trade-off between the size of the hypersphere and the errors. In fact, it is the upper bound of the fraction of target data located outside the hypersphere.

The above problem can be solved by constructing a Lagrangian. Introducing constraints (7.18) to cost function (7.17), we have the following dual problem:

$$\max_{\beta} \left\{ \sum_{i=1}^{N} \beta_i (M(x_i) \cdot M(x_i)) - \sum_{i,j=1}^{N} \beta_i \beta_j (M(x_i) \cdot M(x_j)) \right\} \tag{7.19}$$

with constraints

$$\sum_{i=1}^{N} \beta_i = 1 \tag{7.20}$$

$$0 \leq \beta_i \leq \frac{1}{\nu N}, \quad i = 1, 2, \cdots, N \tag{7.21}$$

Define function $K(\cdot,\cdot)$ as

$$K(x_i, x_j) = M(x_i) \cdot M(x_j) \tag{7.22}$$

Then, Eq. (7.19) becomes

$$\min_{\beta} \left\{ \sum_{i,j=1}^{N} \beta_i \beta_j K(x_i, x_j) - \sum_{i=1}^{N} \beta_i K(x_i, x_i) \right\} \tag{7.23}$$

with the same constraints as (7.19). The cost function of the dual problem (7.23) is convex and quadratic in terms of the unknown parameters β_i. This problem can be solved by quadratic programming for which some standard algorithms such as sequential minimization optimization can be employed [43, 47].

Through quadratic programming, the Lagrangian (7.23) is optimized with respect to β. The center of the hypersphere c and multiplier γ_i can be calculated using the optimal solution β. Because $\| M(x_i) - c \|^2 = R^2$ holds for all the unbounded support vectors (USVs), the radius R can be calculated by choosing any of the USVs x_s

$$R = \left[K(x_s, x_s) + \sum_{i,j=1}^{N} \beta_i \beta_j K(x_i, x_j) - 2 \sum_{i=1}^{N} \beta_i K(x_i, x_s) \right]^{-2} \tag{7.24}$$

Given a new pattern z, the decision function is

$$\begin{aligned} f(z) &= R^2 - \| M(z) - c \|^2 = R^2 - K(z, z) \\ &\quad - \sum_{i,j=1}^{N} \beta_i \beta_j K(x_i, x_j) + 2 \sum_{i=1}^{N} \beta_i K(z, x_i) \end{aligned} \tag{7.25}$$

If the value of the decision function is greater than zero, the new sample lies inside the hypersphere and hence is classified as a target. Otherwise, it is classified as an outlier.

Similar to binary SVMs, kernel function can be used in SVDD. Although nonlinear mapping has been used to improve the effectiveness of the hyperspherical description, neither does the explicit nonlinear mapping $M(.)$ appear in the dual problem of SVDD (7.23), nor in the decision function (7.25). They are expressed completely in terms of $K(x_i, x_j)$, which is the advantage of kernel method. In fact, since the problem is stated completely in terms of the inner products of the vectors, the inner products of the patterns can be replaced by a kernel function (7.22), provided that this kernel $K(x_i,x_j)$ satisfies the Mercer's theorem [47].

The Gaussian RBF kernel in Eq. (7.16) provides a very flexible description, which has been proven in [48]. Because it only depends on $x_i - x_j$, $K(x,x)$ is constant 1. Therefore, Eq. (7.23) becomes

$$\min_{\beta} \sum_{i,j=1}^{N} \beta_i \beta_j K(x_i,x_j) \tag{7.26}$$

subject to the same constraints as (7.23). The decision function (7.25) can be reformulated as follows using (7.24),

$$f_d(z) = \sum_{i=1}^{N} \beta_i [K(x_i,z) - k(x_i,x_s)] = \sum_{i=1}^{N} \beta_i K(x_i,z) - b \tag{7.27}$$

where the bias term b is

$$b = \sum_{i=1}^{N} \beta_i K(x_i,x_s) = \sum_{i=1}^{N} \beta_i e^{-\frac{\|x_i - x_s\|^2}{\sigma^2}} \tag{7.28}$$

Here, the SVDD decision function behaves as a template-matching detector in the mapped feature space. Since $\beta_i \neq 0$ holds only for those USVs and bounded support vectors (BSVs), these patterns form a known template. Given a new pattern, it is compared with only the USVs and BSVs in the mapped feature space. A pattern similar to all of the USVs and BSVs tends to have a large negative value in (7.27) and it is more likely to be an outlier. A pattern different from all of the USVs and the BSVs tends to have a large positive value in (7.27) and it is more likely to be a target.

7.2.2.3 ν-Support Vector Classifier

Another way of estimating the support of a data distribution in the kernel feature space is the νSVC. The kernel mapping is different from that of SVDD. The target data are mapped into a higher-dimensional space called feature space $M(x)$ in which the dot product can be computed using some kernel function. The mapped target data are away from the origin as shown in Fig. 7.4, which can be found by solving the following problem

$$\min_{w,S_i,b} \frac{\| w \|^2}{2} + \frac{1}{\nu N} \sum_{i=1}^{N} S_i - b \tag{7.29}$$

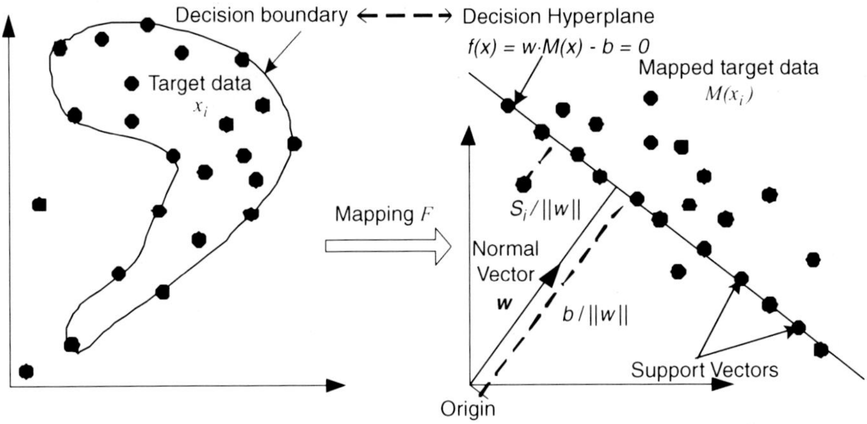

Fig. 7.4 Illustration of the kernel mapping of $vSVC$

subject to

$$w \cdot M(x_i) - b + S_i \geq 0, \quad \gamma_i \geq 0, \quad i = 1, 2, \cdots N \tag{7.30}$$

where S_i are slack variables. $v \in (0, 1]$ is a regularization parameter to control the effect of outliers and allows for target samples falling outside the decision boundary. The decision function corresponding to the hyperplane is

$$f(x) = w \cdot M(x) - b \tag{7.31}$$

By similar analysis as in SVDD, this problem can be solved as a quadratic programming problem which is exactly the same as the dual problem (7.26) in SVDD when the Gaussian kernel is used. Hence, $vSVC$ and $SVDD$ are equivalent to each other [43].

7.3 The Imbalanced Data Problem

The imbalanced data problem has received considerable attention in recent years in the machine learning community. This is the problem when the size of the training set from one class is significantly larger than that of the other class in a two-class classification setting. An example is illustrated in Fig. 7.5. This problem is often encountered in real applications such as in medical screening for abnormalities, image retrieval, and oil spill in satellite images [5, 23]. In applications such as medical screening for abnormalities, the data of the normal class can be easily obtained. On the other hand, the data of the abnormal class are more difficult to be collected than the normal ones. Therefore, the data from the abnormal class (usually

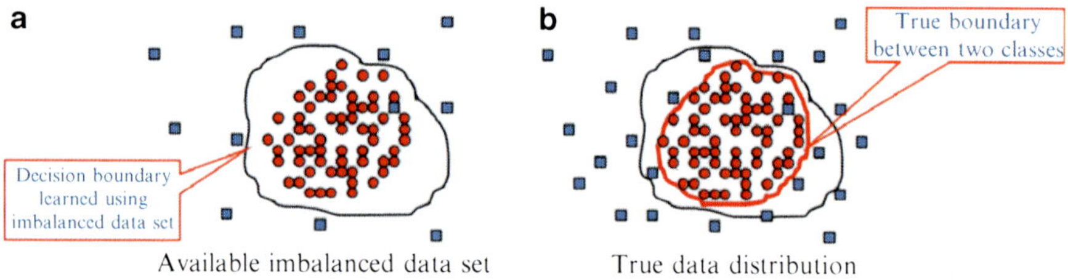

Fig. 7.5 Example of the imbalanced data problem. (**a**) is the decision boundary learned using the available (imbalanced) dataset which is different from the true boundary between the two classes in (**b**)

the minority class) cannot represent its true distribution well compared to the other class (usually the majority class, i.e. the normal class). It can be assumed that the minority class is positive and the majority class is negative in this chapter and it is formulated as a simple binary classification problem. Hence, it is logical to use discriminative binary classifiers such as binary SVMs. However, such classifiers are designed to minimize the overall misclassification rate on the training set, their classification performance degrades if they are trained with a highly imbalanced dataset.

Some attempts have been reported to deal with the imbalanced data problem, which can be classified into three approaches [17, 19] presented in the following subsections.

7.3.1 Resampling

The first approach is resampling the training dataset to make it balanced, such as in [10, 24]. Resampling is probably the most extensively studied approach, which consists of two main techniques:

1. *Undersampling:* The data from majority class are down-sampled so that the size of the majority class matches the size of the minority class. The sampling can be either done randomly [19] or based on some rules [24]. But the problem is that some of the information may be lost if down-sampling is not done properly.
2. *Oversampling:* The data from minority class are over-sampled so that the size of minority class matches the size of the majority class. Similar to random under-sampling, random oversampling has been shown to be effective in improving the classification [19]. There are also some attempts to improve the performance of oversampling. For example, Chawla et al developed a Synthetic Minority Oversampling Technique (SMOTE) by generating artificial data (the nearest neighbors of the original minority data) [4].

It is unclear which of these two is more effective in solving the imbalanced data problem [8, 10]. Therefore, some attempts have also been made to combine these two approaches [4, 10].

7.3.2 *Using Different Costs to Two Classes*

The second approach is to compensate for the class imbalance by altering the costs of the minority and majority classes in the training of classifiers. For example, Karakoulas et al proposed an algorithm called ThetaBoost, which is a boosting algorithm with unequal loss functions [20]. Some attempts have also been made to compensate the class imbalance by using different costs to the two classes in the training of SVMs [53]. Raskutti et al used different penalizing factors for two classes and resampling for SVM in [41]. Wu et al proposed the class-boundary alignment algorithm to deal with imbalanced data problem in SVM [56].

7.3.3 *Recognition-Based Approach*

The third approach is to use recognition-based instead of discrimination-based learning strategy by leaving one of the two classes totally unused (usually the minority class). The recognition-based method resembles that of a density estimation without finding the true density explicitly. This is an extreme case where only the data from one class are used to construct the learning model. For example, Japkowicz proposed to use an autoencoder to solve the imbalanced data problem [18]. This method works well when the majority class can be well modelled by a novelty detector such as an autoencoder. However, a recognition-based method is usually outperformed by a discrimination-based one due to the exclusion of the information from the minority class in training the model [41].

7.4 Hybrid Kernel Machine Ensemble

It has been discussed in the previous sections that discriminative two-class SVMs have problems in dealing with imbalanced datasets and the recognition-based one-class SVM cannot always do better than two-class SVMs. Basically, a two-class classifier *BSVC* benefits from the information from two classes while suffering from inadequate representation of the minority class. But, a one-class classifier *OSVC* benefits from more precise representation of the majority class but is not highly discriminative. There is a need to develop a classifier which is in-between the one-class classifier and the two-class classifier. Such a classifier can be named as ***one and half (1.5) classifier***. By exploiting the different properties of the two types of kernel

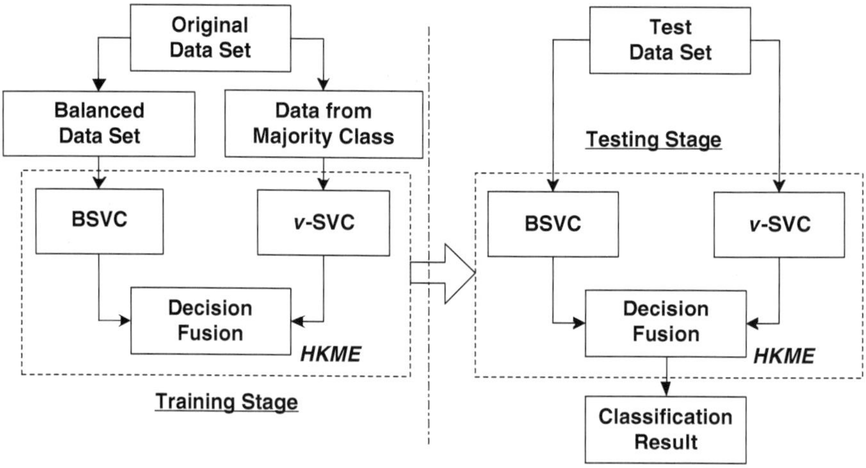

Fig. 7.6 Illustration of hybrid kernel machine ensemble (*HKME*) framework [26]

machines, an ensemble can be constructed by combining these two types of kernel machines. Such an ensemble is called Hybrid Kernel Machine Ensemble (*HKME*). *HKME* is designed to benefit from both discriminative *BSVC* and recognition-based *OSVC* and such an ensemble is expected to perform better in some applications such as in imbalanced datasets or in other cases where there is a need to combine the information from these two types of kernel machines. Most of the material presented in the following subsections has been published in [26].

7.4.1 Hybrid Kernel Machine Ensemble Framework

The hybrid kernel machine ensemble (*HKME*) framework is illustrated in Fig. 7.6. A *HKME* consists of two different types of *SVMs*, i.e. a discriminative *BSVC* and a non-discriminative recognition-based νSVC (or *SVDD*). Hence, the *HKME* is expected to benefit from the strength of both *BSVC* and νSVC.

HKME is designed for problems where *BSVC* does not perform well or costly to construct while νSVC shows good performance. For example, there is a type of imbalanced data problem in which the majority class is compactly clustered and the minority class is scattered in the input space. One example is in heart patient monitoring using ECG. The ECG signal morphologies from normal activities (normal class) are similar and the data from this class can be easily collected (majority class), while those from abnormal activities (abnormal class) may exhibit various morphologies and are more difficult to collect (minority class). A discriminative model, such as a *BSVC*, can be trained by manually balancing the data or compensating the imbalance using different costs to the two classes. Thus, the discriminative model uses the information from both majority class and minority class. However, its performance can still be poor due to the poorly represented minority class. A recognition-based one-class *SVM* may do better than the discriminative *BSVC* in this situation by modeling the well-represented majority class only. Since the majority class satisfies the assumption of one-class classification where the majority class is well represented and compactly clustered, it avoids the problem faced by the binary *SVM* due to the inadequate representation of the minority class. However, as a descriptive model, such a recognition-based model is not highly discriminative because the information from the minority class is left totally unused. Hence, there is a need to incorporate the information from the minority class to the recognition-based model or exploit the well-represented majority class further in the discriminative model. Exploiting the complementary nature of these two different types of models, a combination of them is expected to perform better than using either of them separately for the classification of this type of imbalanced dataset. Hence, constructing a *HKME* by integrating these two types of kernel machines in an ensemble is presented here to address this type of imbalanced data problem.

In this framework, a *νSVC* can be trained using only the data of majority class, so it can avoid the problem of poor representation of the minority data. On the other hand, a *BSVC* can be trained using balanced dataset using oversampling or undersampling, so it benefits from the information from both classes. The outputs of the two *SVM*s can be integrated using some fusion rules. Since the *νSVC* and *BSVC* are trained using different datasets, the training sets of such two kernel machines can be considered diverse. Furthermore, the different nature of the two *SVM*s can further help to increase the diversity. Therefore, the ensemble of such two kernel machines is expected to improve the classification compared to using either of the two types of *SVM*s.

7.4.2 *Binary* SVM *Training*

Performance of the classifiers is closely related to the parameters used by the classifiers. There are two hyper-parameters to be tuned in *BSVC* when using the Gaussian RBF kernel, the width parameter σ of the RBF kernel and the regularization parameter C which is used to control the trade-off of errors. The hyper-parameters of *BSVC* can be optimized using cross validation on the training set. The use of cross validation is able to avoid over-fitting [3]. The values of the hyper-parameters are chosen so that the errors of both classes on the validation set are minimized.

Another problem in *BSVC* is its training using imbalanced datasets. It has been shown that balanced dataset generally leads to results which are no worse than or superior to those of using natural class distribution, although it does not always produce the optimal results [54]. Since *BSVC* suffers from the imbalanced data problem, the original dataset can be balanced first using oversampling, undersampling or *SMOTE* algorithms aforementioned. The trained *BSVC* using the balanced dataset can then be integrated with the one-class *SVM* to form the *HKME*.

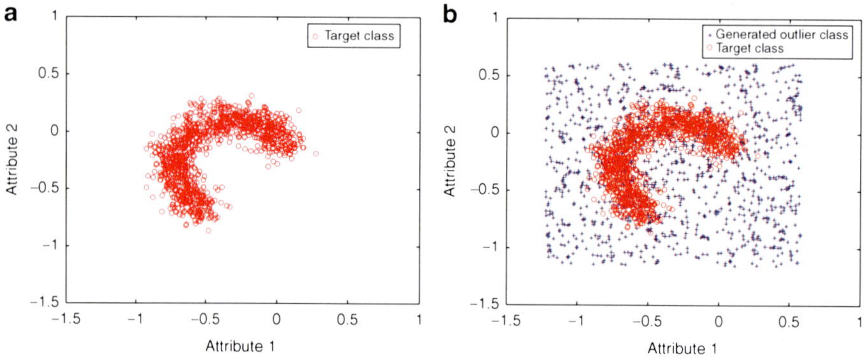

Fig. 7.7 (**a**) Original target dataset and (**b**) Generated artificial outliers around the target class in a toy problem in 2-D space

7.4.3 *One-Class* SVM *Training*

The hyper-parameters of *νSVC* or *SVDD* using Gaussian RBF kernel are the same as those of the BSVCs, i.e. the width parameter of the RBF kernel σ and the regularization parameter ν are used to control the trade-off of errors. The parameters of two-class classifiers can be optimized using cross validation on the training set. However, the information about the outlier class is assumed to be unavailable for one-class classifiers, hence the hyper-parameters can only be estimated using the data from target class or be chosen heuristically. This problem can be solved by generating artificial outliers [50]. Given a set of target samples, some outlier samples are generated randomly with the assumption that the outliers are uniformly distributed around the target class. The union of targets and generated outliers is used as a validation set to optimize the hyper-parameters of one-class *SVM*. A toy dataset and generated artificial outliers are illustrated in Fig. 7.7.

As for the imbalanced data problem in question, there are still some outlier samples, i.e., data from the minority class. The hyper-parameters may be tuned to minimize the training error on the whole training set which consists of both majority and minority classes. But, this might be undesirable if the minority class is not well represented by the sampled data. This problem will be discussed further in the experimental section.

7.4.4 *Fusion Rules for Integration of Hybrid* SVMs

Integrating two *SVM*s in a hybrid is posed as a decision level fusion problem. It is nontrivial to properly combine the two sources of information from these two types of *SVM*s.

Many ensemble learning methods have been developed. In this subsection, several ensemble methods are reviewed to understand how the imbalanced data problem can be handled by them and these include Decision Template (*DET*), Stacking, Average (*AVG*), Maximum (*MAX*), Minimum (*MIN*), Product (*PROD*) [22, 25].

Let $C_i(x) = \{C_{i1}(x), C_{i2}(x), \cdots, C_{ik}(x)\}$ be a set of individual classifiers, called an ensemble, each of which gets an input feature vector $\mathbf{x} = [x_1, x_2, \cdots, x_d]^T$ and assigns it to a class label y_i from $Y = \{-1, +1\}$, the goal of the ensemble is to find a class label L_{ens} for $\mathbf{x}$ based on the outputs of k classifiers $C_1(x), C_2(x), \cdots, C_k(x)$ corresponding to labels $L_1(x), L_2(x), \cdots, L_k(x)$. $C_i(x)$ is often an estimate of the posterior probability $P(y_i|\mathbf{x})$.

- *Decision template:* The decision template DET_j for class $y_j \in \{-1, +1\}$ is the average of the outputs of individual classifiers with respect to the training set for class y_j [25]. The ensemble *DET* assigns the input x with the label given by the individual classifier whose Euclidean distance to the decision template DET_j is the smallest.
- *Stacking (Stacked generalization):* Taking the output of individual classifiers $C_i(x)$ as input to an upper layer classifier and the final decision is determined by the upper layer classifier [55].

$$L_{ens}(x) = F(C_1(x), C_2(x), \cdots, C_k(x)) \tag{7.32}$$

  The upper layer classifiers used here include linear discriminant classifiers (*LDC*s) and quadratic discriminant classifiers (*QDC*s) assuming normally distributed classes. Because the covariance matrices for the classes are near singular, *QDC*s may fail when trying to estimate and invert the covariance matrices [25].
- *Average:*

$$L_{ens}(x) = \arg\max_j \left(\sum_{i=1}^{k} \frac{C_{ji}(x)}{k} \right) \tag{7.33}$$

  where $j \in \{-1, +1\}$. The *AVG* rule calculates the average of the outputs of the k individual classifier and assigns the input x to the class with the largest posterior probability.
- *Maximum:*

$$L_{ens}(x) = \arg\max_j \left(\max_i C_{ji}(x) \right) \tag{7.34}$$

  where $j \in \{-1, +1\}$. The *MAX* rule takes the maximum value of the outputs from the k individual classifier for each class and assigns the input x to the class with the largest posterior probability.

- *Minimum:*

$$L_{ens}(x) = \arg\max_j \left(\min_i C_{ji}(x) \right) \tag{7.35}$$

where $j \in \{-1, +1\}$. The *MIN* rule takes the minimum value of the outputs from the k individual classifier for each class and assigns the input x to the class with the largest posterior probability.

- *Product:*

$$L_{ens}(x) = \arg\max_j \left(\prod_i C_{ji}(x) \right) \tag{7.36}$$

where $j \in \{-1, +1\}$. The *PROD* rule calculates the product value of the outputs from the k individual classifier for each class and assigns the input x to the class with the largest posterior probability.

The problem here is to fuse the outputs of two classifiers. The generally used majority voting is not suitable here. Furthermore, it can be proved that Maximum, Minimum, Averaging, Product rules are equivalent to each other when they are used to combine two classifiers with posterior probability outputs for a two-class classification task. It has been proved that Maximum and Minimum are equivalent when combining multiple classifiers for two-class classification in [46]. Due to the equivalence of *MAX*, *MIN*, *AVG*, and *PROD* rules for the two-class problem using two classifiers with posterior probability as outputs, only *AVG* is investigated in the following subsection.

7.4.5 Estimating the Posterior Probability for Outputs of SVMs

The outputs of *SVM*s are not posterior probabilities and are in different ranges, and hence are not comparable directly. Thus, their outputs have to be normalized for use in this hybrid. It is observed that the outputs of *SVM*s show similar forms. One can estimate the posterior probabilities $P_i(y_j|\mathbf{x})$ of the i-th *SVM* using a sigmoid function by minimizing the negative log likelihood of the training data [40]

$$P_i(y_j|\mathbf{x}) = \frac{1}{1 + e^{p_i f_i(x) + q_i}} \tag{7.37}$$

where p_i is a coefficient to control the shape of sigmoid function and q_i is a coefficient to control the shift along the horizontal axis ($f_i(x)$). Thus, the ensembles can be constructed using these estimated posterior probabilities.

When estimating the posterior probability of *BSVC*, the training set of the *BSVC* has to be balanced. Otherwise, it may lead to biased fitting of a sigmoid to outputs of nonlinear *SVM*s [40, 52]. The balancing of the training set can be done using oversampling such as *SMOTE* [4].

To our best knowledge, this is the first attempt to estimating the posterior probability of *νSVC* or *SVDD*. Since there is only the target data for training a one-class *SVM*, a set of artificial data can be generated whose sample size is the same as that of targets [50]. The union of target data and artificial data can be used to estimate the posterior probability of output from one-class *SVC*.

The posterior probability generated here is only an estimation of the true posterior probability. Bias is unavoidable. This may create some problems to the fusion rules such as *MAX* or *MIN*.

7.5 Experimental Results on Artificial Dataset

The performance of the proposed *HKME* is evaluated on an artificial dataset. The evaluation measure is as follows.

7.5.1 Evaluation Measure

The decision table of two-class classification outcome for calculating the evaluating criteria used in this study is illustrated in Table 7.1.

Four classification outcomes are considered, i.e. true positives (TP), true negatives (TN), false positives (FP), and false negatives (FN). Let A_+ and A_- denote the classification accuracy rates for positive class and negative class, respectively.

$$A_+ = \frac{TP}{TP+FN} \tag{7.38}$$

$$A_- = \frac{TN}{TN+FP} \tag{7.39}$$

The most commonly used measure is the Average Classification Rate (*ACR*) which is the fraction of all correctly classified samples among all the samples, regardless of the classes:

$$ACR = \frac{TP+TN}{TP+TN+FP+FN} \tag{7.40}$$

Table 7.1 Decision table of two-class classification outcome for calculating the evaluating measure

Ground truth	Classification outcome: Positive	Negative
Positive	True positive	False negative
Negative	False positive	Truth negative

In imbalanced datasets, the negative (majority) class dominates. The generally used *ACR* is not valid for evaluating the performance of the classifiers in such an imbalanced dataset. For example, if a classifier classifies all the data as negative samples, it has $A_- = 100\,\%$ and $A_+ = 0\,\%$, but the ACR is still high. Hence, another measure called the Balanced Classification Rate (*BCR*) is used in this study. *BCR* is the algebraic mean of A_+ and A_-:

$$BCR = \frac{A_+ + A_-}{2}. \tag{7.41}$$

This measure has been used for evaluating the performance of classifiers on imbalanced datasets [11, 45]. This measure is more suitable for evaluating the performance of the classifiers here than the generally used *ACR* for the following reason. Only when both A_+ and A_- have large value *BCR* can have a large value. Therefore, the use of *BCR* can give a balanced assessment of the classifiers for the imbalanced datasets as the *BCR* favors both lower false positives and false negatives.

7.5.2 *Artificial Dataset*

To investigate the effects the imbalanced data have on BSVC and OSVC and HKME, experiments were conducted using a checkerboard dataset similar to the one in [56]. The checkerboard data are shown in Fig. 7.8. The negative samples (majority class) occupy two diagonal squares of the checkerboard in the center and the positive samples (minority) surrounds the negative samples. The data are uniformly distributed and it is in agreement to the assumption that the data of the majority class is compactly clustered and the data of the minority class is scattered in the input space.

7.5.2.1 Influence of Class Imbalance to Discriminative *BSVC*s

In order to show the influence of imbalanced dataset on the performance of discriminative *BSVC*s, the following experiments were conducted.

In the first experiment, the size of negative training data in the 2×2 checkerboard data was fixed at 128, the size of the positive training data was reduced from 128 to 4, with increasing imbalance ratio (majority to minority) from 1:1 to 32:1. The test data consists of 1,000 positive samples and 1,000 negative samples. *BSVC*s with RBF kernel were trained using these data. The hyper-parameters of the *BSVC*s were optimized using threefold cross validation on the training set. The experiment was repeated ten times and the average value and standard deviation of the *BCR*s achieved by *BSVC*s are plotted in Fig. 7.9.

It can be observed that *BSVC* performs well when the training dataset is balanced which is expected. But its performance deteriorates gradually as the imbalance ratio

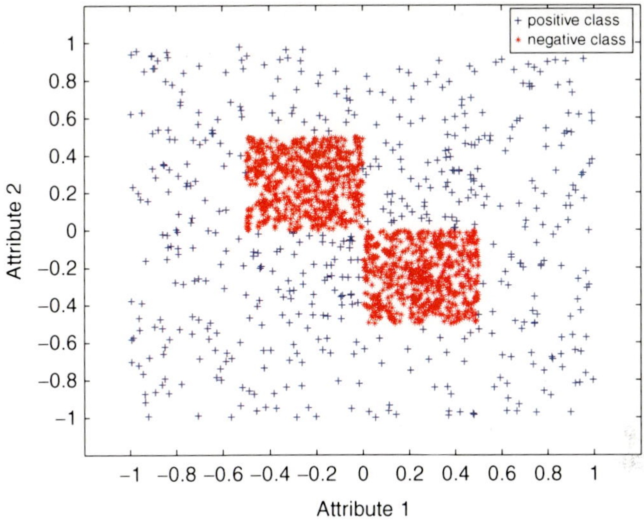

Fig. 7.8 2 × 2 checkerboard dataset

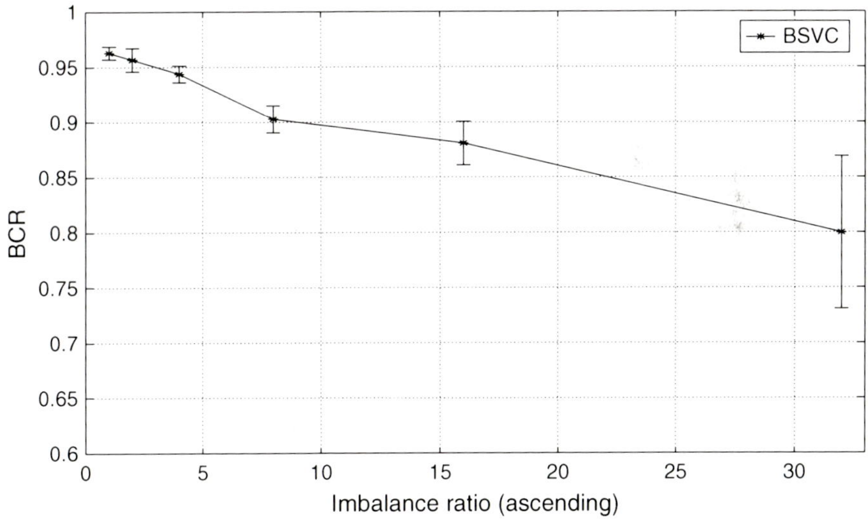

Fig. 7.9 The influence of class imbalance on the performance of *BSVC* using 2 × 2 checkerboard dataset (with negative samples as majority class)

increases. It indicates that discriminative *BSVC* suffers from the class imbalance. When the number of minority samples is very small, the data from this class cannot represent its true distribution well. This can be observed from the larger variation in the performance of the *BSVC* when the imbalance ratio is large.

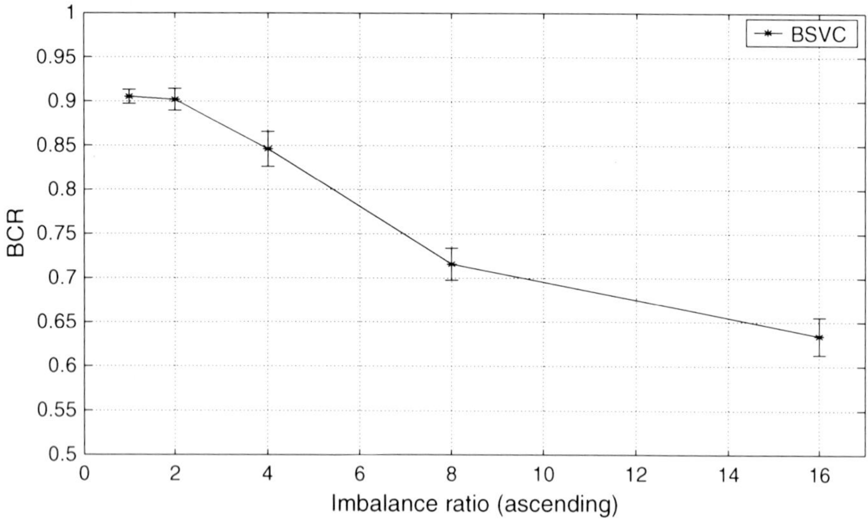

Fig. 7.10 The influence of class imbalance on the performance of *BSVC* using 2×2 checkerboard dataset (with positive samples as majority class)

It is unclear whether *BSVC* also suffers from the imbalanced data problem when the two classes are adequately represented while the samples from two classes are imbalanced. Therefore, the size of negative training data was still fixed at 128 in the second experiment, while the size of the positive training data was increased from 128 to 2,048, with corresponding imbalance ratio (minority to majority) increased from 1:1 to 1:16. Other settings are the same as the first experiment. The experimental result is illustrated in Fig. 7.10.

It can be observed that the result is similar to that in the first experiment. It shows that the discriminative *BSVC* also suffer from the class imbalance when the data are more precisely represented while the two classes are highly imbalanced. In summary, it has been shown that the discriminative *BSVC* suffer from the class imbalance problem. Hence, some measures have to be taken to alleviate this problem.

7.5.2.2 The Performance of Recognition-Based *OSVM*s

It has been mentioned that one approach to address the class imbalance is to use recognition-based model instead of discriminative model by training a one-class classifier using the data from the majority class only. However, one-class classifiers seldom outperform two-class classifiers when the data from two class are available. One reason is that the one-class classifier is designed for describing the majority class rather than for discrimination purpose, leaving the information from another class totally unused. Another reason may be that the concept to be

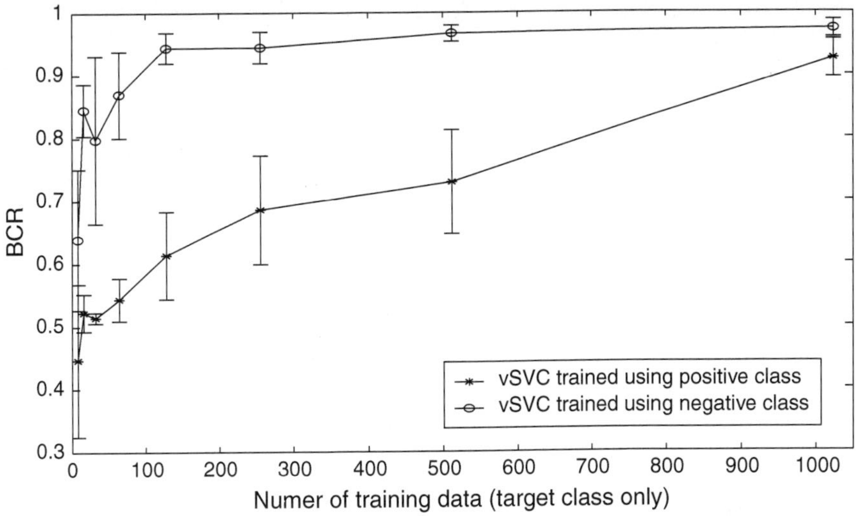

Fig. 7.11 The performance of *νSVC* in terms of *BCR* using 2×2 checkerboard dataset, with negative samples or positive samples as target class for training, respectively

learned is not suitable for description by the one-class classifiers. For example, in patient monitoring, the concept of "normal" is suitable for description by a one-class classifier while the concept of "abnormal" is not. This is because the "normal" class is usually compactly clustered in the input space, while the "abnormal" class is usually scattered. Furthermore, there is no clear boundary between "normal" and "abnormal" class. If a one-class classifier is to enclose the scattered "abnormal" data, it will also include some "normal" data. It may be better to construct a one-class classifier to enclose the compactly clustered "normal" data. The negative class of the checkerboard data in Fig. 7.8 seems more suitable to be described by a one-class classifier than the positive class since it is compactly clustered. This can be ascertained in the following experiment.

In the experiment, the negative samples and positive samples in the checkerboard dataset were taken as target class, respectively, to train a *νSVC*. The number of training data was varied from 8 to 1,024. The test data consists of 1,000 positive samples and 1,000 negative samples. The hyper-parameters of the *νSVC* were optimized using artificially generated dataset described in Sect. 7.4.3. The experiment was repeated ten times and the average value and standard deviation of the *BCR*s achieved by *νSVC*s are reported in Fig. 7.11.

It can be observed that the *νSVC* trained using compactly clustered negative samples outperforms that of using scattered positive samples. This supports the earlier claim that compactly clustered negative class is more suitable for training one-class classifiers than scattered positive class. Furthermore, the performance of *νSVC* is directly related to number of training samples. It seems that $128 \sim 256$ negative samples have been quite good to train a *νSVC* in this dataset, whose

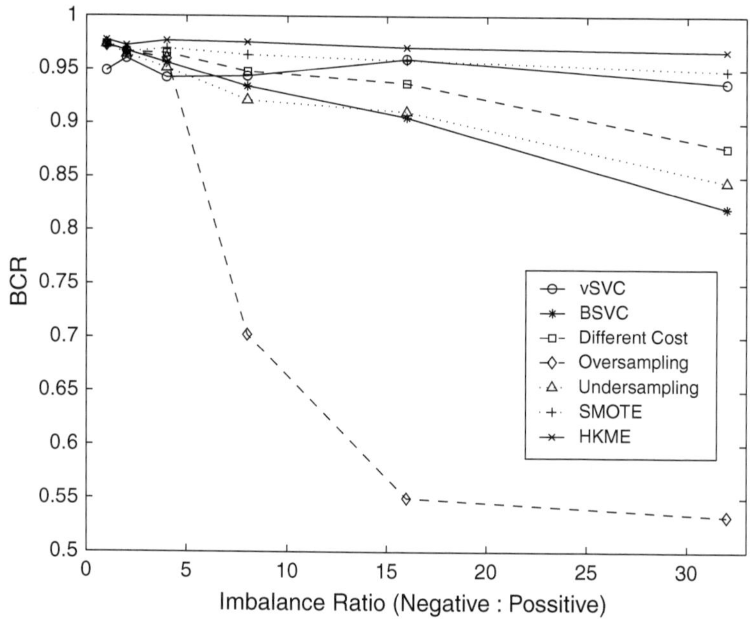

Fig. 7.12 The comparison of different schemes using 2×2 checkerboard dataset with different imbalance ratio [26]

performance is even better than the *νSVC* trained using 1,024 positive samples. It has been pointed out that more data is needed in one-class classification than a two-class classification [47]. So the number of training samples of *νSVC* should be large enough to have a good description of the target class. In imbalanced datasets, a compactly clustered majority class is more suitable for the one-class classifiers to learn.

7.5.2.3 The Performance of HKME

The proposed *HKME* is compared to other commonly used methods to deal with class imbalance using the artificial dataset, including oversampling, down-sampling, *SMOTE* and *BSVC* using different costs to the two classes. The number of negative samples was fixed at 256, the number of positive samples was decreased so that the imbalance ratio (negative to positive) is increased from 1:1 to 32:1. When the imbalance ratio increases to 32:1, the number of positive samples is only eight. The positive samples are too sparse to represent the true distribution. It is thus meaningless to decrease the number of positive samples further. The test data consists of 1,000 positive samples and 1,000 negative samples. The experiment was repeated ten times and the average value of the *BCR*s achieved by different schemes are plotted in Fig. 7.12. The comparison includes the following:

- *Oversampling:* The positive class was randomly oversampled (duplication) so that the training set is balanced.
- *Undersampling:* The negative class was randomly under-sampled so that the training set is balanced.
- *SMOTE:* A balanced dataset was created by adding some artificially generated data in-between the three nearest neighbors of each data point in the original dataset.
- *Using different costs to the two classes:* The *BSVC* was trained using different costs to the two classes. The primal problem in (7.9) becomes

$$\min_{w,b,\theta}\{\frac{1}{2}\|w\|^2 + +C_+\sum_{i=1}^{N_+}\theta_{i+} + C_-\sum_{i=1}^{N_-}\theta_{i-}\} \tag{7.42}$$

where N_+ and N_- are the numbers of positive and negative samples, respectively. $(N_+ < N-)$ and θ_{i+} and θ_{i-} are the errors of positive and negative samples, respectively. The regularization parameter becomes:

$$C_+ = \frac{C}{2N_+}, \quad y_i = +1 \tag{7.43}$$

and

$$C_- = \frac{C}{2N_-}, \quad y_i = -1 \tag{7.44}$$

Hence the error of minority negative class is penalized more than for the majority positive class in order to compensate for the class imbalance.

The parameters of all the *BSVC*s are optimized using threefold cross validation. The parameters of the νSVC are optimized using artificially generated outlier data aforementioned. The *BCR* achieved by *HKME* using *AVG*, *DET*, *LDC*, and *QDC* fusion rules are shown in Fig. 7.13.

It can be observed from Fig. 7.12 that discriminative *BSVC* (trained using original dataset) perform well when the imbalance ratio is not very high, but its performance deteriorates with the increasing imbalance ratio. *HKME* using *AVG* rule performs the best among all the approaches. The *BSVC* trained using different costs to the two classes perform quite well compared to the *BSVC* trained using the same cost to the two classes. Undersampling performs better than original *BSVC*, but is outperformed by using different costs. SMOTE performs reasonably well. It is better than both original *BSVC* and νSVC. Oversampling performs the worst among all the approaches.

The good performance of *HKME* may come from the fact that it benefits from the strength of both of its individual classifiers in the ensemble, the discriminative *BSVC* and recognition-based νSVC. This can be explained using their decision boundaries as illustrated in Fig. 7.14. νSVC performs well due to its ability to model compactly

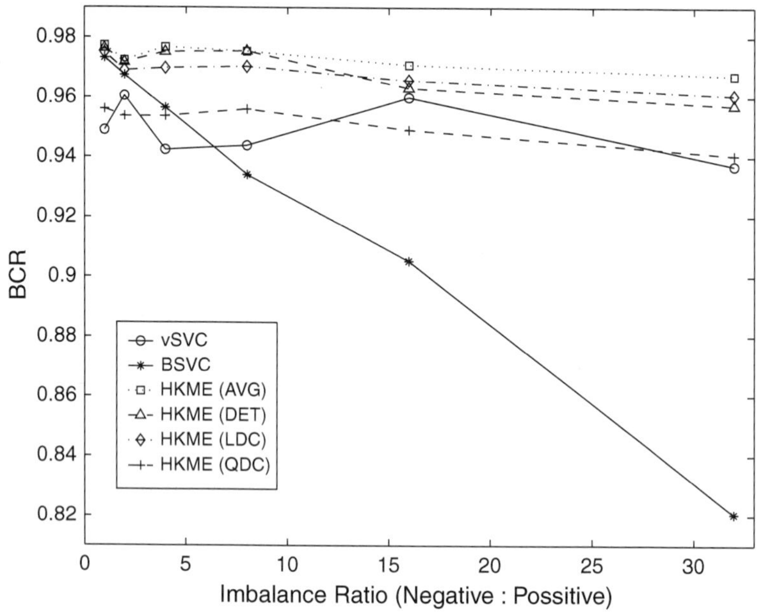

Fig. 7.13 The comparison of different fusion rules for *HKME* using 2×2 checkerboard dataset with different imbalance ratio

clustered target class. But it has to reject some target samples to form a tighter boundary as mentioned in Sect. 7.2.2, so it tends to push the decision boundary towards the majority (negative) class. However, discriminative *BSVC* tends to push the decision boundary toward the minority positive class. The ensemble of these two *SVM* tends to compensate these two different trends and strike a balanced compromise. As shown in the figure, the decision boundary of *HKME* is located in-between two classifiers, which is closer to the ideal decision boundary (two squares in the checkerboard).

The *HKME* using four different fusion rules are compared in Fig. 7.13. *AVG*, *DET*, and *LDC* performs well. But *QDC* does not perform well in some cases. This is because the covariance matrices for the classes are nearly singular in these cases, *QDC*s failed when trying to estimate and invert the covariance matrices [25]. However, it still performs quite well when it is properly trained.

The performance of other methods in Fig. 7.12 may also be explained using their decision boundaries on a checkerboard dataset with 256 negative samples and 16 positive samples, as shown in Fig. 7.15. The *BSVC* tends to push the decision boundary toward the minority class as aforementioned. Using different costs to the two classes, the decision boundary tends to be closer to the majority negative class as shown in Fig. 7.15a. So this approach performs better than *BSVC* trained using original dataset. The artificially generated positive data using *SMOTE* seems

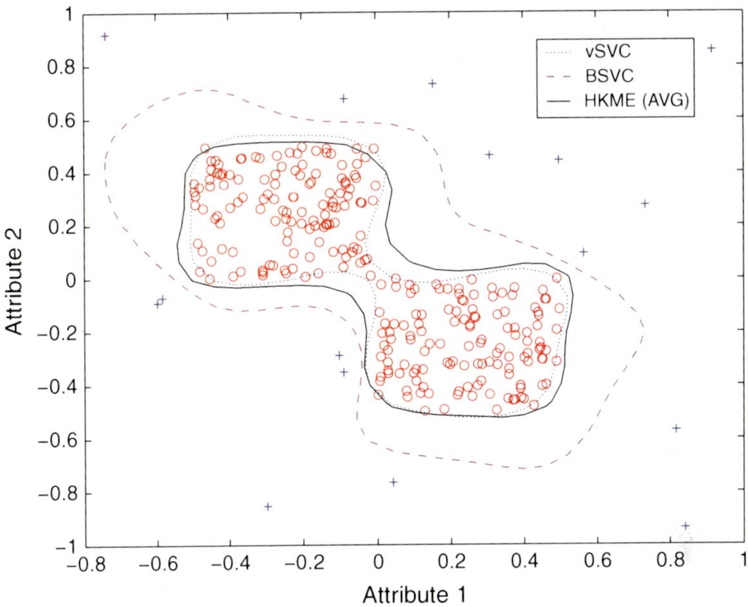

Fig. 7.14 The decision boundaries of *vSVC*, *BSVC*, and *HKME* using 2 × 2 checkerboard dataset (256 negative samples and 16 positive samples) [26]

closer to its original distribution in Fig. 7.15b, which makes the *BSVC* trained using *SMOTE* performs much better than the others. But it must satisfy the assumption that the samples between the nearest neighbors of a sample are from the same class. Due to the duplication of the minority samples in oversampling, the *BSVC* overfits the minority positive class, which is clearly illustrated in Fig. 7.15c. This leads to poor performance of oversampling shown in Fig. 7.12, especially when the imbalance ratio is high. Therefore, random oversampling the minority data is not suitable in *BSVC* training for imbalanced datasets. Undersampling the majority class seems to produce better decision boundary than that using oversampling. But the shape of the decision boundary is quite different from the ideal one as shown in Fig. 7.15d. This may be because that some useful information is lost when some samples from the majority class are removed from the training set. This detrimental effect is especially obvious when the size of the minority class is very small.

To sum up, *HKME* performs well in the checkerboard dataset. SMOTE and using different costs to the two classes seem quite efficient for this dataset. Random undersampling is better than the *BSVC* trained using original imbalanced dataset. Random oversampling is not suitable for *BSVC* when the imbalance ratio is high.

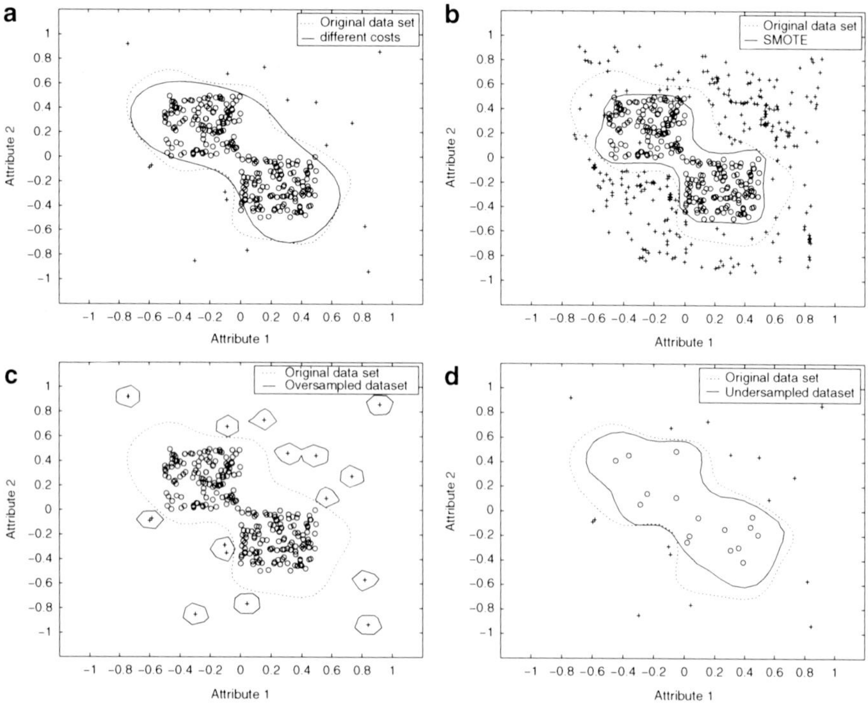

Fig. 7.15 The decision boundary of *BSVC* training using original dataset and those of *BSVC*s trained using (**a**) Different costs to two classes, (**b**) SMOTE, (**c**) Over-sampled dataset, and (**d**) Under-sampled dataset

7.6 Application of HKME in ECG Annotation and Colonoscopic Image Analysis

Since *HKME* is seen to perform well on the artificial dataset, it is deployed in two biomedical applications in which the problem of class imbalance exists and sharing similar properties in the distribution of the class samples to the artificial dataset.

7.6.1 Abnormal ECG Beat Annotation for Long-Term Monitoring of Heart Patients

7.6.1.1 Abnormal ECG Beat Annotation

ECG is a recording of the heart's electrical activity obtained from electrodes attached on the body surface of a patient [57]. Different segments of the ECG signal characterize different cardiac activities. A typical normal ECG beat is illustrated in Fig. 7.16.

Fig. 7.16 A typical normal ECG beat

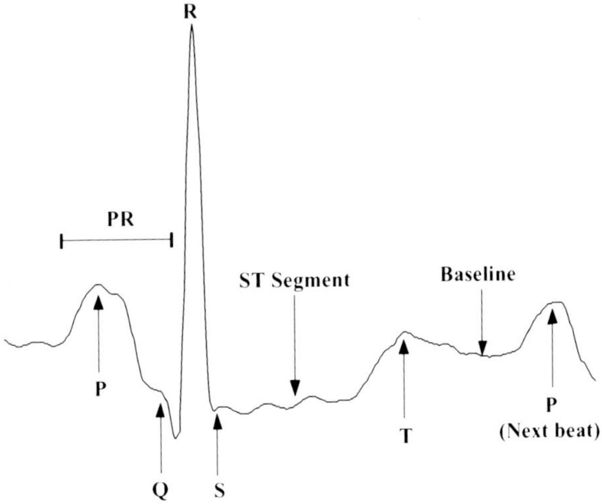

The analysis of heart beat cycles in ECG signal is very important for long-term monitoring and diagnosis of the patients' heart conditions in an intensive care unit or at patients' homes through a telemedicine network. However, it is very costly for the physicians to analyze the ECG recordings beat by beat since the ECG recordings may last for hours. Therefore, it is significant to develop a computer-assisted technique to examine and annotate the ECG recordings automatically, so to facilitate review by medical experts. This computer annotation will assist physicians to select only the informative (abnormal) beats for further analysis.

7.6.1.2 Generalization and Imbalanced Data Problem in ECG Beat Annotation

Generalization Problem

A fundamental assumption in the field of pattern recognition is that the underlying distribution of the training samples is the same as that of the test samples. However, such assumption may not hold in practical application. The abnormal ECG beat annotation problem is one of the examples. Figure 7.17 illustrates the distribution of the first two principal components of the original $181 - dimensional(D)$ feature vector of ECG beats obtained by using Karhunen–Loeve transform (PCA) from 4 recordings of MIT/BIH arrhythmia database [13], preserving 69 % of the total variance, where the circles indicate normal ECG beats and the cross signs are abnormal ones. Although some discriminative information may be lost using PCA, it can be observed that the distributions of "normal" ECG beats are different among patients.

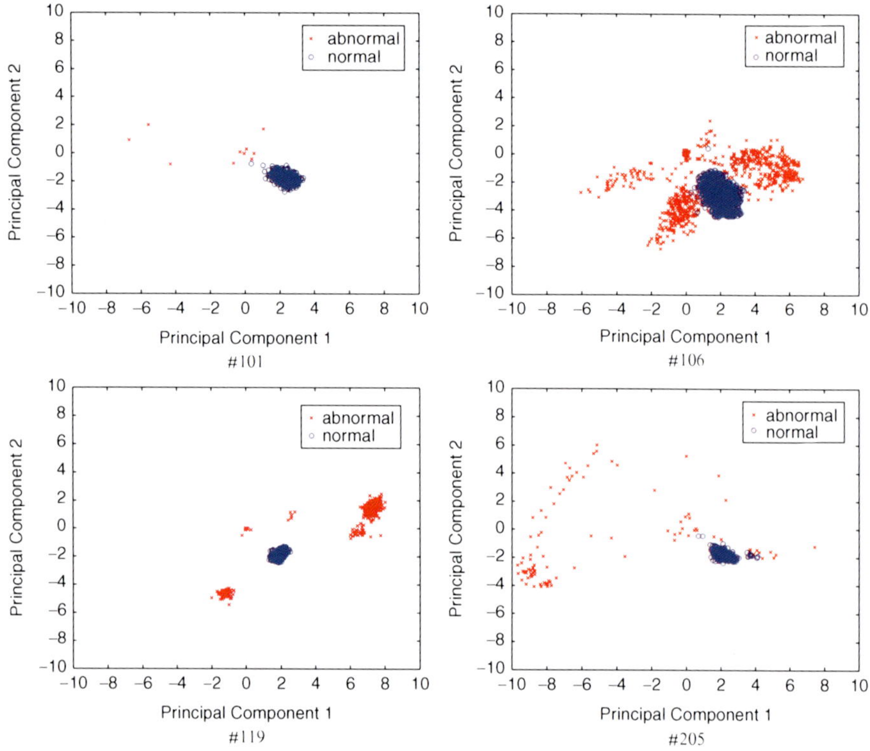

Fig. 7.17 Scatterplot of ECG data of four patients in MIT/BIH arrhythmia database [13] showing the first two principal components of PCA projection

Figure 7.18 illustrates the distribution of the ECG data from 44 recordings of MIT/BIH arrhythmia database using same PCA projection. Although an ECG detector can be finely trained using the ECG beats from a large database which consists of the ECG beats from different patients, it may perform poorly in annotating the ECG beats of other patients who are not in the database. This is the problem of poor generalization.

The solution to such generalization problem lies in the incorporation of local information of a specific patient to the ECG annotator. Since the distribution of the training samples is not the same as that of the test samples, some information about the true distribution of samples from each patient has to be added to train the ECG annotator properly.

In long-term monitoring of patients suffering from cardiovascular diseases, the normal ECG beats usually dominate the ECG recordings such as in patients suffering from or suspected to suffer from asymptomatic heart failure, congestive heart failure, cardiac dysfunction, and cardiac arrhythmias, etc., i.e. the number of abnormal ECG beats is far less than that of the normal ones. It may take a long time

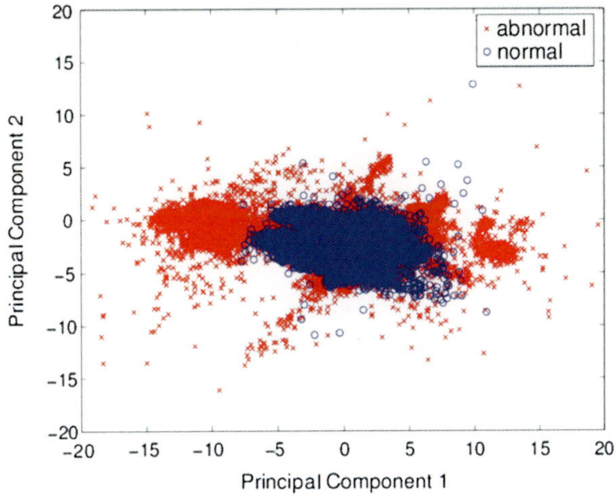

Fig. 7.18 Scatterplot of ECG data of all the ECG data from 44 patients in MIT/BIH arrhythmia database showing the first two principal components of PCA projection

to collect sufficient and balanced normal and abnormal ECG data to construct a good classifier; otherwise, the classifier may suffer from the imbalanced data problem [19].

7.6.1.3 HKME for ECG Annotation

One-Class Classification-Based Approach

A straight way to solve the generalization and imbalanced data problem aforementioned is recognition-based approach. A one-class classifier, νSVC can be trained using only about 5 min of normal ECG beats from a patient to adapt to the specific reference value of the patient. The trained model can then be used to determine whether the other ECG beats from the same patient belong to the "normal beats." Hence the abnormal ECG beats can be annotated automatically for further analysis. Such model has been proposed in [28].

As there is an innate difference between the normal range of each patient and that of a group of patient, there is a need to incorporate the local information of each patient to improve the generalization of the ECG beat annotator. Since the distribution of training data of the νSVC model is more similar to the ECG beats of the same patient than those of the data from a large group of patients, the νSVC model trained properly using about 5 min of the "normal" ECG data from a patient is expected to perform better than the classifiers trained using the data from a large group of patients.

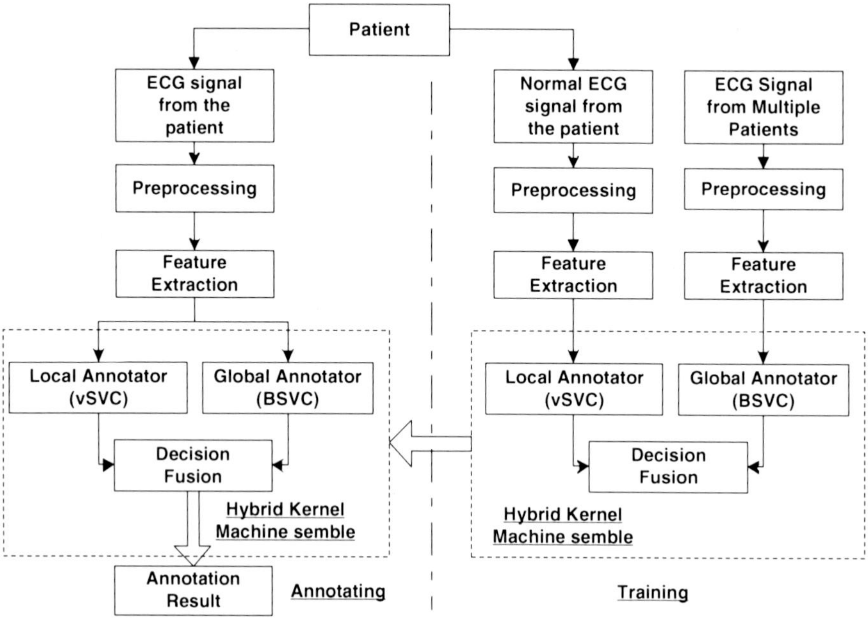

Fig. 7.19 Flowchart of the proposed framework for abnormal ECG beat annotation

HKME-Based Approach

Although the one-class classification-based approach proposed in the previous section performs well in the abnormal ECG beat annotation, it may be possible to be further improved. When a physician examines the ECG recordings of a patient, the physician considers not only the specific reference from each patient to be examined but also the standard reference from the patient-group due to the innate difference between the normal range of each patient and that of a group of patient aforementioned. That is to say, the diagnosis made by the physician is based on the information from both the patient-group and each specific patient. Motivated by this, a *HKME*-based approach is proposed for the ECG beat annotation problem for long-term monitoring heart patients. The following is based on authors' earlier publication in [27].

Figure 7.19 illustrates the flowchart of the proposed Hybrid Kernel Machine Ensemble (*HKME*)-based ECG beat annotator. This *HKME* consists of two base classifiers, one is a binary *SVM* trained using the ECG data from a large group of patients, the other is a one-class classification model, *νSVC* trained using only about 5 min of normal ECG beats from each patient to be monitored. The final decision is determined by a fusion rule. The recognition-based *νSVC* has been described in the previous section. It represents the specific reference value of the patient. The discriminative binary *SVM* is incorporated the global information of a large group of people and thus it can be regarded as the reference values based on the general patient population. Due to different information learned by these two *SVM*s, they usually perform differently in classifying the ECG beats in the long-term ECG recording of the patient. Furthermore, *νSVC* is a non-discriminative recognition-based model and *BSVC* is a discriminative model. Due to the complementary nature of such two types of *SVM*s, integration of the two types of kernel machines using an ensemble is expected to perform better than using either of them separately.

In this study, the raw amplitude of the time domain ECG signals after noise suppression and baseline shift removal was investigated as feature vectors to represent the ECG beats. After the R-peak is detected, the ECG signal in a window of 500 ms is taken as an ECG beat. The lengths of the signal before and after the R-peak in each beat are 167 and 333 ms, respectively, such that the window covers most of the characterization of the ECG beat (for an ECG signal sampled at 360 Hz, 180 samples around each R-peak are taken in a window, with 59 samples before the R-peak and the other 120 samples behind the R-peak). The amplitude of sampled signal in each window is then taken to form a feature vector of 180-dimensions. It has been shown that R–R interval (the interval between two consecutive R-peaks) is useful in recognition of some abnormal ECG beats [6, 16]. Therefore, it is also included in this study by appending it to the 180-dimensional(D) feature vector. The length of the feature vector to represent the ECG beat is then 181.

Normalization

There are some variations in the amplitude ranges of ECG signals among the human beings. Hence a normalization procedure to the ECG feature vectors is necessary; otherwise, the ECG beats may not be comparable. The feature vectors are divided by the mean value of R-peaks in the training data of each patient, such that the maximum amplitude in each ECG beat window is around 1. The normalized ECG feature vectors are then used for the annotation process using the trained HKME models.

7.6.1.4 Experiments

Experimental Setting

The proposed *HKME*-based patient-adaptable ECG beat annotator was evaluated on MIT/BIH arrhythmia database [13]. The experimental setting is as follows.

1. 22 recordings are selected as local training sets and test sets, in which the number of abnormal beats is significantly less than that of the normal beats, to be in agreement to the scenario in long-term monitoring of patients suffering from cardiovascular diseases. These recordings include # 100, 105, 106, 108, 114, 119, 121, 200, 203, 205, 208, 209, 210, 213, 215, 221, 222, 223, 228, 230, 233, and 234. Each of the 22 recordings is split into two sets.

Table 7.2 Results (average ± standard deviation) of abnormal ECG beat annotation (in percentage)

Classifiers	BCR	SEN	SPE	ACR
BSVC	80.3 ± 16.9	81.3 ± 24.5	79.3 ± 29.6	80.2 ± 26.1
νSVC	83.6 ± 14.7	87.5 ± 22.5	79.7 ± 21.3	81.7 ± 18.5
MAX	86.2 ± 16.4	87.0 ± 21.9	85,4 ± 20.6	85.8 ± 19.5
LDC	86.5 ± 16.2	82.6 ± 27.3	90.5 ± 16.1	90.1 ± 14.1
QDC	83.1 ± 17.3	74.6 ± 35.1	91.6 ± 12.2	90.3 ± 10.3
DET	87.5 ± 15.0	83.3 ± 26.6	91.6 ± 13.7	91.2 ± 11.5

- The first 200 normal ECG beats in each of the 22 recordings (about 3 min) are used as the local training set to construct the *νSVC*s.
- The first 350 normal ECG beats in each of the 22 recordings (about 5 min) are used as the training set to train the ensembles.
- The second $1/2$ of each of the 22 recordings (about 15 min or 1,000 beats) is used as test set to evaluate the performance of the ECG annotators.

2. 10,000 ECG beats (with half normal beats and half abnormal beats) from 22 recordings are used as global training set (DB_G) to train some classical binary classifiers for comparison with the proposed *HKME*-based patient-adaptable ECG beat annotator. These recordings are # 101, 103, 109, 111, 112, 113, 115, 116, 117, 118, 122, 123, 124, 201, 202, 207, 212, 214, 220, 222, 231, and 232.

Results and Discussion

The annotation results of using the proposed *HKME* with different fusion rules, the global binary *SVM* and the local *νSVC* are given in Table 7.2.

The reported results are averaged over 22 test ECG recordings as shown in Fig. 7.20. Test sets #1 – #22 correspond to recording 100, 114, 119, 121, 200, 203, 205, 208, 209, 210, 213, 215, 221, 222, 223, 228, 230, 233, 234, 105, 106, and 108 in MIT/BIH arrhythmia database, respectively.

- *Overall Performance and Generalization*

  It is observed from Table 7.2 that all the *HKME*s except using QDC rule outperforms both the global *BSVC* trained using the ECG data from a large patient-group and the local *νSVC* trained using some normal ECG data from each patient. The generalization of both the global *RSVC* and the local *νSVC* can be improved when the information from these two sources are integrated properly by the *HKME*.

  The best BCR achieved by *HKME* is using DET rule, whose BCR is 7.2 and 3.9 % higher than the global *RSVC* and the local *νSVC*, respectively. DET-based *HKME* outperforms both *SVM*s in more than 80 % of the test sets as reported in Table 7.3. The second best BCR achieved by *HKME* is using LDC-based stacking rule, which outperforms both *SVM*s in about 72 % of the test sets.

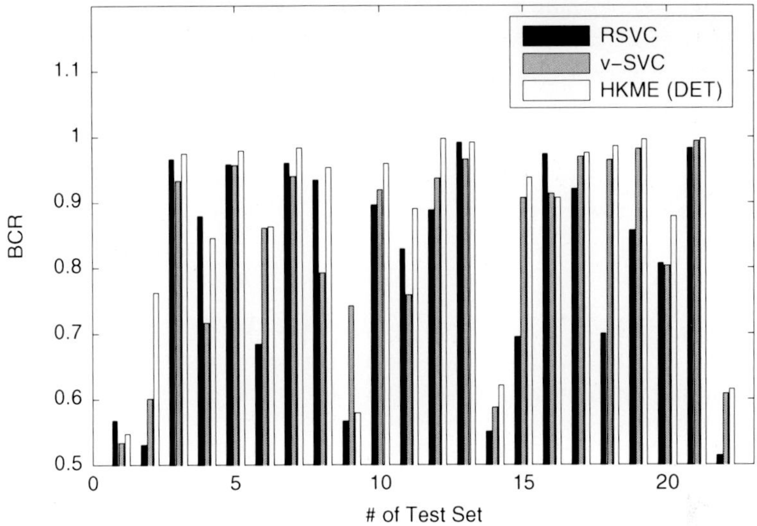

Fig. 7.20 Comparison of the annotation results of the global *BSVC*, local *vSVC* and *HKME* (with DET fusion rule) in each recording of the test sets in terms of BCR

Table 7.3 Number and percentage of test sets in which *HKME* outperforms both global *BSVC* and local *vSVC* among all 22 test sets

Fusion rule	*MAX*	*LDC*	*QDC*	*DT*
In number	14	16	11	18
In percentage	63.6	72.7	50.0	81.8

The performance improvement in terms of BCR using MAX rule is observed in about 64 % of the test sets. Its average of BCR is 5.9 and 2.6 % greater than the global *BSVC* and the local *vSVC*, respectively. The only exception is QDC rule. This may be resulted by the fact that the covariance matrices for the classes are near singular sometimes, these QDC classifiers may fail when trying to estimate and invert the covariance matrices [25]. It is expected that its performance may be better if it is properly trained.

It can be observed that the performance of trained *HKME*s such as DET and LDC is better than that of the non-trained *HKME*, the MAX rule. It shows that the proper training of the fusion rules is helpful to the improvement of the ensemble over the base classifiers. These findings are similar to those in previous section.

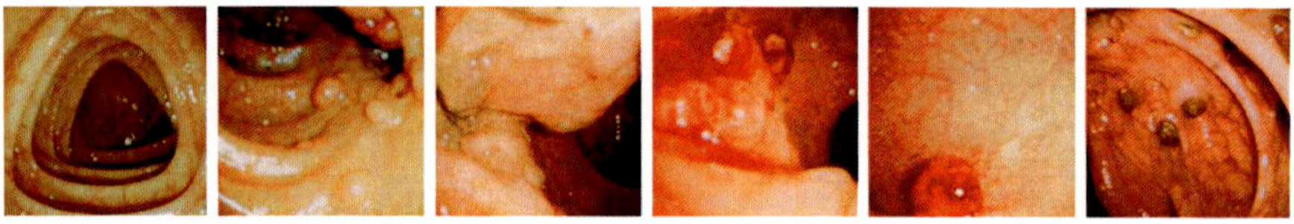

Fig. 7.21 A normal colonoscopic image and five colonoscopic images with different types of abnormalities

7.6.2 Application to Colonoscopic Image Analysis

7.6.2.1 Colonoscopic Image Analysis

In this section, an application of the HKME method to the detection of abnormal region in colonoscopic images is presented. This work has been published in [29]. Colonoscopy is a minimal invasive procedure of screening the colon and rectum using a colonoscope. The procedure is used to look for signs of cancer in the colon and rectum and diagnose the causes of unexplained changes in the bowel such as inflamed tissue, abnormal growths, ulcers, and bleeding. Analyzing colonoscopic images for clinical diagnosis of abnormalities relies on the experience and expertise of the medical experts, which need years of training to acquire. It is thus significant to develop a computer-assisted technique to help the screening process of these potentially lethal diseases by the health-care provider.

Previous research on colonoscopic image analysis focused on the classification between normal tissues and tumors. However, few work has been done to discriminate normal tissues from different kinds of abnormalities including tumors in colonoscopic images, which is more significant for screening purpose. In fact, many categories of abnormalities can be seen in colonoscopic images, such as polyps, tumors, inflammation, bleeding, ulceration, and diverticula (Fig. 7.21) and their image content shows large variations. The abnormal regions usually do not occupy the whole image and vary in color, size, and shape, which add more difficulties to the discrimination of the normal regions from the abnormal ones in colonoscopic images. In addition, this leads to an good example of imbalanced data problem.

The patch-based approach seems to be a good representation model for image segmentation in which the full image is cropped into a set of image patches and these patches can be classified into different categories corresponding to different types of segments. This approach has been used extensively in many applications, such as face detection [15], object detection [7], and image segmentation [42].

In this section, a *HKME*-based approach using multi-size patches is presented for detecting abnormal regions in colonoscopic images. Multiple sizes of patches provide multiple level visual cues of the image regions, which can help produce better perceptually agreeable segmentation. Represented as multi-size patches, the abnormal region detection in colonoscopic images turns into a binary classification problem to discriminate the patches from normal regions (normal class) and those from abnormal ones (abnormal class). Each pixel in a given image can be categorized as normal or abnormal using a trained patch-based classifier. Using multiple sizes of patches, multi-labels can be given to a pixel, the final label of the pixel can be obtained using the ensemble of these multiple classifiers based on different patch sizes.

The performance of the ensemble depends on the individual classifiers used. A set of individual classifiers have to be trained for the binary classification problem to discriminate the normal patches from those abnormal ones. This problem can be solved using a discriminative model, such as *BSVCs* . Such a *BSVC*-based abnormal region detection approach in colonoscopic images using multiple-size patches was published in [30] (A preliminary work by the authors).

Rather than a typical binary classification problem, the abnormal region detection in colonoscopic images can also be treated as a one-class classification problem. A lot of patterns from abnormal regions in colonoscopic images for each categories of abnormalities have to be collected for training a reliable classifier, which means the concept "abnormal" is not easy to learn. On the other hand, the normal patterns show smaller variations than those of the abnormal ones and are much easier to be obtained. This means the concept "normal" can be easier to learn. Therefore, the concept "normal" can be learned using a one-class classifier, such as νSVC or *SVDD*. Such a one-class classification-based abnormal region detection approach in colonoscopic images was published in [31] (Another preliminary work by the authors). Trained using only the data from one class, νSVCs try to find a decision boundary around the training data—called targets, which is different from the decision boundary of *BSVC* trained using the data from both normal and abnormal classes. As explained in the previous section, νSVC tries to represent of target samples rather than for discrimination purpose. On the one hand, multi-size patches produce multi-level cues of image content, which in turn produce a diverse feature set. On the other hand, the combination of the two different types of kernel machines νSVC and *BSVC* can produce more diversity to the ensemble, which may further improve the abnormal detection in colonoscopic images. Experimental results show that the multi-size patch-based hybrid kernel machine ensemble method is superior to that of using single patch size only for the abnormal region detection in colonoscopic images and can produce more perceptual agreeable image segmentation.

7.6.2.2 HKME for Detecting Abnormal Regions in Colonoscopic Image Analysis

Figure 7.22 illustrates the flowchart of the proposed *HKME*-based approach using multi-size patches for abnormal region detection in colonoscopic images. The detail is as follows.

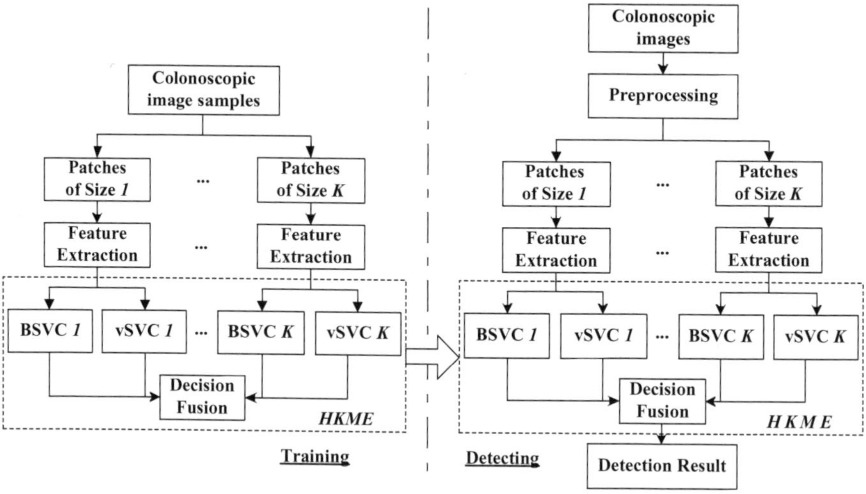

Fig. 7.22 The flowchart of *HKME*-based approach for abnormal region detection in colonoscopic images

Image Region Representation Using Multiple-Size Patches

As illustrated in Fig. 7.21, the abnormal regions in colonoscopic images come from different categories and they vary in location, shape, color, and size. The representation of these regions has to be considered carefully. It is similar to object detection in which the abnormal regions are the objects to be detected.

Patch-based approach turns the abnormal region segmentation into a binary classification problem [21]. As illustrated in Fig. 7.23, each colonoscopic image can be cropped into a set of overlapping image patches and these image patches can be categorized as abnormal region class or normal region class by a classifier. The abnormal regions can thus be segmented from the normal ones.

An open problem in patch-based approach is what patch-size to choose. Compared to large ones, small-size patches (the extreme of a small-size patch is single pixel) can represent the image regions more precisely, but it contains less information of the image content than large-size patches (the extreme of large-size patch is the full image) and usually lead to larger classification error. The large-size patches contain more information of the object, but small abnormal regions in these patches may be missed. This is why it is very difficult to determine the appropriate patch-size to use.

In this section, a multi-size patch-based representation is presented in which multi-size patches are used simultaneously to represent the image regions in colonoscopic images. Using patches of multiple sizes aims at overcoming the scale problem, i.e, an abnormal region may appear at different sizes in different images. Multi-size patches provide multiple-level representation of the image contents. At least some among all the patch sizes can better characterize the object. Hence, the

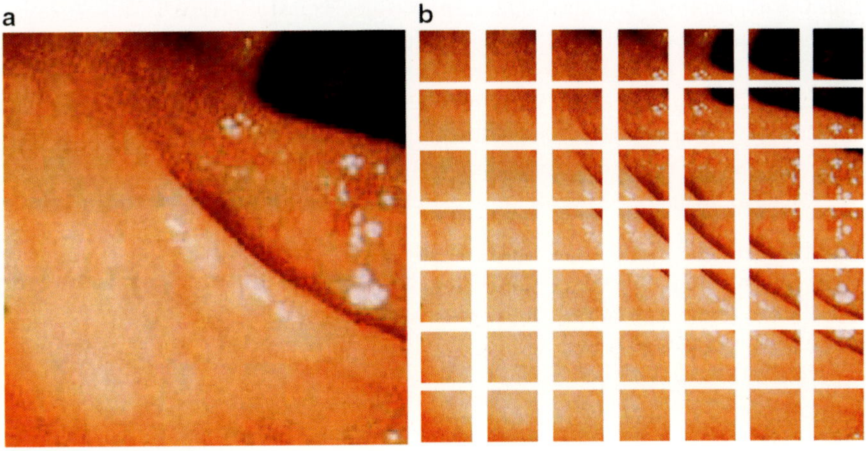

Fig. 7.23 Patch-based image region representation. (**a**) Original colonoscopic image, (**b**) Overlapping image patches

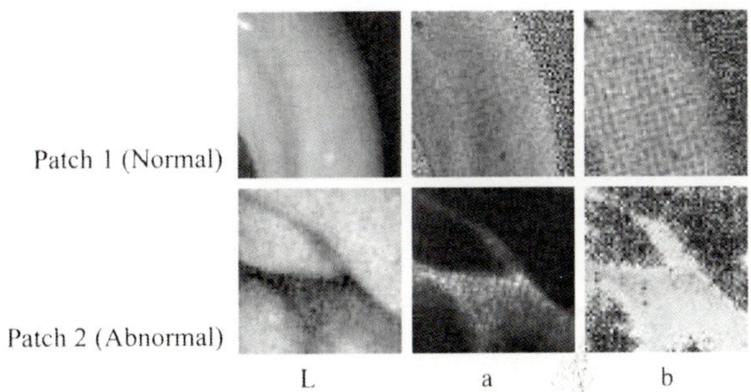

Fig. 7.24 Two examples of patches from colonoscopic images. The component L, a and b of the image patches are illustrated listed in a row

integration of the detection result based on multi-size patches is expected to detect the abnormal regions more precisely than those based on single-size patches only. This is the novelty of this method.

The colonoscopic images in RGB color space are transformed into three bands in *CIELab* color space through which the color and luminance component can be processed and analyzed individually. The image is scanned across and cropped into a set of fixed sizes of patches, respectively. The patches are overlapped by 50 % to ensure that no abnormal region is missed. Here, 3 sizes of the image patches are investigated for abnormal region detection in the colonoscopic images, namely, 48×48, 32×32 and 16×16 (pixels). Two samples of the image patches are illustrated in Fig. 7.24. Feature can be then extracted from these image patches for classification.

Both color and texture features are extracted.

- *Color features:* The color features are Two-dimensional(D) histograms of the components a and b in *CIELab* color space. The number of bins of the histogram is 8 for 2-D histograms.

- *Textural features:* Two-level Discrete Wavelet Transform (DWT)-based statistical features and 1-D histograms of luminance (the number of bins of the 1-D histogram is 16) are employed as textual features. The image patches are processed using two-level DWT. The mean and standard deviation of the absolute value of the approximate and detailed coefficients from the two-level DWT decomposition of the image patches in the three channels of the *CIELab* color space are calculated as the textural features.

Altogether 128 features are extracted, giving rise to a feature vector of 128-D. Then the feature vectors from patches can be used to form the dataset for classification. A set X of N feature vector x_i, $X = \{x_i \in R^{128} | i = 1, 2, \cdots, N\}$ for N patches, are labelled as $y_i \in \{+1, -1\}$ to indicate whether it is a normal patch or a patch containing abnormalities.

Learning *SVMs* for Image Patch Classification

BSVC Learning

Using overlapped image patches, each pixel in the patch can be classified as normal or abnormal by an *SVM* classifier corresponding to the patch size. Thus each pixel in the original image can have at least one label. If a pixel is classified differently by overlapped patches, the label of the patch that has the largest absolute decision value (confidence) is chosen as the label of that pixel.

νSVC Learning

The classification between normal and abnormal patches can also be solved by a one-class classifier, such as *νSVC*. A *νSVC* can be trained using the data from normal image patches for each patch size. Using overlapped image patches, each pixel in the patch can be classified as normal or abnormal by the trained *νSVC* classifier corresponding to the patch size. Thus each pixel in the original image can have at least one label. If a pixel is classified differently by overlapped patches, the label of the patch that has the largest confidence is chosen as the label of that pixel.

7.6.2.3 Decision Fusion Using *HKME*

Since there are many kinds of abnormalities in colonoscopic images showing large variation, many patterns from abnormal regions in colonoscopic images have to be collected for training a reliable classifier and it is difficult to collect. This leads to an imbalanced data problem. One class—"normal" has many training samples and is easier to model, while the other class—"abnormal" is difficult to model because it has more diverse distributions than the normal class. Therefore, *νSVC* is very suitable for this problem. As a recognition-based model, *νSVC* tries to describe the target data rather than for discrimination purpose, it can handle the problem of missing information. However, *vSVC* is often inferior to *BSVC* for discrimination purpose. There is a need to combine these two types of kernel machines for this problem.

A set of 2-*SVC*s can be constructed for the classification, while *ν*-*SVC*s can be used to provide further decision information. The classification results of the two kernel machines can be aggregated using an ensemble. The different natures of the two types of *SVM*s adds more diversity to the ensemble, which may further improve the performance of the ensemble.

7.6.2.4 Experimental Results and Discussions

Data Preparation

The proposed approaches were evaluated using a database which consists of 58 clinically obtained colonoscopic images. There are 12 normal images and 46 images with abnormal regions. The abnormal regions mostly occupy only some parts of the whole image and the abnormalities include polyps, tumors, inflammation, bleeding, ulceration, and diverticula, etc. The images are RGB images with the resolution of 256×256 pixels. The pixels in the original images were manually labeled to provide the ground truths. The detection results were compared with the ground truth and evaluated.

In the experiment, the numbers of collected image patches for training of 48×48, 32×32, and 16×16 (pixels) patches are 2,002, 2,090, and 2,126, respectively. The pixels in the original image are manually labeled as the ground truth for comparison. The patches containing mostly abnormal region were labeled as a positive sample, otherwise, a negative one. A leave-one-out experiment was performed to evaluate the performance of the proposed method for abnormal region detection in colonoscopic images. In each round, one of the colonoscopic images was selected for testing and the patches from other 57 images were used for training. The experiment was repeated 58 times, the detected results were compared to the ground truth image and the average value of the total 58 results was taken as the final result.

Evaluation Measure

The evaluation criteria are specificity (SPE), sensitivity (SEN), and Balanced classification rate (*BCR*). Where SPE is the fraction of normal regions detected among all the normal regions, SEN is the abnormal regions detected among all the abnormal regions and BCR is the weighted average of SPE and SEN.

$$BCR = \lambda SPE + (1 - \lambda) SEN \tag{7.45}$$

Table 7.4 Results of abnormal region detection using single patch sizes

Patch size	Classifier	*BCR*	SPE	SEN
48×48	2-SVC	0.744	0.675	0.813
48×48	ν-SVC	0.539	0.991	0.088
32×32	2-SVC	0.738	0.675	0.802
32×32	ν-SVC	0.546	0.998	0.094
16×16	2-SVC	0.745	0.668	0.822
16×16	ν-SVC	0.538	0.946	0.094

where $\lambda \in [0, 1]$ can be tuned to favor *SPE* or *SEN*. Smaller λ favors more on *SEN*, which means that the error on the abnormal class is punished more seriously. On the contrary, larger λ favors more on *SPE*, which means that the error on the normal class is taken more seriously. At the extreme case, only *SEN* or *SPE* will be considered when λ is 0 or 1, respectively. $\lambda = 0.5$ is used here, so that *SPE* and *SEN* are treated as equally important. Other values might be selected with respect to the requirement of the medical experts.

νSVC vs *BSVC* Using Single Patch Size

In Table 7.4, it is observed that *BSVC*s outperform *νSVC* in all the cases which agrees with the postulate that discriminative models are superior to that of recognition-based models. *BSVC*s achieved *BCR* around 74 %, while *νSVC*s achieved only 55 %. The *νSVC*s have a very high SPE, but almost completely fail for SEN. This may be resulted that the training set size used for *νSVC* was too small and it also suffered from the curse of dimensionality. Compared to *νSVC*s, *BSVC* have higher SEN while much less SPE, which may be good for adding more diversity to the ensembles. The best *BCR* is 74.5 % which was achieved using patches of size 16×16.

Multi-Size Patch Ensemble of *νSVC*s or *BSVC*s

Table 7.5 illustrates the detection results of three patch size ensembles using *νSVC*s or *BSVC*s separately. Obviously, the best ensembles outperform that of the best *SVM*s using single patch size, which supports the claim that multi-size patch-based *SVM* ensemble can achieve better abnormal region detection in colonoscopic images. Due to the poor performance of individual *νSVC*s, the improvement of their ensemble is limited although there are still some.

HKME Using Single-Size Patches

Table 7.6 shows the detection results of the ensemble of a *νSVC* and a *BSVC* based on single-size patches. Only *DET* and LDC achieved *BCR* comparable to the best

Table 7.5 Detection results (in terms of *BCR*) of abnormal region detection using different patch sizes and ensemble schemes

Ensemble	MAX	AVG	PROD	MV	*DET*	LDC	QDC
BSVC	0.737	0.751	0.745	0.751	0.753	0.763	0.765
νSVC	0.532	0.551	0.535	0.551	0.538	0.533	0.536

The ensembles are constructed using same types of *SVM*s, *BSVC* or *νSVC*

Table 7.6 Detection results (in terms of *BCR*) of abnormal region detection using same patch sizes by *HKME*

Patch size	MAX	AVG	PROD	MV	*DET*	LDC	QDC
48×48	0.551	0.551	0.551	–	0.744	0.746	0.540
32×32	0.556	0.556	0.556	–	0.738	0.741	0.548
16×16	0.563	0.563	0.563	–	0.745	0.745	0.736

single classifier and the performance of other ensembles did not outperform the best single classifier. This may be due to the fact that the *νSVC* and *BSVC* are trained using the same features, which limit the performance of this scheme.

HKME Using Multi-Size Patches

Table 7.7 illustrates the detection results of the *HKME* ensemble of *BSVC*s using all three patch sizes plus 1 to 3 *νSVC*(s) trained using 1 to 3 patch size(s). Most of the ensembles show improvement over the best single *SVM* based on single-size patches. The performance of LDC and AVG outperforms others. Figure 7.25 illustrates the result of the ensemble of *BSVC*s using all 3 patch sizes and a *νSVC*(s) trained using patches with size of 48×48. Obviously, the detection results by the *HKME* ensemble is closer to the ground truth compared to those using single-size patches.

The results of detection of abnormal region using learned *HKME* for four colonoscopic images are illustrated in Fig. 7.25.

7.7 Conclusion

In this chapter, we briefly reviewed the one-class SVM and two-class SVM. Their principles of classification are discussed and the strengths and weaknesses for dealing with imbalanced datasets are illustrated with the checkerboard dataset. The imbalanced data problem is also discussed and the various ways of handling such a problem are also presented. The chapter shows that the one-class SVM and two-class SVM can be integrated into an ensemble classifier to form what we call the Hybrid Kernel Machine Ensemble—*HKME*. This ensemble classifier has been evaluated with artificial dataset. The evaluation results show the benefit

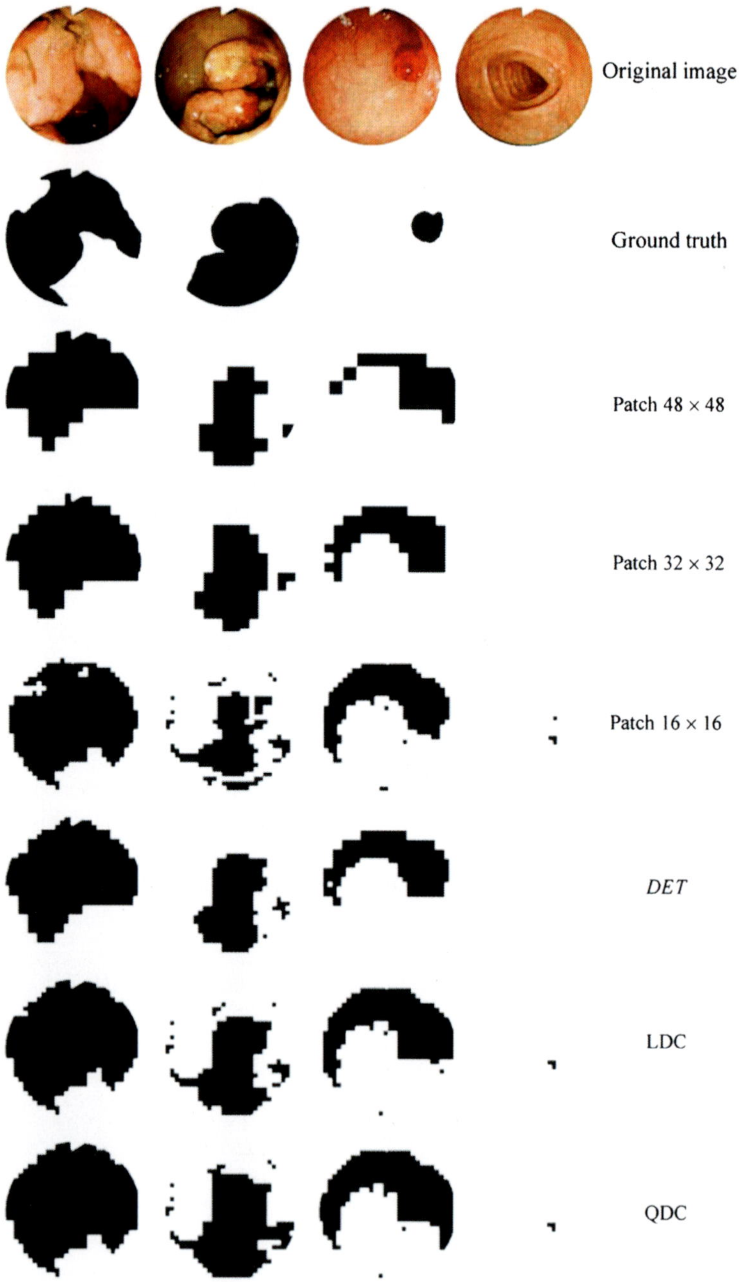

Fig. 7.25 Detection results of four colonoscopic images. The regions in *white* are normal regions detected and the regions in *black* are abnormal ones

Table 7.7 Detection results (in terms of *BCR*) of abnormal region detection using different patch sizes and *HKME*

Ensemble	MAX	AVG	PROD	MV	*DET*	LDC	QDC
A+1	0.705	**0.761**	**0.768**	0.751	0.753	**0.765**	**0.766**
A+2	0.541	**0.761**	0.659	0.751	0.753	**0.764**	0.667
A+3	0.588	**0.754**	0.730	**0.751**	**0.753**	**0.756**	0.730
A+1+2	0.539	**0.769**	0.598	**0.765**	0.753	**0.765**	0.542
A+1+3	0.587	**0.765**	0.656	**0.765**	0.753	**0.763**	0.743
A+2+3	0.538	**0.762**	0.572	**0.765**	0.753	**0.763**	0.559
ALL	0.537	0.565	0.548	0.704	0.751	**0.764**	0.549

Row $A + *$ are the results of ensembles using all 3-size patches learned by *BSVC*s plus 1-size or 2-size patches learned by νSVC(s) (1 for 48×48, 2 for 32×32, 3 for 16×16). Row *ALL* are the ensemble results using all 3-size patches and both *BSVC* and νSVC

of using such an ensemble to handle an imbalanced dataset. It has been shown that the *HKME* can achieve better performance than using either one-class SVM or two-class SVM alone. Discussions are given on the possible reasons for its better performance. Since such imbalanced data problem exists in many biomedical applications, and encouraged by the good performance of *HKME*, it is deployed in two biomedical applications, namely, abnormal ECG beats annotation and abnormal region detection in colonoscopic images. Experimental results further confirm the superiority of using *HKME*.

References

1. The Kernel-Machine.org http://www.kernel-machines.org/
2. Bach, F., Jordan, M.: Kernel independent component analysis. J. Mach. Learn. Res. **3**(1), 1–48 (2003)
3. Bishop, C.M.: Neural Networks for Pattern Recognition. Clarendon, Oxford (1995)
4. Chawla, N., Bowyer, K., Hall, L., Kegelmeyer, W.: SMOTE: synthetic minority over-sampling technique. Artif. Intel. Res. **16**, 321–357 (2002)
5. Chawla, N.V., Japkowicz, N., Kotcz, A.: Editorial: special issue on learning from imbalanced data sets. SIGKDD Explorations **6**(1), 1–6 (2004)
6. de Chazal, P., O'Dwyer, M., Reilly, R.: Automatic classification of heartbeats using ECG morphology and heartbeat interval features. IEEE T. Bio-Med. Eng. **51**(7), 1196–1206 (2004)
7. Deselaers, T., Keysers, D., Ney, H.: Discriminative training for object recognition using image patches. In: Proceedings of the 2005 IEEE Computer Society Conference on Computer Vision and Pattern Recognition (CVPR'05), vol. 2, pp. 157–162 (2005)

8. Drummond, C., Holte, R.C.: C4.5, class imbalance, and cost sensitivity: why undersampling beats over-sampling. In: Proceedings of the ICML 2003 Workshop on Learning from Imbalanced Data Sets II, vol. 11, Washington, DC (2003)
9. El-Naqa, I., Yang, Y., Wernick, M.N., Galatsanos, N.P., Nishikawa, R.M.: A support vector machine approach for detection of microcalcifications. IEEE T. Med. Imaging **21**(12), 1552–1563 (2002)
10. Estabrooks, A., Jo, T., Japkowicz, N.: A multiple resampling method for learning from imbalances data sets. Comput. Intell. **20**(1), 18–36 (2004)
11. Gal-Or, M., May, J.H., Spangler, W.E.: Assessing the predictive accuracy of diversity measures with domain-dependent asymmetric misclassification costs. Inform. Fusion J. (Special issue on Diversity in Multiple Classifier Systems) **6**(1), 37–48 (2005)
12. Gokturk, S.B., Tomasi, C., Acar, B., Beaulieu, C.F., Paik, D., Jeffrey, B.J., Yee, J., Napel, S.: A statistical 3D pattern processing method for computer aided detection of polyps in CT colonography. IEEE T. Med. Imaging **20**(12), 1251–1260 (2001)
13. Goldberger, A.L., Amaral, L.A.N., Glass, L., Hausdorff, J.M., Ivanov, P.C., Mark, R.G., Mietus, J.E., Moody, G.B., Peng, C.K., Stanley, H.E.: PhysioBank, PhysioToolkit, and PhysioNet: components of a new research resource for complex physiologic signals. Circulation **101**(23), e215–e220 (2000)
14. Hojjatoleslami, A., Sardo, L., Kittler, J.: An RBF based classifier for detection of microcalcifications in mammograms with outlier rejection capability. In: International Conference on Neural Networks, vol. 3, pp. 1379–1384 (1997)
15. Hsu, R.L., Abdel-Mottaleb, M., Jain, A.K.: Face detection in color images. In: Proceedings of 2001 International Conference on Image Processing, vol. 1, pp. 1046–1049 (2001)
16. Hu, Y.H., Palreddy, S., Tompkins, W.J.: A patient-adaptable ECG beat classifier using a mixture of experts approach. IEEE T. Bio-Med. Eng. **44**(9), 891–900 (1997)
17. Japkowicz, N.: The class imbalance problem: significance and strategies. In: Proceedings of the 2000 International Conference on Artificial Intelligence (IC-AI'2000), vol. 1, pp. 111–117 (2000)
18. Japkowicz, N., Myers, C., Gluck, M.: A novelty detection approach to classification. In: Proceedings of the Fourteenth International Joint Conference on Artificial Intelligence, pp. 518–523. Morgan Kaufmann, San Francisco, CA (1995)
19. Japkowicz, N., Stephen, S.: The class imbalance problem: a systematic study. Intell. Data Anal. **6**(5), 429–450 (2002)
20. Karakoulas, G.J., Shawe-Taylor, J.: Optimizing classifiers for imbalanced training sets. In: Proceedings of the 1998 conference on Advances in Neural Information Processing Systems II, pp. 253–259 (1999)
21. Karkanis, S.A., Iakovidis, D.K., Maroulis, D.E., Karras, D.A., Tzivras, M.D.: Computer aided tumor detection in endoscopic video using color wavelet features. IEEE T. Inf. Technol. B. **7**(3), 141–152 (2003)
22. Kittler, J., Hatef, M., Duin, R., Matas, J.: On combining classifiers. IEEE T. Pattern Anal. **20**(3), 226–239 (1998)
23. Kubat, M., Holte, R., Matwin, S.: Detection of oil-spills in radar images of sea surface. Mach. Learn. **30**, 195–215 (1998)
24. Kubat, M., Matwin, S.: Addressing the curse of imbalanced training sets: one-sided selection. In: Proceedings of the Fourteenth International Conference on Machine Learning, pp. 179–186. Morgan Kaufmann, Nashville, Tennessee (1997)
25. Kuncheva, L.I., Bezdek, J., Duin, R.: Decision templates for multiple classifier fusion: an experimental comparison. Pattern Recogn. **34**(2), 299–314 (2001)
26. Li, P., Chan, K.L., Fang, W.: Hybrid kernel machine ensemble for imbalanced data sets. In: 18th International Conference on Pattern Recognition (ICPR'06), vol. 1, pp. 1108–1111 (2006)
27. Li, P., Chan, K.L., Fu, S., Krishnan, S.M.: An abnormal ECG beat detector approach for long-term monitoring of heart patients based on hybrid kernel machine ensemble. In: International Workshop on Multiple Classifier Systems (MCS 2005), Lecture Notes in Computer Science, vol. 3541, pp. 346–355. Springer (2005)

28. Li, P., Chan, K.L., Fu, S., Krishnan, S.M.: Neural networks in healthcare: potential and challenges. In: A Concept Learning-Based Patient-Adaptable Abnormal ECG Beat Detector for Long-Term Monitoring of Heart Patients, pp. 105–128. Idea Group Publishing, Hershey, PA (2006)
29. Li, P., Chan, K.L., Krishnan, S.M.: Learning a multi-size patch-based hybrid kernel machine ensemble for abnormal region detection in colonoscopic images. In: IEEE Conference on Computer Vision and Pattern Recognition (CVPR), vol. 2, pp. 670–675 (2005)
30. Li, P., Chan, K.L., Krishnan, S.M., Gao, Y.: Detecting abnormal regions in colonoscopic images by patch-based classifier ensemble. In: 17th International Conference on Pattern Recognition (ICPR), vol. 3, pp. 774–777. Cambridge, UK (2004)
31. Li, P., Krishnan, S.M., Chan, K.L., Gao, Y.: Abnormal region detection in colonoscopic images using novelty detection technique. In: Proceedings of 7th International Workshop on Advanced Imaging Technology (IWAIT'2004). Singapore pp. 139–154, MIT, Cambridge, USA (2004)
32. Manevitz, L.M., Yousef, M.: One-class SVMs for document classification. J. Mach. Learn. **2**, pp. 139–154. MIT, Cambridge, USA (2001)
33. Markou, M., Singh, S.: Novelty detection: a review-part 1: statistical approaches. Signal Process. **83**(12), 2481–2497 (2003)
34. Markou, M., Singh, S.: Novelty detection: a review-part 2: neural network based approaches. Signal Process. **83**(12), 2499–2521 (2003)
35. Mika, S., Rätsch, G., Weston, J., Schölkopf, B., Müller, K.R.: Fisher discriminant analysis with kernels. In: Hu, Y.H., Larsen, J., Wilson, E., Douglas, S. (eds.) Neural Networks for Signal Processing IX, pp. 41–48. IEEE (1999)
36. Mika, S., Schölkopf, B., Smola, A., Müller, K.R., Scholz, M., Rätsch, G.: Kernel pca and de-noising in feature spaces. In: Proceedings of the 1998 Conference on Advances in Neural Information Processing Systems II, pp. 536–542. MIT, Cambridge, MA (1999)
37. Osowski, S., Hoai, L., Markiewicz, T.: Support vector machine-based expert system for reliable heartbeat recognition. IEEE T. Bio-Med. Eng. **51**(4), 582–589 (2004)
38. Peng, J., Heisterkamp, D., Dai, H.: Adaptive quasiconformal kernel nearest neighbor classification. IEEE T. Pattern Anal. **26**(5), 656–661 (2004)
39. Platt, J.C.: Fast training of support vector machines using sequential minimal optimization. In: Advances in Kernel Methods: Support Vector Learning, pp. 185–208. MIT, Cambridge, MA (1999)
40. Platt, J.C.: Probabilities for SV Machines. In: Smola, A.J., Bartlett, P.J., Scholkopf, B., Schuurmans, D. (eds.) Advances in Large Margin Classifiers, pp. 61–74. MIT, Cambridge, MA (2000)
41. Raskutti, B., Kowalczyk, A.: Extreme re-balancing for SVMs: a case study. SIGKDD Explorations **6**(1), 60–69 (2004)
42. Rothganger, F., Lazebnik, S., Schmid, C., Ponce, J.: Segmenting, modeling, and matching video clips containing multiple moving objects. In: Proceedings of the 2004 IEEE Computer Society Conference on Computer Vision and Pattern Recognition, vol. 2, pp. 914–921 (2004)
43. Schölkopf, B., Platt, J.C., Shawe-Taylor, J., Smola, A.J., Williamson, R.C.: Estimating the support of a high-dimensional distribution. Neural Comput. **13**(7), 1443–1471 (2001)
44. Schölkopf, B., Smola, A.J.: Learning with Kernels Support Vector Machines, Regularization, Optimization, and Beyond. MIT, Cambridge, MA (2002)
45. Shin, H., Cho, S.: How to deal with large dataset, class imbalance and binary output in SVM based response model. In: Proceedings of the Korean Data Mining Conference, pp. 93–107 (2003)
46. Shipp, C.A., Kuncheva, L.: Relationships between combination methods and measures of diversity in combining classifiers. Inform. Fusion **3**(2), 135–148 (2002)
47. Tax, D.: One-class classification:concept-learning in the absence of counter-examples. Asci dissertation series, Delft University of Technology (2001)
48. Tax, D., Duin, R.: Support vector data description. Pattern Recogn. Lett. **20**(11–13), 1191–1199 (1999)

49. Tax, D., Duin, R.: Image database retrieval with support vector data description. In: Proceedings of the Sixth Annual Conference of the Advanced School for Computing and Imaging, ASCI Delft (2000)
50. Tax, D., Duin, R.: Uniform object generation for optimizing one-class classifiers. J. Mach. Learn. Res. **2**, 155–173 (2002)
51. Vapnik, V.: The Nature of Statistical Learning Theory. Springer, New York (1995)
52. Vapnik, V.: Statistical Learning Theory. Wiley, New York (1998)
53. Veropoulos, K., Cristianini, N., Campbell, C.: Controlling the sensitivity of support vector machines. In: Proceedings of the International Joint Conference on Artificial Intelligence, (IJCAI99), Stockholm, Sweden (1999)
54. Weiss, G., Provost, F.: The effect of class distribution on classifier learning: an empirical study. Tech. Report ML-TR-44, Department of Computer Science, Rutgers University, August 2001
55. Wolpert, D.H.: Stacked generalization. Neural Networks **5**, 241–259 (1992)
56. Wu, G., Chang, E.Y.: Class-boundary alignment for imbalanced dataset learning. In: The Twentieth ICML Workshop on Learning from Imbalanced Datasets, pp. 49–56. Washington, DC (2003)
57. Yanowitz, F.G.: The Alan E. Lindsay ECG learning center in cyberspace, http://medlib.med.utah.edu/kw/ecg/ (2003)

Chapter 8
Soft Biometrics from Face Images Using Support Vector Machines

Guodong Guo

Abstract Soft biometrics, such as age, gender, and ethnicity, are useful for many applications in practice. For instance, in business intelligence, it is helpful to automatically extract and compute the statistics of potential customers, such as the number of males and females; the number of young, adult, and senior people; or the number of Caucasian, African American, or Asian people. It is also helpful to use soft biometrics to improve the performance of traditional biometrics for human identification, such as face recognition. Different methods can be developed to recognize the soft biometric characteristics from face images. In this chapter, we present the application of the support vector machines (SVM) to learn an estimator or recognizer to extract these soft biometrics. We will mainly focus on age estimation, while the gender and ethnicity classification will also be discussed. Both classification and regression will be considered. The combination of regression and classifiers based on the SVM will also be described which is useful especially for age estimation.

8.1 Introduction

Support vector machines (SVM) [41] have shown many successful applications in a variety of areas, including computer vision, pattern recognition, image analysis, biometrics, bioinformatics, etc. The SVMs are graceful in theory (e.g., the large margin optimization and mathematical programming solver) and have good performance in

---

G. Guo (✉)
Department of Computer Science and Electrical Engineering, West Virginia University, Morgantown, WV, USA
e-mail: guodong.guo@mail.wvu.edu

Y. Ma and G. Guo (eds.), *Support Vector Machines Applications*,
DOI 10.1007/978-3-319-02300-7_8,

© Springer International Publishing Switzerland 2014

practice (e.g., good generalization capability and high discriminative power). In this chapter, we show the use of the SVMs and their extensions to extract soft biometric characteristics from face images.

The typical soft biometric characteristics include age, gender, and ethnicity. These measures cannot be used to identify a person uniquely. For instance, different subjects can share the same age, gender, or even the same ethnicity. However, when two subjects are confused by a face recognizer, the age, gender, and/or ethnicity might be used to help the discrimination, assuming the two subjects have some differences in those measures. For example, we have shown that the soft biometrics, such as gender, ethnicity, weight, and height can help to improve the face recognition performance [27]. So those characteristics are called "soft biometrics," while the traditional biometric cues, such as face, iris, and fingerprints, are assumed to be unique for each individual.

In addition to helping human identification, the soft biometrics themselves are also useful for other applications. A typical case is business intelligence, where there is no need to know the identities of the customers. The real care is the statistics of the group of customers, such as the number of males and females, young, adult or senior people, and Caucasian or Asian. These soft biometric characteristics can help the business owners or managers to know more about the potential customers, do a better advertisement to the related customers, or introduce commercial products to the appropriate customers who might be interested in those products.

Among the three soft biometric characteristics, age estimation is probably the most challenging problem. Our primary focus here is the age estimation, while we will also consider gender and ethnicity classification. Further, age estimation is a very special problem. The age labels, e.g., 1, 2, 3 in years, can be considered as regression values, thus age estimation can be taken as a regression problem. On the other hand, each age label can also be considered as a separate class, thus age estimation can also be taken as a classification problem [21, 23]. We study the performance of the SVM-based classification and regression for age estimation on different databases. We also present a scheme to combine the regression and classifiers for an improved performance on age estimation [21, 23]. Further, a probabilistic fusion is also presented to make the combination automatic without much parameter adjustment [22].

For gender and ethnicity classification, we show the performance of the SVM classifiers on large databases. We also present a study of whether the gender and ethnicity classification is affected by age or not [15, 24].

Soft biometric characteristics have other measures, in addition to age, gender, and ethnicity. For instance, we have recently developed a computational approach to body mass index (BMI) prediction in face images [43]. We believe that more and more soft biometric cues can be extracted along with practical applications. In this chapter, we just study the most popular soft biometrics, i.e., age, gender, and ethnicity.

In the following, we briefly introduce the support vector regression (SVR) in Sect. 8.2 and the SVM in Sect. 8.3. Then in Sect. 8.4 we present a method, called locally adjusted robust regression (LARR), to combine the SVR and SVM for an improved age estimation. In Sect. 8.5 we describe a probabilistic fusion to combine the SVM and SVR. Some simple introduction of the face image representation is presented in Sect. 8.6. The experiments are conducted in Sect. 8.7, and finally, we draw conclusions.

8.2 Support Vector Regression

The basic idea of SVR is to find a function $f(\mathbf{y})$ that has most ε deviation from the actually obtained target z_i for the training data $\mathbf{y}_i$, and at the same time is as flat as possible [41]. In other words, we do not care about the errors as long as they are less than ε. This property determines the SVR to be less sensitive to outliers than the quadratic loss function. In comparison with the conventional quadratic loss function shown in Fig. 8.1a, the ε-insensitive loss function of SVR is shown in Fig. 8.1b. Given the same input, the ε-insensitive loss function is more robust than the quadratic function in dealing with outliers.

8.2.1 *Linear SVR*

Consider the problem of approximating the set of data $\mathscr{D} = \{(\mathbf{y}_1, z_1), \ldots, (\mathbf{y}_n, z_n)\}$, $\mathbf{y}_i \in \mathbb{R}^d, z_i \in \mathbb{R}$, with a linear function,

$$f(\mathbf{y}) = \langle \mathbf{w}, \mathbf{y} \rangle + b. \tag{8.1}$$

The optimal regression function [41] is given by

$$\min_{\mathbf{w},\xi} \frac{1}{2} \parallel \mathbf{w} \parallel^2 + C \sum_{i=1}^{n} (\xi_i^+ + \xi_i^-)$$

$$z_i - \langle \mathbf{w}, \mathbf{y}_i \rangle - b \leq \varepsilon + \xi_i^+$$

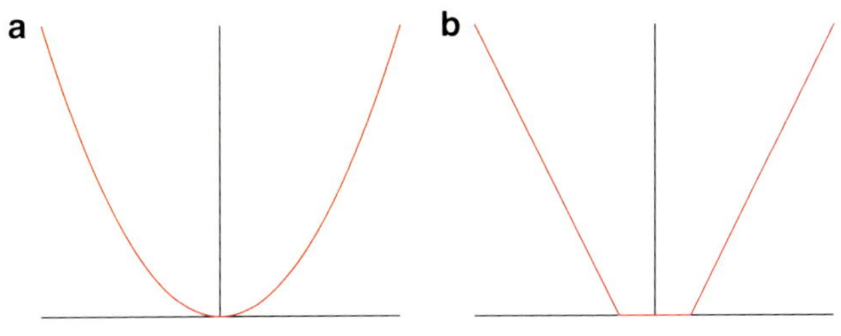

Fig. 8.1 Regression criteria. (**a**) Quadratic regression loss function. (**b**) ε-insensitive loss function which is less sensitive to outliers than the quadratic loss function. Another benefit from this function is a sparse set of support vectors to represent the regression function, i.e., only points outside the ε zone contribute to the regression function. The horizontal and vertical axes are $\mathbf{y}$ and $f(\mathbf{y})$, respectively

$$\text{subject to} \quad \langle \mathbf{w}, \mathbf{y}_i \rangle + b - z_i \leq \varepsilon + \xi_i^-$$
$$\xi_i^+, \xi_i^- \leq 0 \tag{8.2}$$

where constant $C > 0$ determines the trade-off between the flatness of f and data deviations, and ξ_i^+, ξ_i^- are slack variables to cope with otherwise infeasible constraints on the optimization problem of (8.2). The ε-insensitive loss function as shown in Fig. 8.1b is

$$L_\varepsilon(\mathbf{y}, z) = \begin{cases} 0, & \text{if } |f(\mathbf{y}) - z| < \varepsilon \\ |f(\mathbf{y}) - z| - \varepsilon, & \text{otherwise} \end{cases} \tag{8.3}$$

The *primal* problem of (8.2) can be solved more efficiently in its *dual* formulation [41] resulting in the final solution given by

$$\mathbf{w} = \sum_{i=1}^{n} (\alpha_i - \alpha_i^*) \mathbf{y}_i, \tag{8.4}$$

and

$$f(\mathbf{y}) = \sum_{i=1}^{n} (\alpha_i - \alpha_i^*) \langle \mathbf{y}_i, \mathbf{y} \rangle + b, \tag{8.5}$$

where α_i, α_i^* are Lagrange multipliers. The value of b in Eq. (8.1) can be determined by plugging Eq. (8.4) into Eq. (8.1) [12].

8.2.2 A Toy Example

To illustrate the SVR idea and see the importance of proper setting of the parameter ε, we use a toy example that contains 30 points in 2D with 10 in a line and the remaining 20 being outliers distributed on both sides of the line [20]. Hence the data contains 67 % outliers. Using the SVR algorithm implemented by Gunn [12] (which provides a user interface) and a linear kernel with $\varepsilon = 0.02$, the result is shown in Fig. 8.2a. Observe that the line was correctly estimated despite the high percentage of outliers.

On the other hand, observe that SVR returns 27 support vectors (90 % of the input data) and seven of them are very close to the boundaries (two dashed lines), but there are actually 20 outliers in the original data. So we cannot simply classify the support vectors (SVs) as outliers. Increasing the ε value might "drag" the seven closest support vectors inside the dashed boundaries, and then only the outliers in the data would be returned as support vectors. However, when we increase ε gradually up to 0.09, there are still 26 SVs returned which are still not the true outliers, as

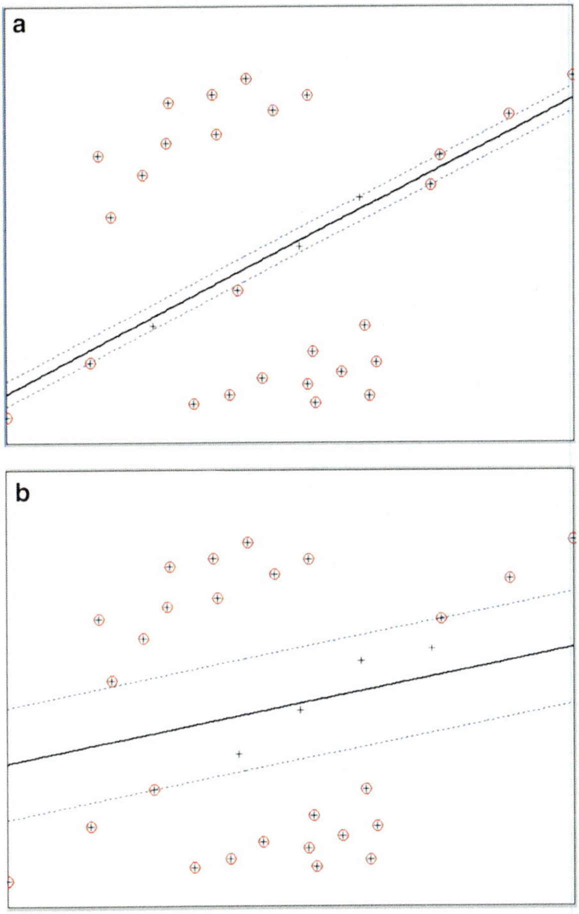

Fig. 8.2 SVR on real 2D data with $\varepsilon = 0.02$ in (**a**) and $\varepsilon = 0.09$ in (**b**). Note that the support vectors (marked by *circles*) are not the true outliers in either case

shown in Fig. 8.2b. And even worse, the slope of the line has changed significantly. This demonstrates that using a large ε is not a good idea because it may degrade the model structure.

Based on this experiment, we observe: (1) the SVR technique can potentially deal with data containing a high percentage of outliers; (2) classifying support vectors as outliers is not workable; (3) using a large value for ε is not a good idea for SVR; and (4) using small ε is preferable, especially when a large number of outliers are present.

This toy example and the above observations were first presented by Guo et al. in [20]. The robust regressor, SVR, was applied successfully for outlier detection and removal in affine motion tracking with the setting of a small ε. Here we adopt the same idea but use it for another application—robust age regression. Instead of using the simple linear regression, we need a nonlinear SVR for the complex aging patterns.

8.2.3 Nonlinear SVR

A nonlinear regression function may be required in practice to adequately model the data. It can be obtained by using kernels, in the same manner as a nonlinear SVM for classification [41]. A nonlinear mapping can be used to map the data into a high dimensional feature space where a linear regression is performed. Different kernels, such as polynomials, sigmoid, or Gaussian radial basis functions, can be used depending on the tasks. For our robust age regression, we found that the Gaussian radial basis function kernel performs much better than the linear regression [21, 23]. The reason is that the linear regression cannot model the complex aging process. A radial basis function is of the form,

$$k(\mathbf{y},\mathbf{y}') = e^{-\gamma\|\mathbf{y}-\mathbf{y}'\|^2}, \tag{8.6}$$

where γ is a constant to adjust the width of the Gaussian function. Given the kernel mapping, the solution of the nonlinear SVR is obtained as [41],

$$\langle \mathbf{w},\mathbf{y}\rangle = \sum_{i=1}^{n}(\alpha_i - \alpha_i^*)k(\mathbf{y}_i,\mathbf{y}), \tag{8.7}$$

and

$$f(\mathbf{y}) = \sum_{i=1}^{n}(\alpha_i - \alpha_i^*)k(\mathbf{y}_i,\mathbf{y}) + b. \tag{8.8}$$

The difference to the linear regression is that $\mathbf{w}$ is no longer given explicitly. Also note that in the nonlinear case, the optimization problem corresponds to finding the *flattest*, or linear regression function in the higher dimensional *feature* space,[1] not in the input space.

[1] Note that the feature space means a higher dimensional space in SVR, which is different from the feature extracted from data in image processing. Actually the extracted features from images are the input data for SVR in our age modelling.

8.3 Support Vector Machine

SVM [41] are a class of classifiers that can learn an optimal separating hyperplane based on the maximum margin criterion. It can use different kernels to make the linear SVM work on a higher dimensional space to improve the separability between two classes. The kernel extension is similar to the SVR learning. In the following, we only briefly introduce the linear SVM. More details on the kernel SVMs can be referred to [41].

8.3.1 Linear SVM

Given a set of training vectors belong to two separate classes, $(\mathbf{y}_1, z_1), \ldots, (\mathbf{y}_n, z_n)$, where $\mathbf{y}_i \in \mathbb{R}^D$, $z_i \in \{-1, +1\}$, the linear SVM learns an optimal separating hyperplane, $\mathbf{w}\mathbf{y} + b = 0$, that maximizes the margin [41]. The SVM learning is to find the saddle point of the Lagrange functional,

$$L(\mathbf{w}, b, \alpha) = \frac{1}{2} \parallel \mathbf{w} \parallel^2 - \sum_{i=1}^{n} \alpha_i \{z_i [(\mathbf{w} \cdot \mathbf{y}_i) + b] - 1\} \tag{8.9}$$

where α_i are the Lagrange multipliers. The Lagrangian has to be minimized with respect to $\mathbf{w}$, b and maximized with respect to $\alpha_i \geq 0$. The optimization is usually transformed to its *dual* problem,

$$\max_{\alpha} W(\alpha) = \max_{\alpha} \left\{ \min_{\mathbf{w}, b} L(\mathbf{w}, b, \alpha) \right\}, \tag{8.10}$$

and the optimal hyperplane is represented by the dual solution, α,

$$\mathbf{w} = \sum_{i=1}^{n} \alpha_i z_i \mathbf{y}_i \tag{8.11}$$

The value of b can be estimated by plugging $\mathbf{w}$ into the original equation, $\mathbf{w}\mathbf{y} + b = 0$.

In testing, the classification is given by

$$f(\mathbf{y}) = \mathrm{sign}(\mathbf{w} \cdot \mathbf{y} + b), \tag{8.12}$$

for any new data point $\mathbf{y}$. If the training data are non-separable, slack variables ξ_i can be introduced. See [41] for more details.

8.4 Locally Adjusted Robust Regression

Age estimation can be considered as a regression problem. Now, a question may be asked, is it "good" enough to use the SVR as a robust regressor for human age prediction? To answer this question, let us look at an estimation result using the SVR [21]. Figure 8.3 shows the predicted ages (red squares) with respect to the ground truth ages (black circles). Note that this is not a regression curve. One thousand data points are sorted in ascending order of the ground truth ages, i.e., from 0 to 91 years for females. The predicted ages are obtained from the SVR method. From this figure, we observe that the SVR method can estimate the global age trend, but cannot predict the ages precisely. By inspecting the result carefully, we find that the SVR predictions give bigger age values for many younger people, and smaller age values for some older people. In some cases, the estimated age values could be far away from the true ages, e.g., more than 40 years. This result was based on a database used in [21].

Why the SVR method cannot show better performance than we expect for age prediction? The reason can be in two aspects: First, the problem of age prediction is really challenging because of the diversity of aging variation. Each individual may age in his/her own way and be affected by external factors, such as health, living condition, and exposure to weather conditions. Second, the SVR method attempts to find a flat curve to approximate the data in order to obtain good generalization capability. As shown in Fig. 8.4, the SVR computes a flat curve within a small ε tube. But the age data may distribute like the (green) irregular curve. One cannot expect the SVR to estimate an irregular curve like this because of the over-fitting

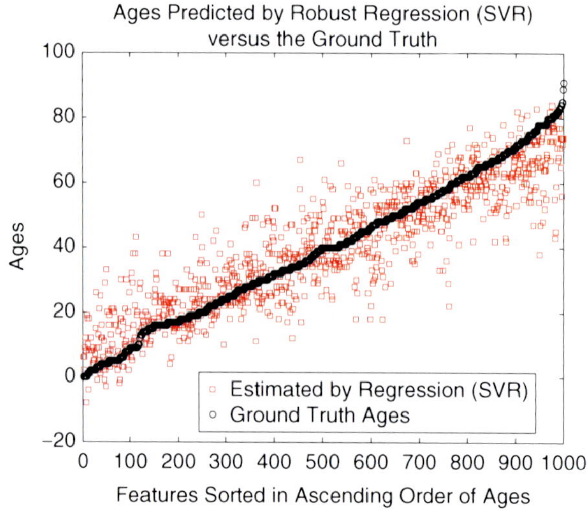

Fig. 8.3 A plot of the true ages (*black circles*) versus the estimated ages (*red squares*) for one thousand female face images. The ages are predicted by the nonlinear SVR with a Gaussian kernel

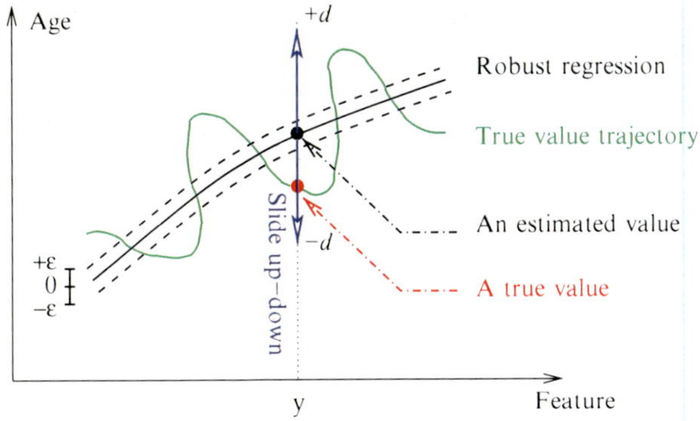

Fig. 8.4 Illustration of the idea of locally adjusted robust regression (LARR)

problem. Further, one cannot assign a large ε to enclose all true data points inside the ε tube, as demonstrated in the toy example in Sect. 8.2.2. So how to model the aging function by allowing the irregular distribution of true ages?

8.4.1 Local Adjustment of the Regression Result

One feasible solution is to adjust the age regression values locally so that the estimated age values can be "dragged" towards the true ages. We call it a locally adjusted robust regression (LARR) [21, 23]. The idea of LARR is illustrated in Fig. 8.4. Suppose the predicted age value by SVR is $f(\mathbf{y})$, corresponding to the input data $\mathbf{y}$. The point $f(\mathbf{y})$ is displayed by the black dot on the regression curve. The estimated age, $f(\mathbf{x})$, may be far away from the true age value, L, shown as the red dot on the true age trajectory curve. The idea of the LARR method is to slide the estimated value, $f(\mathbf{y})$, up and down (corresponding to greater and smaller age values) by checking different age values, $t \in [f(\mathbf{y}) - d, f(\mathbf{y}) + d]$, to see if it can come up with a better age estimation. The value d indicates the range of ages for local search. Hopefully the true age value, L, is also within this range, i.e., $L \in [f(\mathbf{y}) - d, f(\mathbf{y}) + d]$.

Therefore the LARR method is a two-step procedure: (1) a robust regression over all ages of the training data by using the SVR method. This step can be considered as a global regression process; (2) a local adjustment within a limited range of ages centered at the regression result.

Now the key issue is how to verify different age values within a specified range for the purpose of local adjustment. Remember our goal is to "drag" the initially estimated age value, $f(\mathbf{y})$, by the global regressor, towards the true age, L, as close as possible. We take a classification approach to locally adjust or verify different ages, considering each age label as one class. Because only a small number of age labels are used for each local adjustment, regression methods cannot work properly.

For our classification-based local adjustment, there are many possible choices of classifiers, but here we adopt a linear SVM for our local age adjustment. The main reason is that the SVM can learn a classifier given a small number of training examples. This has been demonstrated by the author previously for learning in the small sample case, such as face recognition [16, 18], image retrieval [19], audio classification and retrieval [14], and face expression recognition [13]. The capability of learning a classifier in the small sample case is also important for human age prediction. Usually the number of training examples, e.g., 50, is smaller than the feature dimension, e.g., 150, in age estimation, even though we perform experiments on a large database (see Sect. 8.7 for details).

8.4.2 Binary Tree Search with Limited Range

The classical SVMs are designed to deal with the two-class classification problems. There are three typical ways to extend it to a multi-class classification application. (1) Learning classifiers for each pair of classes, and taking a binary tree search in testing; (2) training SVMs for each class against all the remaining classes; and (3) training SVMs for all classes simultaneously. The last two schemes are not appropriate for our purpose, because in the local adjustment only partial classes of age data are involved. If the last two schemes are used, the SVMs have to be re-trained dynamically for each adjustment, which is computationally expensive. The first scheme is feasible to fulfill our task since there is no need to retrain the SVMs online. Therefore, all pair-wise SVM classifiers can be trained off-line. Only a limited number of classes are involved in the binary tree search for test.

The binary tree structure for multi-class SVM classification has been used successfully in previous research, e.g., face recognition [16]. In general, the number of pair-wise comparisons is $n_c - 1$ for each test in an n_c-class classification problem. Here the number of pair-wise comparisons is limited to $m_c - 1$ when only m_c classes are involved in each local adjustment, and $m_c < n_c$. Each age corresponds to one class label.

8.4.3 Local Search Range Determination

The local search range, m_c, is determined by several factors, such as the scale of the data (large versus small scale) and the performance of the robust regressor (here the kernel SVR).

It is not trivial to determine the local search range. There are some guidelines for choosing local search ranges. The larger the search range, the bigger the chance to contain the true ages within that range. If the search range is too small, the true age label might not be reached and the local search may find an arbitrary age label.

On the other hand, if the search range is too big, it also increases the possibility to obtain an adjusted age that is far away from the true age, because the local classification is just a locally optimal search.

In our experiments, we specify different ranges and demonstrate the effects of different local search ranges on the results [21]. The main goal is to show that the local adjustment can really improve the performance over the robust regressor for human age estimation.

8.5 Probabilistic Fusion of the SVR and SVM

As presented above, the combination of the SVR and SVM can take advantage of both classifiers and regression for age estimation. Our first scheme is a LARR proposed by Guo et al. in [21, 23]. It has been shown that the age estimation performance can be improved significantly by using the LARR method.

However, the LARR method cannot determine the range of local search for the classifier. It has to heuristically try different ranges, such as 4, 8, 16, 32, and 64, and requires the user to choose a best solution among those results. For practical use of the age information, e.g., in multimedia content analysis and understanding, it is important to develop an age information extractor automatically without the user involvement. In other words, the system has to determine the combination parameters *automatically* in a *data-driven* manner. Towards this goal, we interpret the regression and classification results probabilistically in order to fuse them automatically [22].

8.5.1 Theoretical Framework

Consider a pattern recognition problem [42] where pattern Z is to be assigned to one of the m possible labels $L = \{l_1, l_2, \cdots, l_m\}$. For the age estimation problem, the labels are human ages (in years), such as $0, 1, \cdots$. Assume we have a regressor R and a classifier C, each representing the given pattern by a distinct measurement vector, denoted by $\mathbf{x}_R$ and $\mathbf{x}_C$, respectively. In the measurement space each label or class l_k is modeled by the probability density function (PDF) $p(\mathbf{x}_R|l_k)$ or $p(\mathbf{x}_C|l_k)$, and the prior probability of occurrence of each label is denoted by $P(l_k)$.

According to the Bayesian theory, given measurements $\mathbf{x}_R$ and $\mathbf{x}_C$, the pattern, Z, should be assigned label l_j when the posterior probability of that interpretation is maximum, i.e.,

$$l_j = \arg\max_{l_k \in L} P\left(l_k|\mathbf{x}_R, \mathbf{x}_C\right) \tag{8.13}$$

The Bayesian decision rule (8.13) states that all the measurements should be considered simultaneously in order to make a decision utilizing all the available information correctly. The computation of the posterior probability functions in (8.13) depends on knowledge of high-order measurement statistics described in terms of joint PDFs $p(\mathbf{x}_R, \mathbf{x}_C | l_k)$, which are generally difficult to obtain. A classical approach to deal with these kinds of joint probabilities is to assume that all the measurements are independent for a given pattern. For example, the mutual independence assumption was used in combining different classifiers in [31].

Here we build a "causal" relation between R and C. Specifically, the classifier C makes decision based on the output of the regressor R, but the regressor R works on the input data directly. Therefore

$$P(\mathbf{x}_R | \mathbf{x}_C) = P(\mathbf{x}_R). \tag{8.14}$$

There are two reasons to have this causal relation assumption: (1) To reduce the measurement space sequentially—the decisions of the first learner could impact or reduce the measurement space of the second learner. This "early" influence might simplify the original complex decision problem into a simpler one, and therefore improve the recognition accuracy of the second learner. As a result, the performance of the whole system can be improved. (2) To consider the internal structure of the learners—a regressor usually takes into account all data points, computing in a "global" style, while some modern classifiers [41] use a pairwise classification scheme, working in a "local" style. Therefore it might be easier to change the measurement space of the classifiers instead of the regressors.

Now let us go back to the Bayesian decision rule (8.13) and rewrite it. Based on the conditioned Bayes' rule (i.e., Bayes' rule conditioned on another variable; see page 10 in [30]), we have

$$P(l_k | \mathbf{x}_R, \mathbf{x}_C) = \frac{P(\mathbf{x}_R | l_k, \mathbf{x}_C) P(l_k | \mathbf{x}_C)}{P(\mathbf{x}_R | \mathbf{x}_C)} \tag{8.15}$$

which holds in general. Substituting (8.14) into (8.15) we obtain

$$P(l_k | \mathbf{x}_R, \mathbf{x}_C) = \frac{P(\mathbf{x}_R | l_k) P(l_k | \mathbf{x}_C)}{P(\mathbf{x}_R)}. \tag{8.16}$$

By Bayes' rule, we have

$$P(\mathbf{x}_R | l_k) = \frac{P(\mathbf{x}_R) P(l_k | \mathbf{x}_R)}{P(l_k)}. \tag{8.17}$$

Plugging (8.17) into (8.16), we get

$$P(l_k | \mathbf{x}_R, \mathbf{x}_C) = \frac{P(l_k | \mathbf{x}_R) P(l_k | \mathbf{x}_C)}{P(l_k)}. \tag{8.18}$$

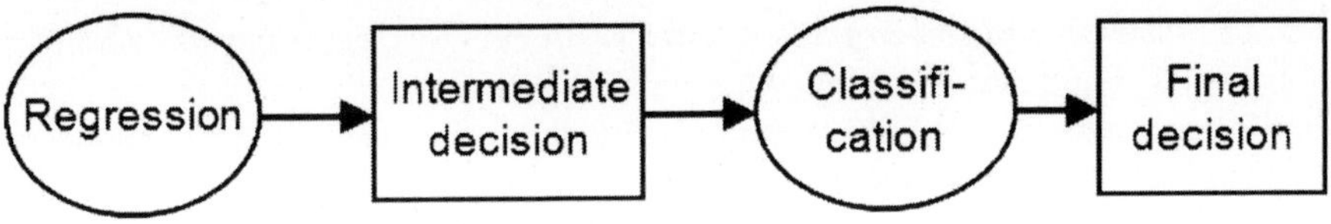

Fig. 8.5 The decision graph of the PFA approach

Now, the decision rule (8.13) becomes:

$$l_j = \arg\max_{l_k \in L} \frac{P(l_k|\mathbf{x}_R)\,P(l_k|\mathbf{x}_C)}{P(l_k)} \tag{8.19}$$

subject to constraints (8.14). Decision rule (8.19) fuses the posterior probabilities computed by the regressor and the classifier sequentially. We call this a *Probabilistic Fusion Approach* (PFA).

In practice, the denominator of (8.19), i.e., the prior probabilities $P(l_k)$, will have equal values if no strong prior knowledge is given for a recognition problem. In this case, the decision rule becomes

$$l_j = \arg\max_{l_k \in L} P(l_k|\mathbf{x}_R)\,P(l_k|\mathbf{x}_C) \tag{8.20}$$

8.5.2 *Fusion Strategy*

Decision rules (8.19) and (8.20) constitute the basic scheme for combining a regression measurement with a classification result in a probabilistic way. Now we develop a specific combination strategy based on decision rule (8.20).

In our sequential probabilistic fusion scheme, the regressor R and classifier C work sequentially so that the output of the regressor, $P(l_k|\mathbf{x}_R)$, is used as an intermediate decision which is then fed to the classifier C to affect the measurement or decision space of the classifier, $\mathbf{x}_C$. The classifier C has no effect on the regression measurement, $\mathbf{x}_R$. This causal relation can be depicted by the decision graph in Fig. 8.5.

To realize the decision process shown in Fig. 8.5, several issues have to be addressed, including (1) which methods to use for the regression and classification modules, (2) how to produce the probabilistic output for each method, and (3) how to alter the measurement space of the classifier based on the regression output.

8.5.2.1 Selection of the Regressor and Classifier

For the regressor, it should have high performance, since its results will influence the decision of the classifier in our sequential fusion strategy. A low performance regressor might "drift" the measurement space badly for the following classifier. The requirement for the classifier is that its measurement space should be able to change (e.g., shrink or expand) easily.

Guided by the above consideration, we chose to use a SVM [41] as the classifier, and the SVR method [41] as the regressor, which were also chosen in [21, 23]. The difference is that there is no probabilistic computation for the SVM and SVR in the LARR method [21,23], while here the results of the SVM and SVR are transformed into probabilities and then fuse them automatically [22] without trying different local ranges and requiring users' selection as in [23].

8.5.2.2 Probabilistic Output for SVMs

Standard SVM provide only an estimated target value, e.g., a category label for classification or a real value for regression. In order to combine the regression and classification measurements probabilistically, probabilities need to be extracted from the standard SVM and SVR results.

For the SVM, some methods have been proposed, mainly in the machine learning literature, to produce probabilistic outputs. For example, Platt [38] proposed a sigmoid training method to post-process standard SVM output, focusing on a two-class classification problem. But it is not clear how to extend this method to a multi-class scenario. In [29], an MAP rule was used on the estimate of the overall posterior probabilities obtained from the outputs of the pairwise classifiers.

Here we adopt a simple yet efficient method to generate a probability estimate for the SVM in a multi-class classification problem, using the counts of occurrences in pairwise comparisons. This simple idea has been used successfully for face recognition [17], for example.

For an n-class classification problem (n could be less than the total number of classes m in the original measurement space), the total number of pairwise comparisons is $n(n-1)/2$. The output of the $n(n-1)/2$ classifiers is used to construct a matrix as shown below:

$$\begin{pmatrix} 0 & \phi_{1,2} & \phi_{1,3} & \cdots & \phi_{1,n} \\ \phi_{2,1} & 0 & \phi_{2,3} & \cdots & \phi_{2,n} \\ \vdots & & \ddots & & \vdots \\ \vdots & & & \ddots & \vdots \\ \phi_{n,1} & \phi_{n,2} & \phi_{n,3} & \cdots & 0 \end{pmatrix}.$$

Each element in the matrix is equal to 1 or 0. $\phi_{i,j} = 1$ if pattern Z is classified as class i in the pairwise competition between classes i and j; otherwise, $\phi_{i,j} = 0$. All elements in the main diagonal are zeros. Based on the measurement matrix, we can create a probability measure for the SVM classifier output as

$$P(l_k|\mathbf{x}_C) = \frac{\sum_{j=1}^{n} \phi_{k,j}}{\sum_{i=1}^{n}\sum_{j=1}^{n} \phi_{i,j}} \tag{8.21}$$

8.5.2.3 Probabilistic Output for the SVR

For the SVR, several methods have been proposed to produce a probabilistic output, but many of them involve either complex computations or modification of the SVR formulation. For example, a Gaussian process is integrated into the SVR to formulate a Gaussian SVM regression model in [10]. A Gaussian (or Laplace with fatter tails) distribution could be used to approximate the probabilistic outputs for SVRs. However, the Gaussian approximation may encounter problems in practice, especially in human age prediction, because of the diversity of aging variations. Each individual may age in his or her own way and be affected by many different external factors.

As pointed out in [23], the ages estimated by the SVR method could be far away from the true age labels. Consequently, a small probability value (possibly close to zero) could be generated for a true age label when a Gaussian model is used for transforming the SVR target values into probabilistic outputs. This would inhibit a correct decision when multiplying the two probabilities in the decision rule (8.19) or (8.20). In order to avoid such undesirable effects, we propose to use a uniform distribution centered at the estimated target value, l_0, obtained from a regressor, i.e., $\mu = l_0$. In fact, we found that the Gaussian model gave much worse results than the uniform distribution in our initial experiment on age estimation which is not shown here.

The uniform distribution model assumes that only a finite range of age labels is possible, each with equal probability. The PDF of the uniform distribution $\mathrm{U}(\mu-\Delta, \mu+\Delta)$ is given by

$$p(x) = \begin{cases} \frac{1}{2\Delta} & \text{for } \mu-\Delta \le x \le \mu+\Delta, \\ 0 & \text{otherwise,} \end{cases} \tag{8.22}$$

where $[\mu-\Delta, \mu+\Delta]$ is the function support. Now the question is how to estimate the range of support for the uniform distribution.

Let us look at the SVR prediction error or residual, ζ_i, with $\zeta_i = l_i - \hat{f}(Z_i)$, where l_i is the true age label for pattern Z_i, and $\hat{f}(Z_i)$ is the regression estimate. Recall that the variance of the uniform distribution satisfies $\sigma^2 = \frac{1}{12}(2\Delta)^2$, i.e., $\sigma^2 = \frac{1}{3}\Delta^2$, so we have $\Delta = \sqrt{3}\sigma$. Thus the function support can be estimated by the sample standard deviation. To compute the sample standard deviation, σ, we can collect the residuals, ζ_i, on a validation data set, and then compute the standard deviation of these residuals. Finally, we have

$$P(l_k|\mathbf{x}_R) \leftarrow U\left(l_0 - \sqrt{3}\sigma,\ l_0 + \sqrt{3}\sigma\right). \tag{8.23}$$

The uniform distribution (8.23) is simple but works well in our experiments. To our knowledge, no previous work uses it to model the probabilistic output of a regressor such as the SVR.

8.5.2.4 Decision Space Deduction

Given the probabilistic outputs, $P(l_k|\mathbf{x}_R)$ and $P(l_k|\mathbf{x}_C)$, for the regressor and classifier, respectively, the next step is to combine the two probabilities together to make a final decision for a given pattern. According to the decision rule (8.20), the two probabilities are multiplied and the label l_j corresponding to the maximum product is selected as the final decision.

Our serial PFA can also be interpreted as a decision space deduction process. The uniform distribution modeling of the probabilistic output of the regressor reduces the original label space (all possible ages) into a smaller decision space, $\left[l_0 - \sqrt{3}\sigma, l_0 + \sqrt{3}\sigma\right]$, by using the cutoff boundaries. The reduced decision space is refined by the classifier to obtain the final decision, l_j. As a result, the probabilistic output of the SVR plays the role of an intermediate decision, as shown in Fig. 8.5, reducing the search space (i.e., less number of classes to compare) for the classifier SVM. The LARR method [21, 23] shares the same spirit as the PFA in terms of decision space deduction, however, it does not address the probabilities for automatic local range determination.

8.6 Soft Biometrics Computation

We have presented the methods of SVM, SVR, and the combinations of them. These methods will be used for age estimation on different databases. For gender and ethnicity classifications, only the SVMs are used, since these problems are typically considered as classifications.

We only use face images for soft biometrics computation. The face images are usually detected, aligned, cropped, and resized into the same size. Various features can be extracted from the face images to characterize the facial appearance. The specific methods for feature extraction will be briefly introduced in the experiments.

8.7 Experiments

We conduct experiments for age, gender, and ethnicity estimation, separately. Different databases might be used for each of the soft biometric measure. Not all databases are proper to study all of the three soft biometric characteristics.

8.7.1 Age Estimation Results

Age estimation experiments are conducted on the FG-NET and Yamaha Aging Databases. The FG-NET Aging Database [7] is a publicly available age database that we adopt for the experiment. The database contains 1,002 color or grayscale face images with variations of lighting, pose, and expression. There are 82 subjects (multiple races) in total with the age ranges from 0 to 69 years, and each face image has 68 labeled points characterizing shape features. The shape features can be combined with appearance features to form a face representation, called active appearance models (AAMs) [5]. The AAMs use 200 parameters to model each face for the purpose of age estimation [11, 46, 47].

The Yamaha Aging database contains 8,000 high-resolution RGB color face images captured from 1,600 different voluntary Asian subjects in an outdoor environment, 800 females and 800 males, in the age range from 0 to 93 years. Each subject has five near frontal images with provided ground truth ages. It has been used in some previous studies, e.g., [8, 9, 46, 47]. The Yamaha database is much larger than the FG-NET.

To evaluate the age estimation performance on Yamaha, a face detector was used to find the face area in each image, and the eye corner locations are labeled for each face subject. Based on the face and eye corner locations, the face images are cropped, scaled, and transformed to 60×60 gray-level patches [21, 23]. The images have significant variances in illumination since the photographs were taken in the outdoor environment. The gray-level values of each face image are normalized to a normal distribution with zero-mean and one standard deviation in order to reduce the effect of out-door illumination changes. The database also contains some facial expression variations and makeup.

The face image patches with the same size of 60×60 are fed into the manifold learning module. The age manifold can be embedded in a low dimensional subspace using different techniques [21]. Some manifold visualizations can be found in [21]. It has been shown in [21] that: (1) The principal component analysis (PCA) method does not show clear manifold trend of ages. The reason is that the PCA is purely unsupervised without using any age label information, which seems to be important for learning the embedded manifold from the complex aging patterns; (2) The manifold learned by the local linear embedding (LLE, a nonlinear embedding method) is approximately an ellipsoid with higher ages in the center and lower ages at periphery; and (3) The OLPP algorithm [4] achieves good visualization of the age manifold with a distinct aging trend. Therefore, we used the OLPP method in our age manifold learning module for age estimation [21].

After the age manifold was learned, each face image can be projected onto the age manifold to extract a feature vector. We used the first 150 features for each face image [21, 23]. The system then learns a robust regression function using the kernel SVR method for females and males separately. Actually the manifold was learned for the female and male independently. As demonstrated in the toy example in Sect. 8.2.2, a small ε value should be chosen for the ε-insensitive loss function in Eq. (8.3). We set $\varepsilon = 0.02$ for our age estimation task. In SVR learning, parameters C and γ are determined on a validation set. Experimentally we found that a good choice is $C = 40$ and $\gamma = 12$, separately. To locally adjust the global regression results, we tried different local search ranges as powers of two, e.g., 4, 8, 16, 32, and 64 classes, and the results from different search ranges are compared to see the effect of local adjustment. The purpose of choosing the powers of two is to simplify the binary search structure. One can observe that the local search range does influence the age estimation results. The pair-wise linear SVM classifiers were used for the local adjustment, centered at the age value (or label) obtained from the global regressor.

We perform a standard fourfold cross validation test to evaluate the accuracy of our algorithms for age estimation on the Yamaha age database. The test was executed on the female and male subsets separately. The females and males age quite differently in the database. For each experiment, about 1/3 of the training data are used as a validation subset to determine the optimal parameter setting such as C and γ. Then the parameters are fixed and the whole training data set is used to learn the robust regression function. The pair-wise linear SVM classifiers are learned using the same training data and used for local adjustment in testing. Finally all performance measures are reported on the unseen test data.

The performance of age estimation can be measured by two different measures: the mean absolute error (MAE) and the cumulative score (CS). The MAE is defined as the average of the absolute errors between the estimated ages and the ground truth ages, $\text{MAE} = \sum_{k=1}^{N} |\widehat{l_k} - l_k|/N$, where l_k is the ground truth age for the test image k, $\widehat{l_k}$ is the estimated age, and N is the total number of test images. The cumulative score [11] is defined as $\text{CS}(j) = N_{e\leq j}/N \times 100\,\%$, where $N_{e\leq j}$ is the number of test images on which the age estimation makes an absolute error no higher than j years.

Table 8.1 shows the experimental results. The first and second columns in Table 8.1 show the MAEs for females and males in the Yamaha aging database, separately. Different ranges, e.g., 4, 8, 16, 32, and 64, were tried for local adjustment of the global regression results. One can see that the local adjustment truly reduces the errors of the global regression. For example, the MAE of the SVR is 7 years for the female (column 1 in Table 8.1), but is reduced to 5.86 (column 1, row 5) when 16 local age classes are used for the LARR method, and so on. Different ranges of adjustment do have different MAEs. For comparison, we also show the results using purely the SVM classifiers in the first row. One can see that the classification scheme has lower errors than the pure regression method for both females and males, but it

Table 8.1 MAEs of the methods: SVM, SVR, and LARR with different settings [21,23]

Various setup	Yamaha (Female)	Yamaha (Male)	FG-NET
SVM	5.55	5.52	7.16
SVR	7.00	7.47	5.16
LARR4	6.83	7.21	**5.07**
LARR8	6.48	6.81	**5.07**
LARR16	5.86	5.95	5.12
LARR32	5.29	**5.30**	6.03
LARR64	**5.25**	5.38	–

The bold fonts indicate the lowest errors in each case.

has higher error rates than some of the locally adjusted results. The best LARR result in terms of MAE is 5.25 years for females when the local search range is 64 classes, while it is 5.30 years for males when the adjust range is 32 classes. The ranges of local adjustment depend on the data and the global regression results. To illustrate the MAEs at each age, two pictures for female and male results are displayed in Fig. 8.6, respectively.

Figure 8.7a, b show the CS measures for females and males separately. We can observe that the LARR methods (with different ranges for local adjustment) improve the score significantly over the pure regression method for lower error levels, e.g., $m_c < 10$ years. For example, in one year error level, most LARRs with proper ranges of local adjustment could improve the accuracy by 175 % and 267 % for females and males separately. This improvement is significant. We also notice that large ranges are required for local adjustment on the Yamaha aging database. For instance, when 16 age classes are used for local adjustment, the CS curve is explicitly lower than 32 or 64 classes. We do not show the cumulative scores for four and eight classes here in order to not mess up the figures. Those two CS curves are even lower than 16 classes. One may also notice that the CS curve of SVM classifiers is close to the LARR32 and LARR64 for both females and males, but the MAEs of the SVM are higher than the LARR16 or LARR32 as shown in Table 8.1. This indicates that we need both MAE and CS measures complementarily to measure the performance of an algorithm in age estimation.

As shown in Table 8.2, we also compare our results with all previous methods reported on the Yamaha aging database. It turns out our LARR method has the MAEs of 5.25 and 5.30 years for females and males separately, which are explicitly smaller than the previous results under the same experimental protocol. Our method brings about 24 % deduction of MAEs over the best result of previous approaches, given in [46].

For age estimation on the FG-NET database, we used the same AAM features as in [11, 46, 47] to evaluate our LARR method [21, 23]. Since the FG-NET database has small size, we do not learn any age manifold but use the AAM features directly. Our focus is then to evaluate the performance of the LARR method for age estimation on the FG-NET database. The popular test strategy, namely leave-one-

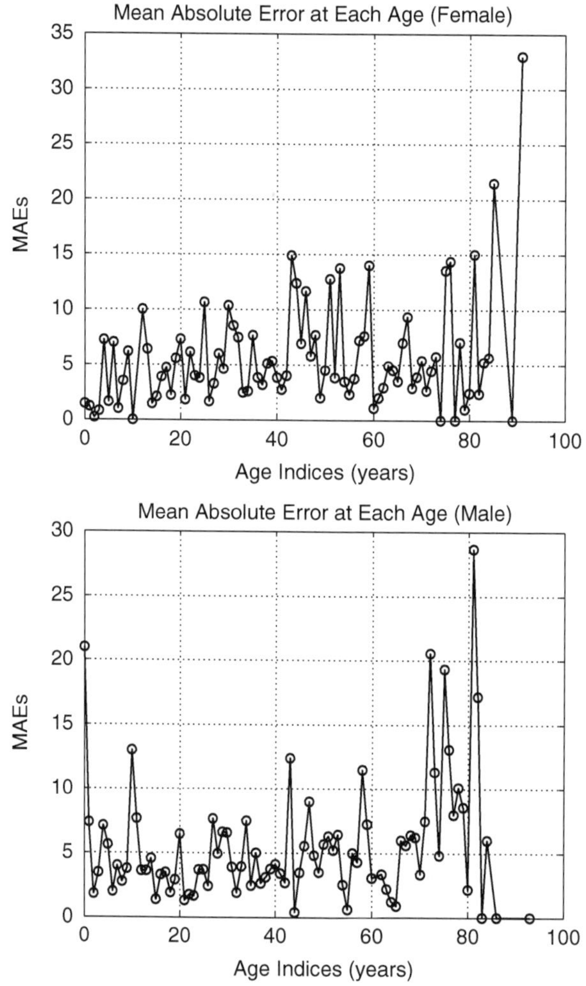

Fig. 8.6 MAEs at each age for females and males on the Yamaha Aging database, obtained by the LARR method [21]

person-out (LOPO), was usually taken for the FG-NET age database, as suggested by the existing work [11, 46, 47]. We follow the same strategy and compare our results with the state-of-the-art methods. The experimental results are shown in the third column of Tables 8.1 and 8.2. One can see that the LARR method has an MAE of 5.07 years which is lower than the previous methods listed in Table 8.2 [21]. The best MAE was obtained using either four or eight classes for local adjustment as shown in Table 8.1. Increasing the local search ranges for the LARR method will make the errors larger. For example, the MAE will be 6.03 years when 32 classes are used for local adjustment. We cannot get the result for 64 classes since there are

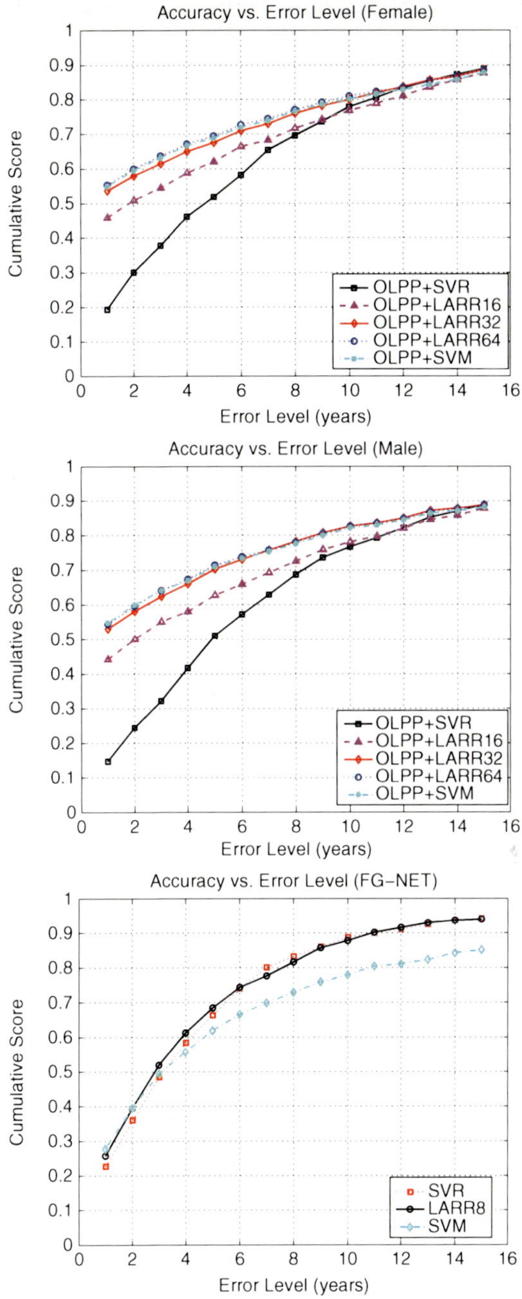

Fig. 8.7 Cumulative scores of the algorithms with different settings for (**a**) *Top*: female age estimation, (**b**) *Middle*: male age estimation on the Yamaha Aging database, and (**c**) *Bottom*: age estimation on the FG-NET database, at error levels from 1 to 15 years [21]

Table 8.2 MAE comparisons of different algorithms [21, 22]

Method	Yamaha (Female)	Yamaha (Male)	FG-NET
WAS [11]	–	–	8.06
AGES [11]	–	–	6.77
QM [32]	9.96	10.51	6.55
MLPs [32]	10.99	12.00	6.98
RUN1 [47]	9.79	10.36	5.78
RUN2 [46]	6.95	6.95	5.33
LARR [21]	5.25	5.30	5.07
PFA [22]	**5.11**	**5.12**	**4.97**

at most 63 or 61 age labels in the LOPO test. In other words, there are missing ages in the FG-NET database. When the pure classifiers, SVMs, are used, the MAE is 7.16, which is much higher than the 5.16 years of the pure regression. One possible reason is that there is not sufficient data for pair-wise SVM training, while the global SVR uses all the data in the model. Another observation is that the robust regression itself (without local adjustment) has an MAE of 5.16 years, which is still lower than all previous methods shown in Table 8.2. The LARR method further reduces the MAE to 5.07 years [21].

Figure 8.7c shows the cumulative scores of the LARR method on the FG-NET database. LARR8 means using eight classes for local adjustment. We do not show LARR4, LARR16, and LARR32 in order to avoid messing up the display. The cumulative scores of those ranges are close to LARR8 with slight differences. LARR8 has higher accuracy than the pure regression by SVR at lower error levels (1–6), but close to it at higher error levels. The cumulative scores of the pure SVM are much lower than the pure SVR for most error levels, which indirectly indicates the significance of constraining the SVM search in a local range. The LARR method performs much better than the QM and MLP methods. The method of RUN1 [47] is close to our LARR in low age error levels, but worse than LARR in high levels. In contrast, the method of RUN2 [46] is close to our LARR in high age error levels, but worse than the LARR in low error levels. Overall, the LARR method has higher accuracy than both the RUN1 and RUN2 on the FG-NET database.

Following the idea of combining the SVR with SVMs in the LARR method [21, 23], we proposed a PFA to combine the classifiers with regression in a probabilistic manner [22]. The PFA method can avoid the search range selection in LARR. To validate the PFA method, we performed age estimation experiments [22] on the Yamaha and FG-NET databases using the same protocol and data. The experimental results are shown in the last row in Table 8.2. The first and second columns in Table 8.2 show the MAEs for females and males in the Yamaha aging database, respectively. The last column shows the MAEs on the FG-NET aging database. From the table, we can see that the PFA method can improve the age estimation accuracies over the LARR method, in addition to the determination of local adjustment automatically.

Table 8.3 Numbers of male and female faces in the three age groups of the Yamaha aging database: young, adult, and senior

	Young (0–19)	Adult (20–60)	Senior (61–93)	All ages (0–93)
Male	1,000	2,050	950	4,000
Female	1,000	2,050	950	4,000
Both	2,000	4,100	1,900	8,000

In summary, we have shown that the SVM and the regression formulations can do well for age estimation. Age estimation can be considered as either a regression or a classification problem. Different results might be obtained in different databases, when classification or regression is applied. We proposed two methods to combine the SVR and SVM in order to improve the performance in age estimation. Both the LARR and PFA methods can take advantage of the SVM classifiers and the SVR for an improved performance.

8.7.2 Gender Classification

Gender classification is an interesting topic in both psychology [3, 44, 45] and computer vision [2, 28, 34, 48]. In computational approaches, various methods have been proposed for gender classification based on different facial image representations and classifier learning. Some typical approaches were listed in [24]. Among the different methods, an earlier work [34] applied the SVM to raw face images for gender classification.

We have studied the influence of age on gender classification in [24], based on several face image representations. The Yamaha database was used for the study. The number of males and females in each age group is shown in Table 8.3. One can see that it is very balanced for males and females in the database.

In addition to the raw face images, the LBP, HOG, and BIF features were used for gender recognition experimentally [24]. The goal was to evaluate the influence of age on gender recognition using several facial representations.

Both LBP and HOG features were extracted from each face image at various patch positions for the three age groups. Face images are of size 60×60, and the patch size is 16×16 with an interval of eight pixels between neighboring patches. The HOG operator has eight directions as in [6], and the LBP operator uses the uniform pattern as in [1]. Since HOG features were initially used with a linear SVM for pedestrian detection [6], we also show gender classification results based on linear SVMs (labeled as L-SVM) in addition to nonlinear SVMs with the RBF kernel (denoted as N-SVM). The LBP operator can be applied to the whole face image (denoted as LBP(W)) or applied to small patches on the face (denoted as LBP(P)). The patch-based LBP is much better than the whole face-based LBP in

Table 8.4 Gender recognition with different representations: raw pixels, LBP, HOG, and BIF, using the linear SVM (L-SVM) or nonlinear SVM (N-SVM) with the RBF kernel as classifiers [24]

Methods	Young (0–19)	Adult (20–60)	Senior (61–93)
Raw + L-SVM	78.59 %	89.91 %	81.17 %
Raw + N-SVM	84.38 %	94.56 %	85.32 %
LBP(W) + L-SVM	68.17 %	72.33 %	63.40 %
LBP(W) + N-SVM	69.65 %	77.08 %	68.40 %
LBP(P) + L-SVM	79.76 %	92.65 %	87.55 %
LBP(P) + N-SVM	81.93 %	94.96 %	90.64 %
HOG + L-SVM	75.83 %	88.00 %	77.13 %
HOG + N-SVM	86.44 %	94.03 %	89.04 %
BIF + L-SVM	83.01 %	94.22 %	91.81 %
BIF + N-SVM	87.13 %	96.03 %	92.34 %
Average(L-SVM)	80.52 %	91.20 %	84.42 %
Average(N-SVM)	84.97 %	94.90 %	89.34 %

gender recognition, as shown in Table 8.4. In each case, the parameters of the SVM are adjusted to optimal values on a tuning set (part of the training data).

8.7.2.1 HOG Feature

When the HOG features are used with nonlinear SVMs, gender recognition accuracies were 86.44 %, 94.03 %, and 89.04 %, for the young, adult, and senior groups, respectively, as shown in row 8 of Table 8.4. When compared with the "Raw+SVM" approach, the accuracies improved from 84.38 % to 86.44 % for the young faces, and improved from 85.32 % to 89.04 % for seniors, while the accuracy of 94.03 % for adults is slightly lower than the 94.56 % based on raw pixel representation. These results demonstrate that the HOG operator can characterize shape and improve recognition accuracies for young and senior faces. However, the two improved accuracies are still much lower than the 94.03 % accuracy for adult faces. On the other hand, the results indicate that the "Raw+SVM" approach is still good for gender recognition on adult faces.

We further explain the result [24] as: (1) adult male and female faces have local shape differences that can be described by the HOG operator and (2) shape changes in young faces and wrinkles in senior faces result in gradient variations that can be encoded by the HOG operator to some extent. However, the HOG performs much better for gender recognition on adult faces than on young and senior faces.

Also notice that linear SVMs performed much worse than kernel SVMs for each age group, as shown in rows 7 and 8 of Table 8.4.

8.7.2.2 LBP Feature

When LBP features were used with kernel SVMs, gender recognition accuracies were 81.93 %, 94.96 %, and 90.64 % for the young, adult, and senior groups, respectively, as shown in row 6 of Table 8.4. Here "P" represents patch-based LBP. When compared with the "Raw+SVM" approach, LBP features improved gender recognition accuracy for seniors (from 85.32 % to 90.64 %), but this is still lower than the accuracy of 94.96 % for adult faces. More interestingly, the accuracy reduced to 81.93 % for young faces, which is even lower than the 84.38 % accuracy of the "Raw+SVM" approach, and much lower than the 94.96 % accuracy for adult faces. Again, gender recognition performance is very different for the three age groups using LBP features: high performance for adult faces, lower performance for senior faces, and very low performance for young faces. Possible reasons for this phenomenon are: (1) adult male and female faces have local texture differences that can be described well by the LBP operator and (2) complex textures (e.g., wrinkles) on senior faces can also be described well by the LBP operator. For young faces, facial textures are not very rich and the main changes are facial shapes where the LBP operator does not work well [24].

It should be mentioned that linear SVMs with LBP features did not perform well for gender as shown in row 5 of Table 8.4. In addition, the LBP operator performed much worse when applied to whole faces, as shown in rows 3 and 4 of Table 8.4, no matter what classifier was used.

8.7.2.3 BIF Feature

For the biologically inspired features [26], we need to find the best structure and setting. To simplify the process, the gender recognition is performed over all ages first. A twofold cross validation was used as the test scheme. The same divisions of training and test data are used for all algorithms here, either over all ages or at separate age groups.

First, we evaluated the C2 features with a nonlinear SVM for gender classification over all ages. The feature extraction process is almost the same as that in [40]. The only difference is the number of prototypes to represent the gender. Since we have 8,000 images for the two-class classification problem, a small number of prototypes cannot work well (not shown here). We let the algorithm randomly select 2,000 prototypes from the female faces for S2 and C2 feature calculations. An accuracy of 81.05 % was obtained. This result is much worse than the 89.28 % using the raw pixel representation, the 88.65 % accuracy of the HOG method, and the 90.53 % of the LBP, shown in Table 8.4. We also randomly selected 2,000 prototypes from the male faces, and the result was 81.00 %—almost the same. Finally, we also let the algorithm randomly select 4,000 prototypes from both males and females, and got an accuracy of 83.00 %—still very low. From this experiment, we believe that C2 features do not work well for gender recognition. We notice that Meyers and Wolf [33] did not use C2 features in their face recognition problem, but they did not show any results when C2 features were used for face recognition. Based on our experience, C2 features are not a good choice for face-based gender classification, although these features demonstrated super performance on object category recognition [35, 40].

When the proper structure is determined for the BIF features, a better result can be obtained. More details about the BIF can be found in [24]. The results of BIF with both linear and nonlinear SVMs are given in Table 8.4. One can see that the kernel SVM performs better than the linear SVM in each age group. The BIF features combined with the kernel SVM can perform better than all other approaches in our comparisons.

8.7.2.4 Summary

We have shown the performance of the SVM for gender classification on a large database. The nonlinear SVM with the RBF kernel can perform significantly better than the linear SVM in all cases. Different methods have been used for facial image representation in the context of gender classification. The BIF features are better than the LBP and HOG features. More interestingly, we have shown that the gender classification is affected by ages. The adult faces can provide a much higher accuracy for gender classification than on young or senior faces. This was discovered quantitatively for the first time [24].

8.7.3 Ethnicity Estimation

We study ethnicity classification under variations of gender and age [15], using the SVM [41] as the classifier. We investigate whether the ethnicity estimation performance is affected by other human attributes, such as gender and age. Towards this goal, we designed experiments under two situations: (1) using female faces to learn an ethnicity classifier and then apply to males, and vice versa, and (2) learning ethnicity classifiers using faces from three age groups, and testing with different age groups.

The data were selected from the MORPH database [39] for this study [15]. The distribution of the selected data is shown in Table 8.5. The BIF features [26] were used for facial image characterization combined with manifold learning techniques [15].

8.7.3.1 Ethnicity w.r.t. Gender

To study whether the performance of ethnicity classification is affected by gender, we learn the ethnic classifiers using female and male faces, separately. Then we test the performance on the same and different gender to observe the difference.

Table 8.5 The distribution of the data selected from MORPH for the study

	Female	Male	Female and male
White	2,570	7,960	10,530
Black	2,570	7,960	10,530
White and black	5,140	15,920	21,060

Table 8.6 A study of ethnicity estimation with respect to gender [15]

		Ethnicity classification concerning gender						
		BIF		BIF+PCA		BIF+OLPP		
Train.	Test	Accuracy	Accuracy decrease	Accuracy	Accuracy decrease	Accuracy	Accuracy decrease	Comments
F1	F2	98.7 %	–	98.9 %	–	99.1 %	–	Same gender
	M1	94.0 %	4.8 %	93.7 %	5.3 %	90.3 %	8.9 %	Female → Male
	M2	94.1 %	4.7 %	93.8 %	5.2 %	90.4 %	8.8 %	Female → Male
F2	F1	98.6 %	–	98.9 %	–	99.3 %	–	Same gender
	M1	91.4 %	7.3 %	90.9 %	8.1 %	92.3 %	7.1 %	Female → Male
	M2	91.4 %	7.3 %	91.1 %	7.9 %	92.2 %	7.2 %	Female → Male
M1	M2	98.8 %	–	98.8 %	–	99.1 %	–	Same gender
	F1	96.8 %	2.0 %	97.2 %	1.6 %	97.7 %	1.4 %	Male → Female
	F2	96.5 %	2.3 %	97.1 %	1.7 %	97.3 %	1.8 %	Male → Female
M2	M1	98.7 %	–	98.7 %	–	98.8 %	–	Same gender
	F1	97.5 %	1.2 %	97.6 %	1.1 %	98.3 %	0.5 %	Male → Female
	F2	97.1 %	1.6 %	97.3 %	1.4 %	97.6 %	1.2 %	Male → Female
F1	F2	98.7 %	–	98.9 %	–	99.1 %	–	Same gender
	$M1_S$	93.9 %	4.9 %	93.1 %	5.9 %	89.6 %	9.6 %	Female → $Males_S$
	$M2_S$	94.1 %	4.7 %	94.1 %	4.9 %	90.5 %	8.7 %	Female → $Males_S$
F2	F1	98.6 %	–	98.9 %	–	99.3 %	–	Same gender
	$M1_S$	90.6 %	8.1 %	90.4 %	8.6 %	91.9 %	7.5 %	Female → $Males_S$
	$M2_S$	91.7 %	7.0 %	91.7 %	7.3 %	92.5 %	6.8 %	Female → $Males_S$
$M1_S$	$M2_S$	98.4 %	–	98.6 %	–	99.0 %	–	Same $genders_S$
	F1	96.2 %	2.2 %	96.8 %	1.8 %	96.9 %	2.1 %	$Males_S$ → Female
	F2	95.8 %	2.6 %	96.3 %	2.3 %	96.6 %	2.4 %	$Males_S$ → Female
$M2_S$	$M1_S$	98.3 %	–	98.7 %	–	98.8 %	–	Same $genders_S$
	F1	96.4 %	1.9 %	97.3 %	1.4 %	97.6 %	1.2 %	$Males_S$ → Female
	F2	96.3 %	2.0 %	97.0 %	1.7 %	97.2 %	1.6 %	$Males_S$ → Female

There are two ethnic groups, white (W) and black (B). So we have four groups with gender: white female (WF), black female (BF), white male (WM), and black male (BM). Each of the four groups is randomly divided into two subgroups for cross validations.

Through comparisons, we can infer the effect of gender difference on ethnicity estimation [15].

For the selected data shown in Table 8.5, we have four groups, black female (BF), white female (WF), black male (BM), and white male (WM). Within each group, the data are randomly divided into two subgroups, labeled as 1 and 2, in order to do cross validations. Suppose we choose one subgroup from the BF and another subgroup from WF to learn the ethnic classifier, labeled as F1 = BF1 + WF1, without any loss of generality. Then we can test the performance on female or male faces. Remember that we have another female data set, denoted as F2 = BF2 + WF2, and two subsets for male faces, M1 = BM1 + WM1, and M2 = BM2 + WM2. For ethnicity estimation with the same gender, the subset F2 is tested, denoted as F1 $\rightarrow$ F2. For different gender evaluation, we use M1 and M2 for testing, denoted as F1 $\rightarrow$ M1 and F1 $\rightarrow$ M2. Similarly, we can use F2, M1, or M2 for training, and use the remaining data for testing.

The experimental results are shown in Table 8.6. We have 16 ethnicity classification experiments for each face representation. So there are 48 experiments in total, using the three face representations. The original dimensionality of the BIF is 4,376. It is reduced to about 500 using PCA, and reduced to about 100 using OLPP. These numbers are kept the same throughout the experiments.

The 48 experiments can be categorized into three kinds of ethnicity classifications: same gender, female $\rightarrow$ male, and male $\rightarrow$ female. From Table 8.6, we can observe that (1) for ethnicity estimation using the same gender, the classification accuracies are very high (from 98.6% to 99.3%) for all three face representations. This demonstrates that our face representations have very good performance for ethnicity estimation; (2) for ethnicity estimation of male $\rightarrow$ female (using male faces to learn and females to test), the classification accuracies are slightly lower than using the same gender, ranging from 96.5 % to 98.3 %, but the accuracy decreases (accuracy difference between the cases of cross-gender and the same gender using the same training data, divided by the accuracy in the same gender case) are relatively small, e.g., from 0.5 % to 2.3 %; and (3) for ethnicity estimation of female $\rightarrow$ male, the classification accuracies range from 90.3 % to 94.1 %, with quite large accuracy decreases, e.g., from 4.7 % to 8.9 %, corresponding to different face representations.

One might notice that the number of female faces is smaller than males in Table 8.5. Do the accuracy differences come from the different sample sizes? To check this issue, we reduce the number of males in Table 8.5 to make the number of males equal to females. Specifically, we randomly chose partial males from M1 (i.e., 1,285 faces) and from M2 (1,285 faces), denoted as $M1_S$ and $M2_S$, respectively. Now, F1, F2, $M1_S$, and $M2_S$ have the same number of faces. Then we use the reduced data set to re-learn the ethnicity classifiers, and re-perform the 48 ethnic classification experiments, with the results shown in the lower part of Table 8.6. One can see that almost the same accuracy decreases can be observed from the equal-sized-data experiments.

As a result, our study demonstrates that ethnicity estimation is influenced by gender significantly when the female faces are used for training while males for testing, i.e., female $\rightarrow$ male. However, the reversed process (male $\rightarrow$ female) has some influence but not very significant. This *unsymmetric* influence is interesting. We are not very clear about how to interpret this phenomenon yet; however, we hope the computational results inspire more psychological studies [37,49,50] to get a reasonable interpretation.

8.7.3.2 Ethnicity w.r.t. Age

To study whether ethnicity estimation is affected by age [15], we divided the data set into three age groups, labeled as A, B, and C. The partition considers the number of face images in different age groups to make them comparable, since the original data in MORPH do not have balanced number of faces at each age. Based on this and the age range (from 16 to 67 years), we determined that age group A contains ages less than or equal to 25 years, group B has ages greater than 25 but less than or equal to 40, and group C contains ages above 40. Remember that we still need two subgroups (1 and 2) within each age group for the purpose of cross validations, and each subgroup has both black and white faces to learn the ethnic classifier. The final distribution of the age groups is that A1 (2,756 faces, $16 \leq age \leq 25$), B1 (4,508 faces, $25 < age \leq 40$), C1 (3,266 faces, $40 < age \leq 67$), A2 (2,756 faces, $16 \leq age \leq 25$), B2 (4,508 faces, $25 < age \leq 40$), and C2 (3,266 faces, $40 < age \leq 67$). Not strictly, we name groups A, B, and C as young, middle, and old to make it easier to interpret the results.

Then we use one age group to train the ethnicity classifier, and the remaining age groups for testing. There are 30 ethnicity estimation experiments for each face representation, and there are 120 experiments in total given the three face representations. The experimental results are given in Table 8.7.

From the table, we can observe that (1) for ethnicity estimation within the same age group, i.e., A1 $\leftrightarrow$ A2, B1 $\leftrightarrow$ B2, C1 $\leftrightarrow$ C2, the ethnic classification accuracies can be very high, ranging from 98.3% to 99.1%, using the three face representations. (2) for ethnicity estimation with different age groups for training and testing, most of the results still have high accuracies, e.g., from 97.6% to 98.7%, for young $\leftrightarrow$ middle, middle $\leftrightarrow$ old, and old $\rightarrow$ young. In comparison with the same age group results, the accuracy decreases are relatively small, ranging from 0.0% to 1.3%, using three face representations. In the case of young $\rightarrow$ old, the accuracy decreases are slightly larger, e.g., from 2.0% to 2.7%, but not so significant as the gender influence on ethnicity estimation in the case of female $\rightarrow$ male.

8.7.3.3 Summary

We can reorganize the above experimental results by averaging over the subcases, so that one can observe the performance more directly. The new results are shown in Table 8.8. From the results, we can easily observe that (1) ethnicity estimation can have very high accuracies if it is performed within the same gender and age groups; (2) our face representations based on biologically inspired features with or without manifold learning show high performance in ethnicity classification; (3) ethnicity estimation can be affected in the cross-gender case of female $\rightarrow$ male, with accuracy decreases of 6$\sim$8 % in average, which is significantly different from the situations of the same gender and male $\rightarrow$ female; (4) ethnicity estimation is not affected very much under the situation of cross-age.

Table 8.7 A study of ethnicity estimation with respect to age [15]

		Ethnicity classification concerning age						
		BIF		BIF+PCA		BIF+OLPP		
Train.	Test	Accuracy	Accuracy decrease	Accuracy	Accuracy decrease	Accuracy	Accuracy decrease	Comments
A1	A2	98.4 %	–	98.5 %	–	99.0 %	–	Same age group
	B1	97.7 %	0.7 %	97.8 %	0.7 %	98.2 %	0.8 %	Young → Middle
	B2	97.7 %	0.7 %	97.8 %	0.7 %	98.4 %	0.6 %	Young → Middle
	C1	95.9 %	2.5 %	96.0 %	2.5 %	96.8 %	2.2 %	Young → Old
	C2	95.7 %	2.7 %	95.8 %	2.7 %	96.3 %	2.7 %	Young → Old
A2	A1	98.5 %	–	98.5 %	–	98.8 %	–	Same age group
	B1	97.7 %	0.8 %	97.9 %	0.6 %	98.3 %	0.5 %	Young → Middle
	B2	98.3 %	0.2 %	98.2 %	0.3 %	98.5 %	0.3 %	Young → Middle
	C1	96.3 %	2.2 %	96.3 %	2.2 %	96.8 %	2.0 %	Young → Old
	C2	95.8 %	2.7 %	95.8 %	2.7 %	96.2 %	2.6 %	Young → Old
B1	B2	98.9 %	–	98.7 %	–	99.1 %	–	Same age group
	A1	98.1 %	0.8 %	98.3 %	0.4 %	98.6 %	0.5 %	Middle → Young
	A2	98.5 %	0.4 %	98.2 %	0.5 %	98.7 %	0.4 %	Middle → Young
	C1	97.7 %	1.2 %	97.9 %	0.8 %	98.1 %	1.0 %	Middle → Old
	C2	97.6 %	1.3 %	97.8 %	0.9 %	98.2 %	0.9 %	Middle → Old
B2	B1	98.7 %	–	98.8 %	–	99.0 %	–	Same age group
	A1	98.3 %	0.4 %	98.3 %	0.5 %	98.4 %	0.6 %	Middle → Young
	A2	98.5 %	0.2 %	98.4 %	0.4 %	98.7 %	0.3 %	Middle → Young
	C1	97.9 %	0.8 %	97.8 %	1.0 %	97.7 %	1.3 %	Middle → Old
	C2	98.0 %	0.7 %	97.8 %	1.0 %	98.2 %	0.8 %	Middle → Old
C1	C2	98.7 %	–	98.7 %	–	98.8 %	–	Same age group
	A1	98.1 %	0.6 %	98.1 %	0.6 %	98.1 %	0.7 %	Old → Young
	A2	98.1 %	0.6 %	98.3 %	0.4 %	98.2 %	0.6 %	Old → Young
	B1	98.4 %	0.3 %	98.3 %	0.4 %	98.6 %	0.2 %	Old → Middle
	B2	98.4 %	0.3 %	98.4 %	0.3 %	98.6 %	0.2 %	Old → Middle
C2	C1	98.3 %	–	98.3 %	–	98.7 %	–	Same age group
	A1	97.8 %	0.5 %	97.7 %	0.6 %	98.0 %	0.7 %	Old → Young
	A2	97.6 %	0.7 %	98.0 %	0.3 %	97.9 %	0.8 %	Old → Young
	B1	98.2 %	0.1 %	98.0 %	0.3 %	98.5 %	0.2 %	Old → Middle
	B2	98.3 %	0.0 %	98.3 %	0.0 %	98.6 %	0.1 %	Old → Middle

The data set is divided into three age groups: Young or A ($age \leq 25$ years), Middle or B ($age \leq 40$), and Old or C ($age > 40$). Each age group is randomly divided into two subgroups for cross validations.

8.7.3.4 Usefulness of the Study

Our study results have applications in many real problems. For example, for a large database containing multiple ethnic groups, one may categorize the ethnic groups before age estimation [25, 36], since the ethnicity estimation is not very sensitive to age variations from our studies. Categorizing into different ethnicity groups may reduce the age estimation errors [36] significantly. For the problem of gender

Table 8.8 A summary of our studies on ethnicity classification versus the changes of gender and age groups [15]

	Ethnicity classification					
	BIF		BIF+PCA		BIF+OLPP	
Versus Gender or age	Average accuracy	Accuracy decrease	Average accuracy	Accuracy decrease	Average accuracy	Accuracy decrease
Same gender	98.7 %	–	98.8 %	–	99.1 %	–
Female → Male	92.7 %	6.1 %	92.4 %	6.5 %	91.3 %	7.9 %
Male → Female	97.0 %	1.7 %	97.3 %	1.5 %	97.7 %	1.4 %
Same gender$_S$	98.5 %	–	98.8 %	–	99.1 %	–
Female → Male$_S$	92.6 %	6.0 %	92.3 %	6.7 %	91.1 %	8.1 %
Male$_S$ → Female	96.2 %	2.3 %	96.9 %	1.9 %	97.1 %	2.0 %
Same age group	98.6 %	–	98.6 %	–	98.9 %	–
Young → Middle	97.9 %	0.7 %	97.9 %	0.7 %	98.4 %	0.5 %
Young → Old	95.9 %	2.7 %	96.0 %	2.6 %	96.5 %	2.4 %
Middle → Young	98.4 %	0.2 %	98.3 %	0.3 %	98.6 %	0.3 %
Middle → Old	97.7 %	0.9 %	97.8 %	0.8 %	98.1 %	0.8 %
Old → Young	97.9 %	0.7 %	98.0 %	0.6 %	98.1 %	0.8 %
Old → Middle	98.3 %	0.3 %	98.3 %	0.3 %	98.6 %	0.3 %

classification [24, 48] on a large database with multiple ethnic groups, one may also perform ethnic classification first, and then gender recognition is performed within each single ethnic group, since in most cases, the ethnicity estimation is not very sensitive to gender variations based on our studies [15]. We believe that multi-ethnic databases will be more and more popular in computer vision research, considering more databases are collected from the Internet, such as in [36]. We expect more research work will be reported on multi-ethnic face image databases in the near future.

On the other hand, our study based on computational analysis may inspire more psychological studies on ethnic grouping [37, 49, 50] related to age and gender variations. Interpretations about our results could be derived from further psychological studies.

8.8 Conclusions

We have presented the applications of the SVM to soft biometrics recognition in face images. The SVM can have very good performance for gender and ethnicity classification, when combined with appropriate features to characterize the facial appearance. For age estimation, we showed the performance of the SVM and SVR on two databases, since age estimation can be considered either a classification or a regression problem. We found that the two approaches can perform quite differently on different databases. A better way is to combine them to take advantage of both.

Two schemes, called LARR and PFA, have been proposed to integrate the SVM with SVR and validated for age estimation. The performance can be improved significantly when these schemes are used for age estimation. Further, we studied the influence of age on gender classification, and also the influence of age and gender on ethnicity estimation, based on the SVM classifiers. Overall, the SVM and their extensions are very useful for learning soft biometric characteristics from face images.

References

1. Ahonen, T., Hadid, A., Pietikainen, M.: Face recognition with local binary patterns. In: European Conference on Computer Vision, pp. 469–481 (2004)
2. Baluja, S., Rowley, H.A.: Boosting sex identification performance. Int. J. Comput. Vision **71**(1), 111–119 (2007)
3. Bruce, V., Burton, A., Hanna, E., Healey, P., Mason, O.: Sex discrimination: how do we tell the difference between male and female faces? Perception **22**, 131–152 (1993)
4. Cai, D., He, X., Han, J., Zhang, H.: Orthogonal laplacianfaces for face recognition. IEEE Trans. Image Process. **15**, 3608–3614 (2006)
5. Cootes, T., Edwards, G., Taylor, C.: Active appearance models. In: European Conference on Computer Vision, pp. 484–498 (1998)
6. Dalal, N., Triggs, B.: Histograms of oriented gradients for human detection. In: IEEE Conference on CVPR, pp. 886–893 (2005)
7. FGNET: The fg-net aging database. http://www.fgnet.rsunit.com/ (2002)
8. Fu, Y., Huang, T.S.: Human age estimation with regression on discriminative aging manifold. IEEE Trans. Multimedia **10**(4), 578–584 (2008)
9. Fu, Y., Xu, Y., Huang, T.S.: Estimating human ages by manifold analysis of face pictures and regression on aging features. In: IEEE Conference on Multimedia and Expo, pp. 1383–1386 (2007)
10. Gao, J.B., Gunn, S.R., Harris, C.J., Brown, M.: A probabilistic framework for svm regression and error bar estimation. Mach. Learn. **46**(1–3), 71–89 (2002)
11. Geng, X., Zhou, Z.H., Zhang, Y., Li, G., Dai, H.: Learning from facial aging patterns for automatic age estimation. In: ACM Conference on Multimedia, pp. 307–316 (2006)
12. Gunn, S.R.: Support vector machines for classification and regression. ISIS Technical Report 14 (1998)
13. Guo, G.D., Dyer, C.: Learning from examples in the small sample case: face expression recognition. IEEE Trans. Syst. Man Cybern. Part B **35**(3), 447–488 (2005)
14. Guo, G.D., Li, S.: Content-based audio classification and retrieval by support vector machines. IEEE Trans. Neural Netw. **14**(1), 209–215 (2003)
15. Guo, G.D., Mu, G.: A study of large-scale ethnicity estimation with gender and age variations. In: IEEE International Workshop on Analysis and Modeling of Faces and Gestures (2010)
16. Guo, G.D., Li, S., Chan, K.: Face recognition by support vector machines. In: Proceedings of Fourth IEEE International Conference Automatic Face and Gesture Recognition (2000)
17. Guo, G.D., Zhang, H., Li, S.: Pairwise face recognition. In: Proceedings of Eighth International Conference on Computer Vision, vol. 2, pp. 282–287 (2001)
18. Guo, G.D., Li, S., Chan, K.: Support vector machines for face recognition. Image Vis. Comput. **19**(9–10), 631–638 (2001)
19. Guo, G.D., Jain, A., Ma, W., Zhang, H.: Learning similarity measure for natural image retrieval with relevance feedback. IEEE Trans. Neural Netw. **13**(4), 811–820 (2002)

20. Guo, G.D., Dyer, C., Zhang, Z.: Linear combination representation for outlier detection in motion tracking. In: Proceedings of IEEE Conference Computer on Vision and Pattern Recognition, vol. 2, pp. 274–281 (2005)
21. Guo, G.D., Fu, Y., Dyer, C., Huang, T.S.: Image-based human age estimation by manifold learning and locally adjusted robust regression. IEEE Trans. Image Process. **17**(7), 1178–1188 (2008)
22. Guo, G.D., Fu, Y., Dyer, C., Huang, T.S.: A probabilistic fusion approach to human age prediction. In: International Workshop on Semantic Learning Applications in Multimedia (2008)
23. Guo, G.D., Fu, Y., Huang, T., Dyer, C.: Locally adjusted robust regression for human age estimation. In: IEEE Workshop on Application of Computer Vision (2008)
24. Guo, G.D., Dyer, C., Fu, Y., Huang, T.S.: Is gender recognition affected by age? In: IEEE International Workshop on Human-Computer Interaction, pp. 2032–2039 (2009)
25. Guo, G.D., Mu, G., Fu, Y., Dyer, C., Huang, T.S.: A study on automatic age estimation on a large database. In: IEEE International Conference on Computer Vision, pp. 1986–1991 (2009)
26. Guo, G.D., Mu, G., Fu, Y., Huang, T.S.: Human age estimation using bio-inspired features. In: IEEE Conference on Computer Vision and Pattern Recognition, pp. 112–119 (2009)
27. Guo, G.D., Mu, G., Ricanek, K.: Cross-age face recognition on a very large database: the performance versus age intervals and improvement using soft biometric traits. In: International Conference on Pattern Recognition (2010)
28. Gutta, S., Wechsler, H.: Gender and ethnic classification of face images. In: International Conference on Automatic Face and Gesture Recognition, pp. 194–199 (1998)
29. Hastie, T., Tibshirani, R.: Classification by pairwise coupling. Ann. Stat. **26**(2), 451–471 (1998)
30. Jensen, F.V., Nielsen, T.D.: Bayesian Networks and Decision Graphs. Springer, New York (2007)
31. Kittler, J., Hatef, M., Duin, R.P.W., Matas, J.: On combining classifiers. IEEE Trans. Pattern Anal. Mach. Intell. **20**(3), 226–239 (1998)
32. Lanitis, A., Draganova, C., Christodoulou, C.: Comparing different classifiers for automatic age estimation. IEEE Trans. SMC-B **24**(4), 621–628 (2002)
33. Meyers, E., Wolf, L.: Using biologically inspired features for face processing. Int. J. Comput. Vis. **76**, 93–104 (2008)
34. Moghaddam, B., Yang, M.H.: Learning gender with support faces. IEEE Trans. Pattern Anal. Mach. Intell. **24**(5), 707–711 (2002)
35. Mutch, J., Lowe, D.: Object class recognition and localization using sparse features with limited receptive fields. In: IEEE Conference on Computer Vision and Pattern Recognition, pp. 11–18 (2006)
36. Ni, B., Song, Z., Yan, S.: Web image mining towards universal age estimator. In: ACM Multimedia (2009)
37. Okazaki, S., Sue, S.: Methodological issues in assessment research with ethnic minorities. Psychol. Assess. **7**(3), 367–375 (1995)
38. Platt, J., et al.: Probabilistic outputs for support vector machines and comparisons to regularized likelihood methods. Adv. Large Margin Classifiers **10**(3), 61–74 (1999)
39. Ricanek, K., Tesafaye, T.: Morph: a longitudinal image database of normal adult age-progression. In: IEEE Conference on AFGR, pp. 341–345 (2006)
40. Serre, T., Wolf, L., Bileschi, S., Riesenhuber, M., Poggio, T.: Robust object recognition with cortex-like mechanisms. IEEE Trans. Pattern Anal. Mach. Intell. **29**(3), 411–426 (2007)
41. Vapnik, V.N.: Statistical Learning Theory. Wiley, New York (1998)
42. Webb, A.R.: Statistical Pattern Recognition, 2nd edn. Wiley, New York (2002)
43. Wen, L., Guo, G.: A computational approach to body mass index prediction from face images. Image Vis. Comput. **31**(5), 392–400 (2013)
44. Wild, H.A., Barrett, S.E., Spence, M.J., O'Toole, A.J., Cheng, Y.D., Brooke, J.: Recognition and sex categorization of adults' and children's faces: examining performance in the absence of sex-stereotyped cues. J. Exp. Child Psychol. **77**, 269–291 (2000)

45. Yamaguchi, M.K., Hirukawa, T., Kanazawa, S.: Judgment of sex through facial parts. Perception **24**, 563–575 (1995)
46. Yan, S., Wang, H., Huang, T.S., Tang, X.: Ranking with uncertain labels. In: IEEE Conference on Multimedia and Expo, pp. 96–99 (2007)
47. Yan, S., Wang, H., Tang, X., Huang, T.: Learning auto-structured regressor from uncertain nonnegative labels. In: IEEE Conference on ICCV (2007)
48. Yang, Z., Ai, H.: Demographic classification with local binary patterns. In: International Conference on Biometrics, pp. 464–473 (2007)
49. Yee, A.H.: Ethnicity and race: psychological perspectives. Educ. Psychol. **18**(1), 14–24 (1983)
50. Zuckerman, M.: Some dubious premises in research and theory on racial differences: scientific, social, and ethical issues. Am. Psychol. **45**(12), 1297–1303 (1990)

Printed by Publishers' Graphics LLC
CAMZ140316.20.05.114